Natural antimicrobials for the minimal processing of foods

Related titles from Woodhead's food science, technology and nutrition list:

Minimal processing technologies in the food industry (ISBN 1 85573 547 4)
The emergence of 'minimal' processing techniques, which have limited impact on a food's nutritional and sensory properties, has been a major new development in the food industry. This book provides an authoritative review of the range of minimal techniques currently available, their applications and safety and quality issues.

Food preservation techniques (ISBN 1 85573 530 X)
Extending the shelf-life of foods whilst maintaining safety and quality is a critical issue for the food industry. As a result there have been major developments in food preservation techniques, which are summarised in this authoritative collection. The first part of the book examines the key issue of maintaining safety as preservation methods become more varied and complex. The rest of the book looks both at individual technologies and how they are combined to achieve the right balance of safety, quality and shelf-life for particular products.

Antioxidants in food (ISBN 1 85573 463 X)
Antioxidants are an increasingly important ingredient in food processing, as they inhibit the development of oxidative rancidity in fat-based foods, particularly meat and dairy products and fried foods. Recent research suggests that they play a role in limiting cardiovascular disease and cancers. This book provides a review of the functional role of antioxidants and discusses how they can be effectively exploited by the food industry, focusing on naturally occurring antioxidants in response to the increasing consumer scepticism over synthetic ingredients.

'An excellent reference book to have on the shelves' *LWT Food Science and Technology*

Details of these books and a complete list of Woodhead's food science, technology and nutrition titles can be obtained by:

- visiting our web site at www.woodhead-publishing.com
- contacting Customer Services (e-mail: sales@woodhead-publishing.com; fax: +44 (0) 1223 893694; tel.: +44 (0) 1223 891358 ext. 30; address: Woodhead Publishing Limited, Abington Hall, Abington, Cambridge CB1 6AH, England)

If you would like to receive information on forthcoming titles in this area, please send your address details to: Francis Dodds (address, telephone and fax as above; e-mail: francisd@woodhead-publishing.com). Please confirm which subject areas you are interested in.

Natural antimicrobials for the minimal processing of foods

Edited by
Sibel Roller

CRC Press
Boca Raton Boston New York Washington, DC

WOODHEAD PUBLISHING LIMITED
Cambridge England

Published by Woodhead Publishing Limited
Abington Hall, Abington
Cambridge CB1 6AH
England
www.woodhead-publishing.com

Published in North America by CRC Press LLC
2000 Corporate Blvd, NW
Boca Raton FL 33431
USA

First published 2003, Woodhead Publishing Limited and CRC Press LLC

British Library Cataloguing in Publication Data
A catalogue record for this book is available from the British Library.

Library of Congress Cataloging-in-Publication Data
A catalog record for this book is available from the Library of Congress.

Woodhead Publishing Limited ISBN 1 85573 669 1 (book); 1 85573 703 5 (e-book)
CRC Press ISBN 0-8493-1753-3
CRC Press order number: WP1753

Cover design by The ColourStudio
Project managed by Macfarlane Production Services, Markyate, Hertfordshire
(e-mail: macfarl@aol.com)
Typeset by MHL Typesetting Limited, Coventry, Warwickshire
Printed by TJ International, Padstow, Cornwall, England

This book is dedicated to the memory of my father, Dr Dragan Roller, who passed away in April 2003.

Contents

Contributor contact details

Chapter 1

Professor S. Roller
Faculty of Health and Human Sciences
Thames Valley University
32–38 Uxbridge Road
London W5 2BS
UK

Tel: +44 (0) 20 8280 5108
Fax: +44 (0) 20 8280 5289
Email: Sibel.Roller@tvu.ac.uk

Chapter 2

Dr M. R. Adams
School of Biomedical and Life Sciences
University of Surrey
Guildford
Surrey GU2 7XH
UK

Tel: +44 (0) 1483 686492
Fax: +44 (0) 1483 300374
Email: m.adams@surrey.ac.uk

Dr E. Smid
NIZO Food Research
Kernhemseweg 2
PO Box 20
6710 BA Ede
The Netherlands

Email: eddy.smid@nizo.nl

Chapter 3

Professor P. L. Dawson
Food Science and Human Nutrition Department
Clemson University
A203J Poole Hall
Clemson
SC 29634-0371
USA

Tel: +1 864 656 1138
Fax: +1 864 656 0331
Email: pdawson@clemson.edu

Dr B. W. Sheldon
Department of Poultry Science
234D Scott Hall
Box 7608
North Carolina State University
Raleigh
NC 27695-7608
USA

Tel: +1 919 515 5407
Email: brian_sheldon@ncsu.edu

Chapter 4

Dr B. Ray and Dr K. W. Miller
Department of Molecular Biology
University of Wyoming
Laramie
WY 82071-3944
USA

Email: labcin@uwyo.edu
kwmiller@uwyo.edu

Chapter 5

Dr J. Stark
DSM Food Specialties
2600 MA Delft
The Netherlands

Tel: +31 (0) 15 2792226
Fax: + 31 (0) 15 2792490
Email: Jacques.Stark@dsm.com

Chapter 6

Dr J. Samelis
National Agricultural Research Foundation
Dairy Research Institute
Katsikas
452 21 Ioannina
Greece

Tel: +30 2651 0 91785
Fax: +30 2651 0 92523
Email: jsam@otenet.gr

Dr J. Sofos
Department of Animal Sciences
Colorado State University
Fort Collins
CO 80523-1171
USA

Email: john.sofos@colostate.edu

Chapter 7

Professor A. S. Naidu
N-terminus Research Laboratory
981 Corporate Center Drive # 110
Pomona
CA 91768-2600
USA

Tel: +1 909 469 9596
Fax: +1 909 469 6936
Email: asnaidu@aol.com

Chapter 8

Professor S. Roller
Faculty of Health and Human Sciences
Thames Valley University
32–38 Uxbridge Road
London W5 2BS
UK

Tel: +44 (0) 20 8280 5108
Fax: +44 (0) 20 8280 5289
Email: Sibel.Roller@tvu.ac.uk

Chapter 9

Professor G. J. E. Nychas and Dr P. Skandamis
Agricultural University of Athens
Department of Food Science and Technology
Iera Odos 75
Athens 11855
Greece

Tel/Fax: +30 10 529 4693
Email: gjn@aua.gr

Dr C. Tassou
National Agricultural Research Foundation
Institute of Technology of Agricultural Products
S. Venizelou 1
Lycovrisi 14123
Greece

Tel: +30 210 2845940
Fax: +30 210 2840740
Email: microlab.itap@nagref.gr

Chapter 10

Professor A. Ippolito and Dr F. Nigro
Università degli Studi di Bari
Dipartimento di Protezione delle Piante e Microbiologia Applicata
Via Amendola 165/A
70126 Bari
Italy

Tel: +39 080 5443053
Fax: +39 080 5442911
Email: ippolito@agr.uniba.it

Chapter 11

Professor S. M. Alzamora and Dr S. Guerrero
Departamento de Industrias
Facultad de Ciencias Exactas y Naturales
Universidad de Buenos Aires
Ciudad Universitaria
1428 Buenos Aires
Argentina

Tel: +54 2320 491822
Fax: +54 2320 491822
E-mail: alzamora@ciudad.com.ar
alzamora@di.fcen.uba.ar

Dr A. López-Malo and Dr E. Palou
Departamento de Ingeniería Química y Alimentos
Universidad de las Américas
Santa Catarina Mártir
Puebla 72820
México

Chapter 12

Dr L. R. Franssen
General Mills Inc
330 University Avenue SE
Minneapolis
MN 55414
USA

Tel: +1 763 764 7535
Fax: +1 763 764 8211
E-mail: lauren.franssen@genmills.com

Dr J. M. Krochta
Department of Food Science and Technology
Department of Biological and Agricultural Engineering
University of California
One Shields Avenue
Davis
CA 95616-8598
USA

Tel: +1 530 752 2164
Fax: +1 530 752 4759
Email: jmkrochta@ucdavis.edu

Chapter 13

Dr B. Ouattara and Dr A. A. Mafu
Meat Industry Section
Food Research and Development Centre
Agriculture and Agri-Food Canada
3600 Casavant Blvd West
St-Hyacinthe
Quebec J2S 8E3
Canada

Tel: +1 450 773 1105
Fax: +1 450 773 8461
Email: ouattarab@agr.gc.ca

Chapter 14

Professor N. Magan, M. Arroyo and D. Aldred
Institute of BioScience and Technology
Cranfield University
Silsoe
Bedford MK45 4DT
UK

Tel: +44 (0) 1525 863539
Fax: +44 (0) 1525 863540
Email: n.magan@cranfield.ac.uk

Chapter 15

Mr P. Berry Ottaway
Berry Ottaway & Associates Limited
Nesscliffe House
1A Fields Yard
Plough Lane
Hereford HR4 0EL
UK

Email: Berry.Ottaway@dial.pipex.com

1

Introduction

S. Roller, Thames Valley University, UK

1.1 The food safety challenge and the need for new preservatives

Despite our improved understanding of infection processes, better methods for controlling microorganisms and much stricter regulation of food production, foodborne diseases are still a major cause of morbidity and mortality worldwide. In the United States, the annual incidence of foodborne disease has been estimated at 76 million cases (affecting more than one-quarter of the total population), resulting in 325 000 hospitalisations and 5000 deaths (Mead *et al.*, 1999). In England, it has been estimated that 9.5 million cases of Infectious Intestinal Disease (IID) occur annually, of which 1.5 million lead to a visit to the family doctor (Food Standards Agency, 2000). Temporal trends in many developed countries show that the number of known cases has increased rapidly between 1980 and 2000, as illustrated in Fig. 1.1. Furthermore, food and waterborne diarrhoeal diseases are the leading cause of illness and death in developing countries, and kill an estimated 2.1 million people annually (WHO, 2001). In addition to the human costs, foodborne diseases are an economic burden on societies. Annual economic losses attributable to foodborne diseases have been estimated at over $6 billion in the USA and £745 million in England alone (Buzby and Roberts, 1997; Food Standards Agency, 2000; Tauxe, 2002). Economic losses due to food spoilage clearly also play an important role but are more difficult to quantify.

Common food preservatives such as nitrite, sodium benzoate and sodium metabisulfite have a long history of safe use (Gould and Russell, 2003). Their annual market size has been estimated at well over £100 million (Tollefson, 1995). However, reports of occasional allergic reactions in sensitive individuals

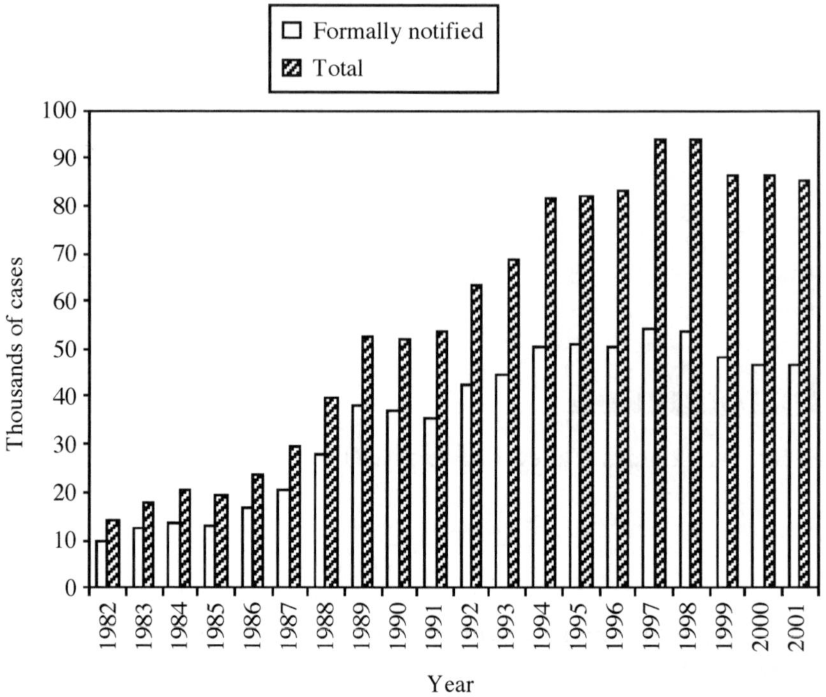

Fig. 1.1 Food poisoning cases in England and Wales, 1982–2001. Data from: UK Public Health Laboratory Service, http://www.phls.co.uk

and the formation of potentially carcinogenic by-products (e.g. nitrosamines from nitrite) have raised concerns about the potential detrimental effects of preservatives on health. Furthermore, demands from increasingly mistrustful consumers (especially in Europe) have led to numerous legislation reviews, which are expected to result in reductions in the permitted levels of use of many traditional preservatives in many countries. Consequently, there has been a resurgence of interest in antimicrobial compounds found in nature and this book is a reflection of that interest. The challenge for food scientists is to develop new preservation systems that will improve the quality and extend the shelf-life of foodstuffs without compromising safety or sensory properties.

1.2 Antimicrobial activity: laboratory results *vs* practical applications

Much has been written about the potential for natural antimicrobial compounds to replace or reduce reliance on synthetic food preservatives. In the last 20 years, hundreds of studies demonstrating antimicrobial activity of natural compounds against pathogenic or spoilage organisms in laboratory media have been published. However, few of these have been translated into real food applications. There has been a tendency to infer from a zone of inhibition on an agar plate that a putative antimicrobial agent has potential for use in real foods. Rarely has the investigator bothered to demonstrate antagonism or biocidal activity in foods. Factors such as temperature, pH, and the presence of fats and proteins, surfactants, minerals and other food components can greatly influence the outcome of antimicrobial testing and render predictions about applicability in foods difficult. For example, the lipophilicity of many phenolic compounds such as cinnamic acid and carvacrol plays an important role in determining their antimicrobial efficacy; yet, in fatty foods, lipophilic compounds can be expected to partition into the lipid phase and thereby lose some of their antimicrobial potency.

When developing or modifying a food product based on natural antimicrobial principles, a series of questions should be posed and answered prior to undertaking laboratory work (Table 1.1). One or more of the answers to the questions in Table 1.1 may well reveal serious obstacles to the application of a particular antimicrobial system in a specific food product. For example, the ease of extraction or production yield of an antimicrobial agent is often of critical importance. A compound that is present in milligram quantities in a rare tropical plant is unlikely to achieve commercial acceptance as a food preservative unless the gene controlling its production can be transferred and expressed in a microorganism such as *Bacillus subtilis* or *Aspergillus niger*, thereby allowing large-scale production by industrial fermentation. The ultimate question that must be asked of any antimicrobial system is: 'Will it function effectively and for a relatively long time in the defence of a food store?' The latter can only be answered by verification in the specific food product that is to be preserved.

1.3 A futile search for the 'magic bullet'

Antimicrobial compounds in nature rarely function in isolation; combination systems such as those found in the hen's egg, are far more common (Banks *et al.*, 1986; Dillon and Board, 1994; Roller and Board, 2003). Like plants and animals, microorganisms synthesise a multiplicity of compounds that act in concert to help individual strains to survive and/or multiply in the presence of competing microorganisms. Therefore, it is unlikely that a single food preservative, whether old or new, synthetic or natural, will provide the 'magic bullet' for the prevention of food spoilage and poisoning. The scientific literature around the turn of the millennium reflects a gradual recognition of this and knowledge-based attempts at developing new combination systems are becoming increasingly evident (Brul *et al.*, 2002).

Table 1.1 Developing food preservation systems based on natural antimicrobial compounds

Question	Possible answers	Examples (Chapter reference in this book)
1. What types of microorganisms need to be controlled?	• Bacteria: Gram-positive, Gram-negative, vegetative/spores • Yeasts • Moulds • Viruses • Prions	Lysozyme inhibits spore-forming spoilage bacteria and *Listeria* spp. (Chapter 7)
2. What are the cellular targets of the antimicrobial (if known)?	• Cell walls • Cell membranes • Metabolism • DNA/RNA	EDTA (ethylenediaminetetraacetic acid) and chitosan permeabilise Gram-negative bacteria (Chapters 2 & 8, respectively)
3. What other antimicrobials could be used in combination?	Combinations of antimicrobials with different cellular targets may lead to synergy or broaden the spectrum of activity of the preservation system	Lysozyme, nisin and EDTA together inhibit both Gram-positive and Gram-negative bacteria (Chapter 2)
4. Does the activity of the antimicrobial match the composition and storage conditions of the food?	Activity may depend on pH, temperature, texture, oil/water content, particle size, presence of enzymes	Nisin may bind to meat particles and/or be degraded by endogenous peptidases in raw meat (Chapters 2 & 3)

5. How stable is the antimicrobial under food processing conditions?	Heat, extremes of acidity or alkalinity, homogenisation, pressure, etc. may degrade some antimicrobials	Many enzymes are denatured by heat (Chapter 7)
6. Any other functional properties not related to antimicrobial activity?	• Desirable: antioxidant activity • Possibly undesirable: high flavour/aroma impact	Essential oils of herbs and spices (Chapters 9–14)
7. Any potential problems with toxicity?	Cytotoxic effects on eukaryotes may preclude regulatory approval	Phytoalexins
8. What is the cost of production?	Agents present at very low levels in natural materials may be too expensive to extract in the near term	Magainins from amphibians (but genetic transfer into industrial microorganisms may solve this problem in the long term)
9. What is the regulatory status of the antimicrobial?	Many newly extracted natural compounds may need regulatory approval to be used as food additives	Many plant phenolic compounds are permitted for use as food flavourings but may require additional clearance if used for preservation (Chapter 15)
10. Has the antimicrobial system been verified in the food?	**Shelf-life and challenge trials are essential to confirm activity in specific food products. Each product is unique with many interactions that cannot be completely predicted from laboratory trials**	

1.4 Beyond the hurdle concept: multifactorial food preservation

The concept of using combinations of preservation factors in foods is not new and has been applied empirically since antiquity in traditional products such as cheese or cured meat. In the early 1980s, the term 'hurdle technology' was coined by Leistner to describe the scientific selection of combined methods in food preservation (Leistner, 1978, 1995, 2000, and numerous references therein). The term 'hurdle technology' is still used widely in the literature. However, the analogy of a hurdle race, often accompanied by a cartoon depicting bacteria as athletes running on a track, gives the unfortunate impression that microorganisms confront each 'hurdle' or 'barrier' in succession. In reality, microorganisms are faced with all the 'barriers' or stresses at once. It has been suggested by Adams (2000) that a brick wall would provide a better way of visualising the effect. Each adverse condition contributes one or more layers of bricks to the wall. The overall height of the wall determines which organisms can climb over it and grow. Therefore, in this book, the term 'hurdle technology' is avoided and 'multifactorial preservation' is used throughout.

1.5 Microbial resistance

In a community of microorganisms, some strains possess innate resistance to a given antimicrobial while others may acquire resistance during continued exposure to the inimical compound. The ability of bacteria to adapt and develop resistance to therapeutic antibiotics is well known and is beginning to compromise humankind's ability to fight infectious diseases (Schlundt, 2002; WHO, 2000). It has also been suggested that some microorganisms may become resistant to therapeutic antibiotics through exposure to common disinfectants (Moken *et al.*, 1997; also reviewed in Maillard, 2002). However, others have argued that many antimicrobials used in food processing (e.g. food preservatives, sanitisers and disinfectants) are too non-specific to make acquired resistance likely (Davidson and Harrison, 2002).

There is currently relatively little direct evidence in the literature for acquired resistance to common food preservatives such as benzoates (when used at traditional levels) or to correctly applied sanitisers such as chlorine. Many of these compounds have been in use in food factories for nearly 100 years without resistance problems (Davidson and Harrison, 2002). Nevertheless, the potential for the emergence of such resistance exists, particularly if the levels of addition of an antimicrobial are sub-lethal, as in minimal processing.

Many foodborne pathogens are capable of mounting an adaptive response to sub-lethal stresses, enabling them to tolerate and survive subsequent exposure to normally lethal levels of the same stress or even a different type of stress (Abee and Wouters, 1999; Rowan, 1999). For example, *Escherichia coli* O157:H7, *Listeria monocytogenes* and *Salmonella enterica* serovar Typhimurium have been adapted to stresses such as acid, heat and osmotic pressure by exposure to relatively mild acidity (reviewed in Chapter 6 (section 6.4); Abee and Wouters, 1999; Rowan, 1999; Davidson and Harrison, 2002). It has also been suggested that adaptation to sub-lethal stresses may lead to enhanced virulence (Rowan, 1999).

'Stress-hardening' and cross-protection could have serious implications for the safety of foods that have been minimally processed where levels of lethality may be very close to the boundaries of microbial control. However, most studies showing cross-protection have been undertaken using pure cultures in laboratory media under conditions that do not fully mimic those in foods. When real foods have been investigated, the evidence for cross-protection has been less convincing. For example, no differences in survival of acid-adapted and non-adapted *E. coli* O157:H7 were reported during the processing of beef jerky and in some cases, acid-adapted cells were inactivated more rapidly than non-adapted cells, depending on the formulation of the jerky marinade (Calicioglu *et al.*, 2002). Clearly, more data relevant to real food processing situations are needed before the significance of cross-protection in real foods can be assessed.

In the dairy industry, susceptibility to bacteriophage attack can be reduced by a planned rotation of starter cultures. It is conceivable that development of resistance in spoilage and pathogenic foodborne organisms may be preventable if the food product developer has at his/her disposal a range of antimicrobial compounds and technologies, which can be rotated. However, while technologically feasible, the implications of continually changing formulations and unit operations on production efficiency, costs and product labelling would probably make this approach impractical.

1.6 How the book is organised

In the past 20 years, the vast literature on natural antimicrobial systems has been thoroughly reviewed by several authors who have attempted to identify agents that may have a role to play in food preservation. The aim of this book is *not* to provide yet another encyclopaedic account of all natural antimicrobial compounds ever described in the scientific literature. Many excellent and comprehensive compilations of natural antimicrobial agents have already been published and will not be reproduced here. For the reader interested in a broad sweep of the subject area, books by Dillon and Board (1994) and Naidu (2000) are recommended. In this book, the accent is on practical food preservation based on combinations of natural antimicrobials and mild processing techniques. Where information on practical applications is not readily available, the focus is on the potential for additive and synergistic combinations that might be developed for use in foods in the future.

Chapters 2–4 are devoted to the antimicrobial peptides from lactic acid bacteria, known as the bacteriocins. In Chapter 2, new applications of nisin, the best-known of all the bacteriocins, in combination with physical treatments such as heat, high hydrostatic pressure (HHP) and pulsed electric fields (PEF), and chemical treatments with plant- and animal-derived antimicrobials are reviewed. Chapter 3 is focused on the application of nisin in the decontamination of animal products including beef and poultry carcases, seafood and egg products; innovative ideas for using nisin in non-edible packaging are also reviewed here. Chapter 4 is focused on bacteriocins other than nisin: many of these have been described in the literature but none was used commercially at the time of publication of this book. Chapter 5 presents another microbial metabolite: natamycin, an established antifungal food preservative particularly suited to the preservation of cheese and salami surfaces, as well as beverages.

A very important class of food preservatives, the organic acids, is reviewed in Chapter 6. Recent studies on the potential consequences of acid tolerance/resistance in foodborne microorganisms are discussed, as are applications in foods. This chapter is followed by two reviews on antimicrobials from animals. The first of these (Chapter 7) is a review of the iron chelators lactoferrin and ovotransferrin, the enzymes lactoperoxidase and lysozyme, and the immunoglobulins from milk and eggs. Chapter 8 is focused on chitosan, a polysaccharide derived from the shells of crab and shrimps, with interesting antimicrobial properties.

Three chapters are devoted to antimicrobials from plants. The first of these (Chapter 9) is focused on applications of essential oils from herbs and spices in dairy and meat products, fish, sauces, and salads and dressings. Postharvest treatment of fruits and vegetables using natural compounds including those extracted from plants, animals and microorganisms is the subject of Chapter 10. Combinations of natural antimicrobials such as vanillin and cinnamon with synthetic preservatives and/or ultrasonic or mild heat treatments in fruit purées are considered in Chapter 11.

Possible future approaches to the application of natural antimicrobial compounds on foods are reviewed in Chapter 12 on edible coatings. The potential for combining organic acids, plant essential oils and antimicrobial-containing edible coatings with gamma irradiation is discussed in Chapter 13. The specific application of natural antifungal agents in bakery products is addressed in Chapter 14. The book concludes with a very important chapter on the international regulatory framework that governs all new introductions of food additives, ingredients and processes. Finally, the information presented in a book of this nature will inevitably require frequent updating. To facilitate the process, an Appendix listing a range of useful web site addresses has been included.

1.7 References

ABBEE, T. and WOUTERS, J.A. (1999) Microbial stress responses in minimal processing. *Intl. J. Food Microbiol.* **50**, 65–91.

ADAMS, M. (2000) *Food Microbiology*, 2nd Edn. The Royal Society of Chemistry,

Cambridge, UK. p. 55.

BANKS, J.G., BOARD, R.G. and SPARKS, N.H.C. (1986) Natural antimicrobial systems and their potential in food preservation in the future. *Biotechnol. Appl. Biochem.* **8**, 103–147.

BRUL, S., COOTE, P., OOMES, S., MENSONIDES, F., HELLINGWERF, K. and KLIS, F. (2002) Physiological actions of preservative agents: prospective of use of modern microbiological techniques in assessing microbial behaviour in food preservation. *Intl. J. Food Microbiol.* **79**, 55–64.

BUZBY, J.C. and ROBERTS, T. (1997) Guillain–Barré syndrome increases foodborne diseases costs. *Food Review* **20**, 36–42.

CALICIOGLU, M., SOFOS, J.N., SAMELIS, J., KENDALL, P.A. and SMITH, G.C. (2002) Inactivation of acid-adapted and non-adapted *Escherichia coli* O157:H7 during drying and storage of beef jerky treated with different marinades. *J. Food Prot.* **65**, 1394–1405.

DAVIDSON, P. M. and HARRISON, M. A. (2002) Resistance and adaptation to food antimicrobials, sanitizers and other process controls. *Food Technology*, **56** (11), 69–78.

DILLON, V.M. and BOARD, R.G. (1994) *Natural Antimicrobial Systems and Food Preservation*. CAB International, Wallingford, UK. 328 pp.

FOOD STANDARDS AGENCY (2000) Report of the study of infectious intestinal disease in England. London, The Stationery Office. Summary also available on Food Standards Agency web site: www.food.gov.uk/science/research/research_archive/scientific/intestinal.

GOULD, G.W. and RUSSELL, N.J. (2003) *Food Preservatives*, 2nd Edn. Kluwer Academic/Plenum Publishers, New York.

LEISTNER, L. (1978) Hurdle effect and energy saving. In: *Food Quality and Nutrition*. Ed. W.K. Downey. Applied Science Publishers, London, UK, pp. 553–557.

LEISTNER, L. (1995) Principles and applications of hurdle technology. In: *New Methods of Food Preservation*. Ed. G.W. Gould. Blackie Academic and Professional (Chapman & Hall), London, UK. pp. 1–21.

LEISTNER, L. (2000) Basic aspects of food preservation by hurdle technology. *Intl. J. Food Microbiol.* **55**, 181–186.

MAILLARD, J.-Y., ED. (2002) Antibiotic and biocide resistance in bacteria: perceptions and realities for the prevention and treatment of infection. *J. Appl. Microbiol. Symp. Suppl.* **92**. 173 pp.

MEAD, P.S., SLUTSKER, L., DIETZ, V., MCCAIG, L.F., BRESEE, J.S., SHAPIRO, C., GRIFFIN, P.M. and TAUXE, R.V. (1999) Food-related illness and death in the United States. *Emerging Infect. Dis.* **5**, 607–625 (also available free-of-charge at http://www.cdc.gov).

MOKEN, M.C., MCMURRY, L.M. and LEVY, S.B. (1997) Selection of multiple- antibiotic-resistant (mar) mutants of *Escherichia coli* by using the disinfectant pine oil: Roles of the mar and acrAB loci. *Antimicrob. Agents Chemother.* **41**, 2770–2772.

NAIDU, A. S. (2000) *Natural Food Antimicrobial Systems*. CRC Press, Boca Raton, FL. 818 pp.

ROLLER, S. and BOARD, R.G. (2003) Naturally occurring antimicrobial systems. In: Food Preservatives, 2nd Edn. Eds. G.W. Gould and N.J. Russell. Kluwer Academic/Plenum Publishers, New York.

ROWAN, N.J. (1999) Evidence that inimical food-preservation barriers alter microbial resistance, cell morphology and virulence. *Trends Food Sci. Technol.* **10**, 261–270.

SCHLUNDT, J. (2002) New directions in foodborne disease prevention. *Intl. J. Food Microbiol.* **78**, 3–17.

TAUXE, R.V. (2002) Emerging foodborne pathogens. *Intl. J. Food Microbiol.* **78**, 31–41.

TOLLEFSON, C. (1995) Stability preserved. *Chem. Mark. Rep.* **29** May, SR28–SR31.

WHO (2000) Overcoming antimicrobial resistance. *World Health Organization Report on Infectious Diseases.* WHO, Geneva.

WHO (2001) Annex Table 2. Deaths by cause, sex and mortality stratum in WHO regions. Estimates for 2000. *The World Health Report* 2001. WHO, Geneva.

2

Nisin in multifactorial food preservation

M. Adams, University of Surrey, UK and E. Smid, NIZO, The Netherlands

2.1 Introduction

Nisin was first discovered in 1928, when its production in milk stored overnight prior to cheese making led to inhibition of a *Lactobacillus* starter culture. Since that time, it has emerged as a compound that combines fascinating properties at the molecular level with significant practical value. It is not the purpose of this chapter to provide a comprehensive account of the history, properties and uses of nisin. These topics have been reviewed periodically and the reader is referred to a particularly useful and thorough account that appeared in 2000 (Thomas *et al.*, 2000). In the present chapter we will confine ourselves to a brief summary of nisin and its activity, and focus primarily on its role as part of a multifactorial approach to food preservation.

2.2 Structure and biosynthesis

Ribosomally synthesised bacterial proteins or peptides which can kill or inhibit other bacteria are described as bacteriocins. Nisin is classified as a Class Ia bacteriocin or lantibiotic (Klaenhammer, 1993). It contains 34 amino acids and is produced by certain strains of the bacterium *Lactococcus lactis*. It is synthesised as a precursor peptide, which is then subject to post-translational modification. This takes the form of dehydration of serine and threonine residues to generate didehydroalanine and didehydroaminobutyric acid followed by the stereospecific addition of cysteine sulfhydryl groups to the α, β-double bonds produced. The unusual thioether-containing amino acids thus formed, lanthionine and β-methyllanthionine, introduce five rings in the final structure.

The modified peptide is then exported from the cell where the C-terminal leader peptide is cleaved, releasing mature nisin. Two natural variants of nisin are known to occur (Mulders *et al.*, 1991). These are designated nisin A and nisin Z and differ in the amino acid at position 27: asparagine in the case of nisin A and histidine in nisin Z.

2.3 Properties

The preponderance of basic amino acids such as lysine and histidine confers a positive charge on the nisin molecule. It is insoluble in non-polar solvents and its water solubility increases as the pH is decreased. Limited solubility is not a practical problem, however, since, even at neutral pH, its solubility exceeds the levels usually necessary for activity in foods. A physical property of some significance is the molecule's considerable heat stability, particularly at acidic pH values. Optimal stability occurs around pH 3.0 where less than 5% activity was reported to be lost after heating at 115 °C for 20 minutes (activity losses at pH 2.0 and 4.0 were 28.5% and 21.4% respectively) (Davies *et al.*, 1998).

In the literature, nisin activity levels have been expressed in several different ways based on IU (international units), weight of pure nisin or weight of the commercial preparation Nisaplin®. To minimise confusion national and international legislators tend to specify addition levels as 'mg pure nisin per kg'. The activity of pure nisin is approximately 40×10^6 IU/g. Nisaplin® has a standardised activity of 1×10^6 IU/g, so that 1 μg Nisaplin® is equivalent to 1 IU.

2.4 Spectrum of activity and mode of action

Nisin differs from most bacteriocins in that its spectrum of activity is not confined to closely related species. Most vegetative Gram-positive bacteria are inhibited to some extent and bacterial endospores show a marked sensitivity. Gram-negative bacteria are resistant to nisin. Their protective outer membrane excludes the free passage of molecules exceeding 700 Da and nisin (MW 3353 Da) is therefore unable to reach its site of action (see below). Yeasts and moulds are also unaffected by nisin.

In vegetative bacteria, nisin acts at the cytoplasmic membrane. It interacts electrostatically with membrane phospholipids to produce transient, non-selective, pores. Nisin has also been shown to exhibit a high affinity for Lipid II, the universal carrier of cell wall peptidoglycan components, and this accounts for its high activity in the nanomolar range when compared with some other cationic polypeptides (Wiedemann *et al.*, 2001). The rapid efflux of ions, amino acids and ATP through the pores results in collapse of the transmembrane protonmotive force and cell death.

Surprisingly in view of its more widespread practical use against spores, the biochemical basis of this activity is less well understood. Its effect, which is normally sporistatic, is enhanced under acidic condition or if the spores have been heat injured. Spore germination is unaffected but post-germination swelling and subsequent spore outgrowth are prevented. The observation that replacement of the dehydroalanine at position 5 with alanine reduced nisin activity against spores but not vegetative cells suggests that dehydroalanine may act as a nucleophile in a reaction with a spore-associated factor essential for outgrowth.

Nisin sensitivity is quite variable. Significant differences have been noted in the sensitivity of *Clostridium botulinum* spores compared with *Cl. butyricum* spores and between mesophilic and thermophilic spores; in both cases the former being markedly more resistant (Ramseier, 1960; Jarvis, 1967). Reduced sensitivity to nisin can be conferred by the production of a specific nisin-inactivating enzyme, nisinase. This has been described for a number of organisms including several *Bacillus* species where it has been shown to reduce the C-terminal dehydroalanyl lysine to alanyl lysine (Jarvis, 1967, 1970; Jarvis and Farr, 1971).

In vegetative bacteria particular attention has focused on nisin's activity against the Gram-positive pathogen *Listeria monocytogenes.* Differences in sensitivity have been noted between strains of this organism (Fereira and Lund, 1996; Ukuku and Shelef, 1997) and acquisition of resistance has been shown to occur (Harris *et al.*, 1991; Ming and Daeschel, 1993; Davies and Adams, 1994). Changes in the cell envelope have been associated with increased resistance. Differences between nisin-sensitive and nisin-resistant strains in the lipid composition of their cytoplasmic membranes have been reported. These tend to confer lower membrane fluidity in resistant strains making it more difficult for nisin to insert and form pores (Ming and Daeschel, 1995; Maisnier-Patin and Richard, 1996; Bennik *et al.*, 1997; Mazzotta and Montville, 1997; Verheul *et al.*, 1997). The cell wall also appears to be involved. Cells of nisin-sensitive strains exhibit greater hydrophobicity, a different sensitivity to cell wall-acting compounds (Ming and Daeschel, 1995; Davies *et al.*, 1996; Crandall and Montville, 1998) and removal of the cell wall has been shown to abolish the difference in nisin sensitivity between strains (Davies *et al.*, 1996).

2.5 Current uses

Since the first production of a commercial preparation in 1953, nisin has been used increasingly widely as a food preservative. It is now approved in more than 80 countries, including the United States, where it has Generally Recognized As Safe (GRAS) status, Malaysia, Saudi Arabia, China, India, Australia, New Zealand and the countries constituting the European Union (Danisco, 2002). Most applications exploit nisin's activity against spore-forming bacteria such as *Bacillus* and *Clostridium* in, for example, clotted cream, canned foods, processed cheeses and pasteurised liquid egg (Table 2.1). Even within this

Table 2.1 Examples of food applications for nisin

Type of food	Nisaplin® addition level (mg/kg or litre)*	Typical target organisms
Processed cheese	100–600	*Bacillus* spp. *Clostridium* spp.
Milk and milk products	10–50	*Bacillus* spp. including *B. sporothermodurans*
Pasteurised chilled dairy desserts	75–200	*Bacillus* spp. *Clostridium* spp.
Liquid egg	50–200	*Bacillus* spp. including *B. cereus*
Pasteurised soups	100–250	*Bacillus* spp.
Crumpets	150–200	*Bacillus cereus*
Fruit juice (pasteurised, stored at ambient temperature)	30–60	*Alicyclobacillus acidoterrestris*
Canned food	100–200	*Bacillus stearothermophilus, Clostridium thermo-saccharolyticum, Cl. botulinum*
Dressings and sauces	50–200	Lactic acid bacteria, *Brochothrix thermosphacta, Listeria mono-cytogenes*
Mascapone cheese	<400	*Clostridium botulinum*
Meat products such as Bologna, frankfurter sausages	200–400	Lactic acid bacteria *Brochothrix thermosphacta,* *Listeria monocytogenes*
Ricotta cheese	100–200	*Listeria monocytogenes*
Beer, wine and spirits production		Lactic acis bacteria e.g. *Lactobacillus, Pediococcus*
Pitching yeast wash	1000–1500	
Reduced pasteurisation	10–50	
During fermentation	25–100	
Post-fermentation	10–50	

* 1 mg Nisaplin® is equivalent to 1000 IU.
Adapted from Danisco (2002).

relatively restricted area, new applications continue to be found in response to developments in the way foods are produced, processed and stored. For example, nisin has been shown to be effective against spores of *Alicyclobacillus acidoterrestris,* a spoilage organism in pasteurised fruit juices first described in 1984 (Komitopoulou *et al.*, 1999; Silva and Gibbs, 2001).

2.6 New applications and the multifactorial approach

Nisin is not toxic for humans and is produced by a food-grade microorganism. It is, therefore, widely perceived as a natural food preservative. Supported by consumer demand for more 'natural', less heavily processed and preserved foods, the natural features of nisin have boosted the interest of industry in extending the use of nisin beyond traditional areas. One potentially fruitful approach to this would be its employment as part of a multiple barrier system of food preservation (Gould, 2000; Leistner, 2000).

The so-called 'hurdle' concept was first described about 30 years ago, although its practical use in a range of traditional products pre-dates this considerably. Essentially, a number of inhibitory/preservative factors are combined to give a useful product shelf-life but without the adverse sensory changes that might result from the use of fewer factors to achieve a similar effect. At the microbial level, individual bacteria in a food matrix encounter several inhibitory factors and must respond to each in order to survive and/or grow. These responses may involve homeostatic mechanisms to maintain internal pH or osmolarity or the repair of damage inflicted by processes such as heat or high pressure. In each case it will require the diversion of cellular energy from growth-associated processes. If this is coupled with additional preservative factors which restrict the organism's ability to generate energy, such as storage at low temperature or under anaerobic conditions, then this reinforces the aggregate effect, further limiting microbial activity. Combining nisin with compounds or processes that eliminate cellular barriers to it can potentiate its activity in cells, which would otherwise be resistant.

Nisin is already employed as part of a multifactorial preservative system in a number of instances, but there is considerable interest in exploring other interactions of potential practical value. Numerous research groups are active in this area. The European Nisin*Plus* project (FAIR CT96-1148) is an example of a consortium that has explored the application of nisin in mild food preservation. The consortium involved five European food companies, four research institutes and two universities joining forces to explore the use of combined applications of nisin with other natural preservatives or mild, mainly non-thermal, physical treatments to develop new preservation strategies for minimally processed vegetables, *sous-vide* processed products, delicatessen salads, marinated meat products, Bologna-type sausages and fermented dairy products. In what follows we will describe examples of the work arising from that project as well as results from the numerous other workers active in this field.

2.7 Physical treatments

2.7.1 Thermal treatments

Thermal processing is by far the most widely used and effective strategy for inactivating foodborne microorganisms and many current applications of nisin employ the additive effect of heat plus nisin. Though primarily added to inhibit the outgrowth of spores surviving the heat process, the presence of nisin during heating has been shown to enhance the process lethality for spores of *B. cereus* and *A. acidoterrestris* (Beard *et al.*, 1999; Komitopoulou *et al.*, 1999; Wandling *et al.*, 1999; Penna and Moraes, 2002).

Similar observations have been made with vegetative cells of Gram-positive bacteria. Nisin increased the heat sensitivity of *L. monocytogenes* in cold pack lobster meat, allowing a milder heat process and reduced loss in drained weight (Budu-Amoako *et al.*, 1999). Nisin has also been shown to act synergistically with mild heat treatments between 48 and 56 °C on *Lactobacillus plantarum* (Ueckert *et al.*, 1998). A marked synergy was seen between nisin and heat against a nisin-resistant mutant of *L. monocytogenes*, although this was absent with the nisin-sensitive wild-type cells (Modi *et al.*, 2000). This suggests that, in nisin-resistant mutants, the heat process overcame the resistance barrier to nisin activity and could thus prove a valuable way of avoiding any problems of acquired nisin resistance in *L. monocytogenes*.

In Gram-negative cells, nisin resistance is conferred by the outer membrane of the cell envelope. Any process or treatment that disrupts this permeability barrier and allows nisin access to its site of action at the plasma membrane will sensitise the cells to its activity. Nisin has been shown to increase the heat sensitivity of a number of Gram-negative bacteria although the effect was variable between organisms and often required relatively high concentrations of nisin (Kalchayanand *et al.*, 1992; Boziaris and Adams, 2001). For *Salmonella* Enteritidis PT4 the effect was small and was much reduced in food systems such as whole egg and egg white (Boziaris *et al.*, 1998). Determination of the extent of outer membrane injury in survivors suggested that the cells succumbing to the presence of nisin during heating were those whose outer membrane was injured but that would otherwise have survived the heating process. Sensitivity to nisin was transient with surviving cells able to rapidly reconstitute the integrity of the outer membrane barrier and display their normal nisin resistance. More prolonged nisin sensitivity has been seen after heating *E. coli* O157:H7 at 55 °C for 15 minutes but not with shorter heating times and lower temperatures (Lee *et al.*, 2002).

Similar transient sensitivity has been seen with low temperature shocks such as chilling and freezing (Kalchayanand *et al.*, 1992; Boziaris and Adams, 2000, 2001). As with heating, cells were only sensitive if nisin was present during stress and resistance recovered rapidly afterwards (Boziaris and Adams, 2001). It appears that thermal shocks produce a transient disruption of the outer membrane, allowing nisin access. After treatment, the permeability barrier is rapidly restored by a process which apparently involves reorganisation rather than biosynthetic repair.

Chill storage (6.5 °C) in peptone water containing nisin (100 IU/ml) over 14 days has been shown to produce an additional 1.5–2.0 log CFU/ml reduction in numbers of *Salmonella* Typhimurium and *Escherichia coli*, suggesting that deterioration in the outer membrane barrier function occurs under these conditions. The cells were stored in peptone water and the effect could be completely reversed by the addition of 25 mM Mg^{2+} or 100 mM K^+ or Na^+, which suggests such effects may not occur in more complex media such as foods (Elliason and Tatini, 1999).

2.7.2 High hydrostatic pressure

The ability of high hydrostatic pressures (>100 MPa) to inactivate micro-organisms has been known since the beginning of the 20th century, although it was not until the 1980s that it was employed in the production of commercial food products. High pressures promote reactions where there is an overall decrease in volume, and they act primarily on non-covalent bonds. Thus in microbial cells, high pressure tends to induce changes such as protein denaturation and the compression and altered permeability of membranes (Adams and Moss, 2000). Such changes may well facilitate the activity of nisin as has been observed for vegetative cells using fluorescent probes for outer membrane permeability (Ganzle and Vogel, 2001).

Both Gram-positives such as *L. monocytogenes* and Gram-negatives such as *E. coli* and *Salmonella* Typhimurium showed increased levels of inactivation when nisin was present during pressure treatment (Kalchayanand *et al.*, 1994) and a more recent study has also reported a slight effect with the yeast *Saccharomyces cerevisiae* at reduced temperature (Ter Steeg *et al.*, 1999). Studies with several members of the Enterobacteriaceae, *Pseudomonas fluorescens* and *Staphylococcus aureus* have confirmed this effect and indicated that the changes that confer nisin sensitivity are transient since none of the organisms tested was nisin-sensitive before or after the treatment (Hauben *et al.*, 1996; Masschalck *et al.*, 2001). This synergistic effect is most pronounced in relatively simple buffer systems. More complex food materials have been shown to exert a strong protective effect, although a synergy between high pressures and nisin was still apparent in trials using milk, goat's cheese and mechanically recovered poultry meat (Garcia-Graells *et al.*, 1999; Capellas *et al.*, 2000; Yuste *et al.*, 2000). Despite the mitigating effect of real food systems, a combination of high hydrostatic pressure and nisin has been shown to achieve 5–6 log CFU/ml reductions in numbers of *Listeria innocua* and *E. coli* in liquid whole egg and offers a non-thermal pasteurisation procedure which will conserve the physical properties of the product (Ponce *et al.*, 1998).

Spores are extremely resistant to elevated pressure, although a synergistic interaction between nisin and pressurisation has been reported (Roberts and Hoover, 1996). This was demonstrated by inoculating media containing nisin with a spore suspension following pressure treatment. In this way it was shown that 400 MPa for 30 minutes at 70 °C in pH 4.0 buffer achieved at least a 6 log CFU/ml reduction in spores of *Bacillus coagulans* when pressurisation was followed by plating onto media containing 0.8 IU/ml nisin. This indicates that the surviving spores had been sub-lethally injured by the pressure treatment, making them more sensitive to nisin. It is possible that even greater lethality may be achieved if nisin is present during the pressure treatment to inactivate spores acquiring more transient injury.

2.7.3 Pulsed electric fields (PEF)

Use of the heat generated by electric fields to inactivate microorganisms is well established in the technologies of ohmic and microwave heating. The use of shorter duration high-voltage pulses for non-thermal inactivation is, however, more recent (Jeyamkondan *et al.*, 1999; Gould, 2000). Inactivation and injury are the result of the permeabilisation of the cell membrane caused by the concentrated voltage gradient which is established across the membrane. At lower voltages this makes cells leaky and is the basis of the electroporation technique used in the laboratory to introduce genetic material into cells.

Since, like high hydrostatic pressure, PEF produces changes in membrane structure and integrity, it is likely that some interaction will be apparent when its use is combined with a membrane-active antimicrobial such as nisin. Synergy has been reported in the inactivation of vegetative cells of *B. cereus* when a low dose of nisin (0.06 μg/ml) and mild PEF (16.7 kV/cm, 50 pulses each of 2 μs) gave a reduction 1.8 log CFU/ml greater than the sum of the reductions obtained with single treatments (Pol *et al.*, 2000). Similarly a combined nisin and PEF treatment of *Micrococcus luteus* cells in buffer gave a reduction 1.4 log CFU/ml higher than the sum of the individual treatments (Dutreux *et al.*, 2000). In this work, similar overall reductions were obtained regardless of whether nisin was present during the PEF treatment or added after. This suggests that residual injury sensitising the cells to nisin persisted after PEF treatment. Although no evidence of sub-lethal injury was apparent from counts on selective media, differences in the osmotic sensitivity of the survivors suggest that survivors carried some membrane injury not detectable by the media used.

Gram-negative bacteria also display nisin sensitivity when combined with PEF. A 4 log CFU/ml reduction in numbers of *E. coli* O157:H7 was obtained with PEF (12.5 kV/cm) alone, but inactivation increased by a further log CFU/ml in the presence of nisin (Kalchayanand *et al.*, 1994). In a simulated milk ultrafiltrate medium, *E. coli* numbers decreased by 4 log CFU/ml in a combination of relatively high levels of nisin (500 and 1000 IU/ml) and three to five pulses of 11.25 kV/cm (Terebiznik *et al.*, 2000). Evidence was found of nisin inactivation during the treatment, which may have been due to interaction of nisin with components leaking from inactivated cells. The same authors have also produced a response surface model of the interaction between PEF, nisin and water activity adjusted with NaCl (Terebiznik *et al.*, 2002).

The effect of the food matrix on the efficacy of the PEF/nisin combination has been investigated using *B. cereus* cells in diluted skimmed milk. In individual treatments, the milk proteins did not appear to interfere with inactivation caused by the PEF, although more nisin was needed to achieve similar reductions to those seen in buffer. The synergy between the two treatments was still apparent, albeit reduced, in milk, presumably due to the reduced effectiveness of the nisin under these conditions (Pol *et al.*, 2001a). Using *E. coli* O157:H7 in apple juice, cell death was found to increase with both temperature and electric field strength over the experimental values used (20, 30 and 42 °C and 60, 70 and 80 kV/cm), achieving a maximum reduction of 5.35 log CFU/ml. This exceeded the 5 log CFU/ml reduction recommended by the US FDA (Food and Drug Administration) for unprocessed packaged juices, but could be improved by a further 3 log CFU/ml when combined with nisin (Iu *et al.*, 2001). A similar study looking at survival of *Salmonella* Typhimurium in orange juice found greater resistance than seen with *E. coli* in apple juice. For example, 90 kV/cm, a pulse number of 20 and 45 °C did not have a notable effect on viability, although at temperatures of 46 °C and above inactivation increased considerably achieving a 5.9 log CFU/ml reduction at 55 °C and 50 pulses at 90 kV/cm (Liang *et al.*, 2002). Addition of nisin increased the level of inactivation, although this effect was found to differ considerably between freshly squeezed orange juice and pasteurised juice, suggesting that relatively subtle differences in physical and chemical properties of the medium can exert a significant effect.

Spores are considerably more resistant to PEF. Germinating spores, however, are sensitive to both nisin and PEF. Spores of *B. cereus* became sensitive to nisin almost immediately after germination but were not inactivated by PEF until 50 minutes after germination had commenced. No additional inactivation was seen when nisin and PEF treatments were combined, which may in part be due to the difference in time scale of sensitisation to the different factors (Pol *et al.*, 2001b).

2.8 Microbiological treatments

The preservative effect of fermentation by lactic acid bacteria (LAB) is well established. The combination of antimicrobial factors which operates in the production of traditional lactic acid fermented foods such as salami or cheese, can itself be regarded as an archetype of the multifactorial concept (Adams, 2001).

Lactic acid is the principal antimicrobial agent produced by LAB. In lactic fermented rice, production of nisin by *Lactococcus lactis* during fermentation did not make any appreciable additional contribution to the inhibition of *Staph. aureus* (Yusof *et al.*, 1993). Where a target microorganism displays particular sensitivity to nisin, then *in situ* production can be of benefit and was described more than 50 years ago as a way of inhibiting the spores of gas-producing clostridia in Gruyere cheese production (Hirsch *et al.*, 1951).

Studies using fluorescent probes have indicated that lactic acid can make the outer membrane of Gram-negative bacteria permeable, suggesting that this could render them sensitive to nisin (Alakomi *et al.*, 2000). Nisin plus lactic acid has been shown to give increased inhibition of a number of bacterial isolates from fish, including the Gram-negative *P fluorescens* and *P. aeruginosa* using a plate inhibition assay (Nykänen *et al.*, 1998). When *E. coli* was added to broths fermented with a nisin-producing *Lactococcus lactis* and with a non-nisin producer, the viable population of *E. coli* declined slightly faster in the nisin-containing broth but still only decreased by about 1 log CFU/ml over 36 hours (Boziaris and Adams, 1999). In vacuum packed beef the addition of nisin (200 IU/g) did not enhance the reduction of *E. coli* O157:H7 when added in combination with lactic acid and polylactic acid (Mustapha *et al.*, 2002; see also Tables 3.2–3.4 for more examples).

The problem posed by the acquisition of nisin resistance in organisms such as *L. monocytogenes* has already been described. One strategy to overcome this would be the use of combinations of bacteriocins, and several authors have reported synergies between nisin and other bacteriocins (Hanlin *et al.*, 1993; Schillinger *et al.*, 1996; Mulet-Powell *et al.*, 1998; Bouttefroy and Millière, 2000). Since nisin is the only bacteriocin explicitly approved for food use, other bacteriocins could only be introduced as a result of their production in a food by bacteriocinogenic cultures (Schillinger *et al.*, 1998). This approach has been described as a means of preventing the outgrowth of nisin-resistant *L. monocytogenes* in tofu (Schillinger *et al.*, 2001). It was found that lower amounts of nisin could be used when combined with bacteriocin-producing strains of *Enterococcus faecium* or *Lactococcus lactis* and these achieved complete suppression of *Listeria* growth in tofu stored at 10°C for 1 week.

Ethanol can be present in fermented foods as a result of the activity of heterofermentative LAB or yeasts – both groups of organisms sharing a common ecological niche. Although ethanol levels of 5% (v/v) had no effect on survival of *L. monocytogenes*, its ability to disrupt biological membranes has been shown to enhance inactivation of *L. monocytogenes* by nisin (Brewer *et al.*, 2002).

The combined use of nisin and a listeriophage in broth considerably reduced the number of listeria cells and no regrowth was seen. However, this effect was not apparent when the same technique was tried in food systems such as raw beef (Dykes and Moorhead, 2002).

2.9 Chemical treatments

There are numerous reports of studies investigating the use of nisin in conjunction with other chemical preservatives. These have focused particularly on preservatives that, like nisin, can be perceived as natural. In general, as with the physical treatments, the best results have been obtained with additives that are membrane active.

2.9.1 Plant-derived antimicrobials

Plants produce a range of chemicals with antimicrobial activity: some are natural constituents of the plant, others are produced in response to physical injury which allows an enzyme contact with its substrate, and some (phytoalexins) are produced in response to microbial invasion. In addition to the protective role they play in the living plant, their potential as antimicrobials in foods has been a longstanding research topic subject to regular review (Shelef, 1983; Beuchat, 1994; Walker, 1994; Nychas, 1995; Smid and Gorris, 1999; see also Chapters 9–11).

When garlic, *Allium sativum*, is crushed, alliin (*S*-allyl-L-cysteine-*S*-oxide) is converted to allicin (2-propenyl-2-propenethiol sulfinate) which has pronounced inhibitory activity against a range of bacteria and fungi (Beuchat, 1994). It readily permeates through phospholipid membranes and is thought to act by reacting with critical thiol groups in the cell, affecting several physiological processes including respiration and RNA synthesis (Feldberg *et al.*, 1988; Miron *et al.*, 2000). Synergy between nisin and an aqueous garlic extract in the inhibition of a number of strains of *L. monocytogenes* has been described (Singh *et al.*, 2001). The garlic extract was not found to be appreciably bactericidal though it enhanced the bactericidal action of nisin. In a food system (hummus), sub-minimum inhibitory concentrations of nisin and garlic were effective at preventing listerial growth and did enhance the slight bactericidal effect of the nisin.

The essential or volatile oils of plants are cocktails of compounds, which play an important role in flavouring foods and beverages and often display appreciable antimicrobial activity. The active components are usually terpenoids and, with the exception of cinnamaldehyde, are largely phenolics such as thymol, carvacrol and eugenol (Fig. 2.1). They act primarily at membranes, with the phenols acting as proton translocators collapsing the protonmotive force (Juven *et al.*, 1994; Helander *et al.* 1998; Ultee *et al.*, 1999, 2002; see also Chapter 9).

A pronounced synergy between nisin and carvacrol has been seen at both bacteristatic and bactericidal levels against *B. cereus* and *L. monocytogenes.* At concentrations higher than its minimum inhibitory concentration (MIC), carvacrol was not bactericidal but considerably enhanced the bactericidal effect of nisin, producing an additional reduction in numbers of *L. monocytogenes* of

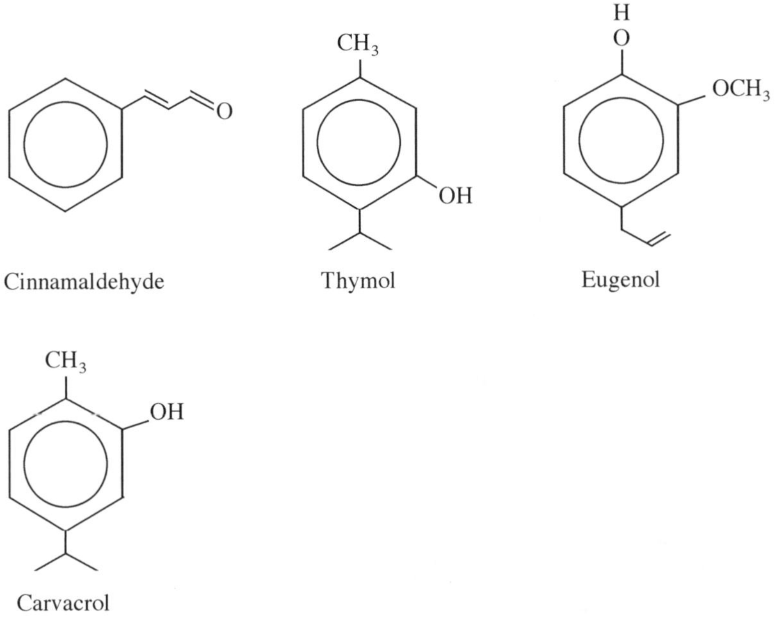

Fig. 2.1 Essential oil components with antimicrobial activity.

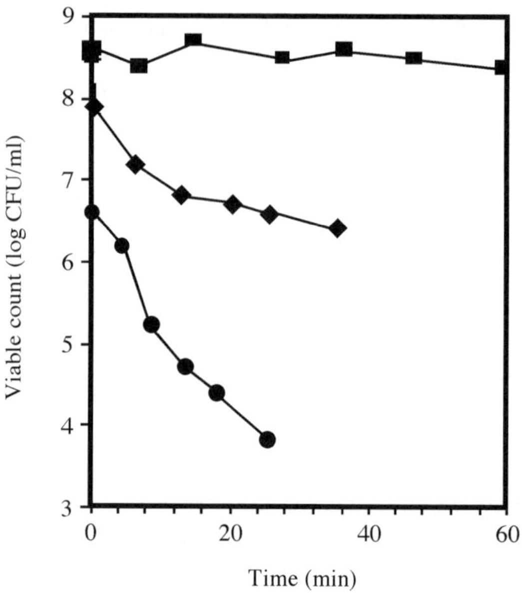

Fig. 2.2 Combined action of nisin and carvacrol on *L. monocytogenes* precultured and treated at 20 °C. Carvacrol (squares) was added to the cell suspension at 1.3 mM and nisin (diamonds) at 5.3 μg/ml. The same concentrations were used in the combined treatment (circles). All data represent means of triplicate measurements. The detection limit was 3.5 log CFU/ml. Data taken from Pol and Smid (1999).

almost 3 log CFU/ml (Fig. 2.2). Thymol and carvone were said to produce similar results (Pol and Smid, 1999). Similar results have been reported subsequently by other workers. The activity of nisin Z against *L. monocytogenes* and *B. subtilis* was greatly potentiated in broth by sub-inhibitory concentrations of thymol (Ettayebi *et al.*, 2000). Similarly, the activity of nisin against heated cells of *B. cereus* was enhanced when combined with thymol or carvacrol, though neither phenol had much effect when applied alone (Periago *et al.*, 2001). Incubation of cells of *B. cereus* with carvacrol sensitised the cells to later exposure to nisin but not to PEF treatment or combined nisin/PEF treatment (Pol *et al.*, 2001b).

In the Gram-negative *Salmonella* Typhimurium and *E. coli* O157:H7, carvacrol and thymol disrupt the outer membrane, thereby releasing outer membrane-associated materials into the medium (Helander *et al.*, 1998). This suggests that these compounds could also help sensitise Gram-negative bacteria to nisin.

Sorbic and benzoic acids both occur naturally in plant materials, but the synthetic materials have found wider application as food preservatives. The use of combinations of nisin with either sorbate or benzoate to preserve spice and dried bean curd has been described (Fang *et al.*, 1997). The results did not record the effect of the individual preservatives although combinations of very high levels of nisin (5000 IU/g) and sorbate (3%) or benzoate (0.12%) did produce significant reductions in the numbers of *B. cereus* and *Staph. aureus* over 14 days' storage at chill temperatures. Sorbate and nisin have also been shown to establish and maintain substantially reduced numbers of *L. monocytogenes* on vacuum-packed and CO_2-packed beef over four weeks' refrigerated storage (Avery and Buncic, 1997; see also Chapter 3).

2.9.2 Animal-derived antimicrobials

Lactoperoxidase, an enzyme found in the secretions of mammalian species, mediates the oxidation of thiocyanate, iodide or bromide by hydrogen peroxide to generate reactive products with a high antimicrobial activity (Ekstrand, 1994; Kussendrager and van Hooijdonk, 2000). Microbial inhibition or inactivation is a result of oxidation of sulfhydryl groups in enzymes and other key proteins, which leads to structural damage to the cytoplasmic membrane and leakage of potassium ions, amino acids and peptides into the surrounding medium.

A US Patent describes the use of a combination of an antimicrobial peptide, such as nisin, with the lactoperoxidase system at pH 3–5 and temperatures of 30–40 °C to inhibit *Salmonella* species (Bycroft *et al.*, 1991). The subsequent published work has, however, largely focused on inhibition of the Gram-positive species *L. monocytogenes*. Combinations of nisin and lactoperoxidase in skim milk showed a synergistic bactericidal effect against *L. monocytogenes*. Nisin alone at 10 or 100 IU/ml had no effect on *L. monocytogenes* counts after 34 hours at 30 °C, while lactoperoxidase produced a reduction of 3 log CFU/ml. In combination, however, a 5.6 log CFU/ml reduction was seen (Zapico *et al.*, 1998). Very similar results were reported by Boussouel and co-workers who produced a response surface model describing the interaction and also noted that the effect persisted for 15 days without the regrowth of bacteria which was seen when either preservative was used alone (Boussouel *et al.*, 1999, 2000). Both groups observed that increased inactivation could be obtained if the nisin and the lactoperoxidase were added at different times, although they differed over the order of the additions and their precise timings.

In situ production of nisin in raw milk enhanced the lethality of the lactoperoxidase system against *L. monocytogenes,* although the effect was small (< 2 log CFU/ml), probably reflecting limited nisin production at the low incubation temperatures (4 and 8 °C) used (Rodriguez *et al.*, 1997).

Lysozyme is a widely occurring enzyme that hydrolyses the glycosidic bond between *N*-acetylmuramic acid and *N*-acetylglucosamine in peptidoglycan. This polymer is present in the cell wall of most Eubacteria and is responsible for its mechanical strength and shape. Consequently lysozyme acts on many Gram-positive bacteria where peptidoglycan is accessible but is inactive against Gram-negative bacteria. Lysozyme is produced commercially from hen's egg white and is used to control late blowing in some cheeses by inhibiting the outgrowth of *Clostridium tyrobutyricum* (Tranter, 1994). Though it does not act at the cell membrane, synergy between lysozyme and nisin has been reported against *L. monocytogenes* and food spoilage lactobacilli (Monticello, 1989; Chung and

Hancock, 2000). This may be the result of lysozyme facilitating access of nisin to the cell membrane or inhibition of the repair of lysozyme damage as a result of nisin draining the cell's energy reserves.

Combinations of nisin and lysozyme have been tried in food systems. Applied as a surface treatment with ethylenediaminetetraacetic acid (EDTA) to ham and bologna, nisin/lysozyme produced an initial reduction in Gram-positive bacteria and controlled their growth over 4 weeks' storage at 8 °C, although results varied with the particular species studied and between meats. Inhibition of *Salmonella* and *E. coli* O157:H7 was also noted under some circumstances (Gill and Holley, 2000a, 2000b; see also Chapter 3). Synergy between nisin and lysozyme was also seen against a *Carnobacterium* species and *Brochothrix thermosphacta* on cores of lean and fatty pork (Nattress *et al.*, 2001).

2.9.3 Fatty acids and their derivatives

The antibacterial activity of free fatty acids (>10 carbons) is restricted to Gram-positive bacteria and is largely associated with the undissociated molecule rather than the anion. Antimicrobial activity is lost when the acid is esterified with a monohydric alcohol but is retained if a polyhydric alcohol is used, when the product retains some hydrophilic character. Fatty acids and their esters are largely inactive against Gram-negative bacteria in normal circumstances, although *Vibrio* species are reported to be sensitive (Kabara, 1993). Owing to their amphiphilic properties, they act at the plasma membrane and therefore would be expected to have some influence on nisin activity against Gram-positive bacteria. This has been confirmed by several studies. A 1989 patent describes how a mixture of nisin and monolaurin was 10 000 times more active against *Streptococcus agalactiae* in a milk-based medium than the individual components (Blackburn *et al.*, 1989). A synergistic bactericidal effect between nisin and monolaurin in milk was also seen against the vegetative cells of four *Bacillus* species (Mansour and Millière, 2001). Sucrose fatty acid esters have also been shown to enhance nisin activity against a number of Gram-positive bacteria but did not sensitise Gram-negatives (Thomas *et al.*, 1998).

In food systems, surfactants may also potentiate nisin activity by restricting its adsorption/inactivation by food components. Jung and coworkers found that nisin activity decreased with the fat content of milk but could be partially counteracted by addition of Tween 80 (Jung *et al.*, 1992). Lauric acid and nisin impregnated film showed a potent synergy against *L. monocytogenes* in broth medium, although results on refrigerated turkey bologna were less encouraging where lauric acid alone produced the greatest reduction in numbers (Dawson *et al.*, 2002; see also Chapter 3).

2.9.4 Carbon dioxide

The preservative effect of carbon dioxide is widely used in vacuum and modified atmosphere packing. Although the mechanism of its action remains to be fully characterised, its ability to interact with the cell membrane appears to be an important factor and one which might lead to a positive interaction with nisin (Stanbridge and Davies, 1998). This has been demonstrated in model systems where CO_2 showed a pronounced synergy with nisin against *L. monocytogenes* and in a number of food systems such as beef, cooked pork, pizza and cold smoked salmon (Fang and Lin, 1994; Nilsson *et al.*, 1997, 2000; Cabo *et al.*, 2001a; Tu and Mustapha, 2002). No interaction was seen between nisin and CO_2 in water-soluble fish muscle extracts (Cabo *et al.*, 2001b).

2.9.5 Chelating agents and nisin against Gram-negative bacteria

A recurring theme in this chapter has been the ability of membrane-active compounds to enhance the effect of nisin against Gram-positives and the relative resistance of Gram-negative bacteria. Procedures that disrupt the outer membrane of the Gram-negative cell, even if only transiently, may permit nisin access and thus render the cells sensitive. Several examples of this effect involving physical treatments have already been noted above.

The stability of the outer membrane is maintained by ionic interactions between Ca^{2+} and Mg^{2+} and phosphate residues on the core polysaccharides of adjacent lipopolysaccharide molecules. Removal of these ions by chelators such as EDTA is known to make the outer membrane leaky. EDTA and nisin combinations produced reductions of 3.2–6.9 log CFU/ml in the number of survivors of several Gram-negative bacteria, including *Salmonella*, whereas individually they had no effect (Stevens *et al.*, 1991). A similar effect was seen with other chelating agents such as citrate (Stevens *et al.*, 1992; Cutter and Siragusa, 1995a; see also Table 3.1). *Arcobacter butzleri* was inhibited by EDTA alone, but addition of nisin gave a significant improvement in some instances (Phillips and Duggan, 2001). Sensitisation of Gram-negatives to nisin in the presence of EDTA was not seen, however, when nisin was generated *in situ* as concurrent acid production by the *Lactococcus lactis* decreased the pH, reducing the chelating power of the EDTA (Boziaris and Adams, 1999).

Addition of bovine serum albumin to EDTA/nisin mixtures did not significantly reduce the inhibitory effect, although Delves-Broughton saw a negation of the effect against *P. fluorescens* by food components (Stevens *et al.*, 1992; Delves-Broughton, 1993).

Food applications of this technique have centred on surface treatment of foods, through procedures such as dipping, absorbent tray pads and impregnated packaging films. Studies with poultry have shown that nisin/EDTA combinations produce appreciable reductions in the surviving numbers of *Salmonella* and an extension of shelf-life through inhibition of Gram-negative spoilage organisms (Shefet *et al.*, 1995; Cosby *et al.*, 1999; Natrajan and Sheldon, 2000a, b).

A similar study of surface contamination of beef with *Salmonella* and *E. coli* found much lower reductions in numbers (< 0.5 log CFU/ml) compared with results in broth systems (Cutter and Siragusa, 1995b), although in later work reductions of more that 3.5 log CFU/ml in the numbers of the Gram-positive *Br.*

thermosphacta were reported. These observations have been supported by other studies on beef where nisin/EDTA failed to have any significant effect on Gram-negatives such as *Salmonella* or *E. coli* but did reduce numbers of *Br. thermosphacta* or *L. monocytogenes* (Zhang and Mustapha, 1999; Hoffman *et al.*, 2001; Tu and Mustapha, 2002). A more detailed review of nisin applications in animal products is given in Chapter 3.

2.10 Conclusions

The numerous investigations described here clearly indicate the potential for using nisin as part of multiple barrier preservation systems. They also bring out the common theme that membrane active compounds or processes show the greatest promise for enhancing nisin activity. Promising laboratory studies in broth media, however, are often disappointing when tried in more realistic food systems and the reliable translation of interesting laboratory results to practical applications requires greater understanding of the factors that inhibit nisin activity as well as those that promote it.

2.11 References

ADAMS, M.R. (2001) Why fermented foods can be safe. In: *Fermentation and Food Safety.* Adams, M. R. and Robert Nout M.J. (eds) Aspen Publishers Inc., Gaithersburg, pp. 39–52.

ADAMS, M.R. and MOSS, M.O. (2000) *Food Microbiology*, 2nd edn, Royal Society of Chemistry, Cambridge pp. 93–95.

ALAKOMI, H.-L., SKYTTÄ, E., SAARELA, M., MATTILA-SANDHOLM, T., LATVA-KALA, K. and HELANDER, I.M. (2000) Lactic acid permeabilizes Gram negative bacteria by disrupting the outer membrane. *Appl. Environ. Microbiol.* **66** 2001–2005.

AVERY, S.M. and BUNCIC, S. (1997) Antilisterial effects of a sorbate-nisin combination *in vitro* and on packaged beef at refrigerated temperature. *J. Food Prot.* **60** 1075–1080.

BEARD, B.M., SHELDON, B.W. and FOEGEDING, P.M. (1999) Thermal resistance of bacterial spores in milk-based beverages supplemented with nisin. *J. Food Prot.* **62** 484–491.

BENNIK M.H., VERHEUL A., ABEE T., NAAKTGEBOREN-STOFFELS G., GORRIS L.G. and SMID E.J. (1997) Interactions of nisin and pediocin PA-1 with closely related lactic acid bacteria that manifest over 100-fold differences in bacteriocin sensitivity. *Appl Environ Microbiol*, **63**(9) 3628–3636.

BEUCHAT, L.R. (1994) Antimicrobial properties of spices and their essential oils. In: *Natural antimicrobial systems and food preservation*, V.M. Dillon and R.G. Board (eds), CAB International, Wallingford, pp. 167–179.

BLACKBURN, P., POLAK, J. , GUSIK, S.A. and RUBINO, S.D. (1989) Nisin composition for use as enhanced broad range bacteriocides. *International Patent* No. PCT/US89/02625; WO 89/12399.

BOUSSOUEL, N., MATHIEU, F., BENOIT, V., LINDER, M., REVOL-JUNELLES, A.-M. and MILLIèRE, J.

B. (1999) Response surface methodology, an approach to predict the effects of a lactoperoxidase system, nisin, alone or in combination, on *Listeria monocytogenes* in skim milk. *J. Appl. Microbiol.* **86** 642–652.

BOUSSOUEL, N., MATHIEU, F., REVOL-JUNELLES, A.-M. and MILLIÈRE, J. B. (2000) Effect of combinations of lactoperoxidase system and nisin on the behaviour of *Listeria monocytogenes* ATCC 15313 in skim milk. *Int. J. Food Microbiol.* **61** 169–175.

BOUTTEFROY, A. and MILLIÈRE, J-B. (2000) Nisin-curvaticin 13 combinations for avoiding the regrowth of bacteriocin resistant cells of *Listeria monocytogenes* ATCC 15313. *Int. J. Food Microbiol.* **62** 62–75

BOZIARIS, I.S. and ADAMS, M.R. (1999) Effect of chelators and nisin produced *in situ* on inhibition and inactivation of Gram negatives. *Int. J. Food Microbiol.* **53** 105–113.

BOZIARIS, I.S. and ADAMS, M.R. (2000) Transient sensitivity to nisin in cold-shocked Gram negatives. *Lett. Appl. Microbiol.* **31** 233–237.

BOZIARIS, I.S. and ADAMS, M.R. (2001) Temperature shock, injury and transient sensitivity to nisin in Gram negatives. *J. Applied Microbiol.* **91** 715–724.

BOZIARIS, I.S., HUMPHESON, L. and ADAMS, M.R. (1998) Effect of nisin on heat injury and inactivation of *Salmonella enteritidis* PT4. *Int. J. Food Microbiology* **43** 7–14.

BREWER, R., ADAMS, M.R. and PARK, S.F. (2002) Enhanced inactivation of *Listeria monocytogenes* by nisin in the presence of ethanol. *Lett. Appl. Microbiol.* **34** 18–21.

BUDU-AMOAKO, E., ABLETT, R.F., HARRIS, J. and DELVES-BROUGHTON, J. (1999) Combined effect of nisin and moderate heat on destruction of *Listeria monocytogenes* in cold-pack lobster meat. *J. Food Prot.* **62** 46–50.

BYCROFT, N.L., BING, G.S. and GOOD S.R. (1991) Synergisitic antimicrobial compositions. US Patent No. 5,043,176. Haarman and Reimer Corp., USA.

CABO, M.L., PASTORIZA, L., BERNARDEZ, M. and HERRARA, J. J. R. (2001a) Effectiveness of CO_2 and Nisaplin® on increasing shelf-life of fresh pizza. *Food Microbiol.* **18** 489–498.

CABO, M.L., PASTORIZA, L., SAMPREDO, G., GONZALEZ, M.P. and MURADO, M.A. (2001b) Joint effect of nisin, CO_2 and EDTA on the survival of *Pseudomonas aeruginosa* and *Enterococcus faecium* in a food model system. *J. Food Prot.* **64** 1943–1948.

CAPELLAS, M., MOR-MUR, M., GERVILLA, R., YUSTE, J. and GUAMIS, B. (2000) Effect of high pressure combined with mild heat or nisin on inoculated bacteria and mesophiles of goat's milk fresh cheese. *Food Microbiol.* **17** 633–641.

CHUNG, W. and HANCOCK, R.E. (2000) Action of lysozyme and nisin mixtures against lactic acid bacteria. *Int. J. Food Microbiol.* **60** 25–32.

COSBY, D.A., HARRISON, M.A., TOLEDO, R.T. and CRAVEN, S.E. (1999) Vacuum or modified atmosphere packaging and EDTA–nisin treatment to increase poultry product shelf life. *J. Appl. Poultry Res.* **8** 185–190.

CRANDALL, A.D. and MONTVILLE, T.J. (1998) Nisin resistance in *Listeria monocytogenes* ATCC 700302 is a complex phenotype. *Appl. Environ. Microbiol.* **64** 231–237.

CUTTER, C.N. and SIRAGUSA, G.R. (1995a) Population reductions of Gram-negative pathogens following treatments with nisin and chelators under various conditions. *J. Food Prot.* **58** 977–983.

CUTTER, C.N. and SIRAGUSA, G.R. (1995b) Treatments with nisin and chelators to reduce *Salmonella* and *E. coli* on beef. *J. Food Prot.* **58** 1028–1030.

DANISCO (2002) *International Acceptance of Nisin as a Food Preservative.* TM 2-2e SPEC, Danisco A/S, Brabrand Denmark, 7 pp.

DAVIES, E.A. and ADAMS, M.R. (1994) Resistance of *Listeria monocytogenes* to the bacteriocin nisin. *Int. J. Food Microbiol.* **21** 341–347.

DAVIES, E.A., FALAHEE, M.B. and ADAMS, M.R. (1996) Involvement of the cell envelope of *Listeria monocytogenes* in the acquisition of nisin resistance. *J. Appl. Bact.* **81** 139–146.

DAVIES, E.H., BEVIS, H.E., POTTER, R., HARRIS, J. and WILLIAMS, G.C. (1998) The effect of pH on the activity of nisin solution during autoclaving. *Lett. Appl. Microbiol.* **27** 186–187.

DAWSON, P.L., CARL, G.D., ACTON, J. C. and HAN, I.Y. (2002) Effect of lauric acid and nisin-impregnated soy-based films on the growth of *Listeria monocytogenes* on turkey bologna. *Poultry Sci.* **81** 721–726.

DELVES-BROUGHTON, J. (1993) The use of EDTA to enhance the efficacy of nisin towards Gram-negative bacteria. *Int. Biodet. Biodeg.* **32** 87–97.

DUTREUX, N., NOTERMANS, S., GONGORA-NIETO, M.M., BARBOSA-CENOVAS, G.V. and SWANSON, B.G. (2000) Effect of combined exposure of *Micrococcus luteus* to nisin and pulsed electric fields. *Int. J. Food Microbiol.* **60** 147–152.

DYKES, G.A. and MOORHEAD, S.M. (2002) Combined antimicrobial effect of nisin and a listeriophage against *Listeria monocytogenes* in broth but not in buffer or on raw beef. *Int. J. Food Microbiol.* **73** 71–81.

EKSTRAND, B. (1994) Lactoperoxidase and lactoferrin. In: *Natural antimicrobial systems and food preservation,* V.M. Dillon and R.G. Board (eds), CAB International, Wallingford, pp. 15–63.

ELLIASON, D.J. and TATINI, S.R. (1999) Enhanced inactivation of *Salmonella typhimurium* and verotoxigenic *E. coli* by nisin at 6.5 °C. *Food Microbiol.* **16** 257–267.

ETTAYEBI, K., EL YAMANI, J. and ROSSI-HASSANI, B-D. (2000) Synergistic effects of nisin and thymol on antimicrobial activities in *Listeria monocytogenes* and *Bacillus subtilis*. *FEMS Microbiol. Lett.* **183** 191–195.

FANG, T.J. and LIN, L.-W. (1994) Growth of *Listeria monocytogenes* and *Pseudomonas fragi* on cooked pork in a modified atmosphere packaging/nisin combination. *J. Food Prot.* **57** 479–485.

FANG, T.J. , CHEN, C.Y. and CHEN, H.H.L. (1997) Inhibition of *Staphylococcus aureus* and *Bacillus cereus* on a vegetarian food treated with nisin combined with either potassium sorbate or sodium benzoate. *J. Food Safety* **17** 69–87.

FELDBREG, R.S., CHANG, S.C., KOTIK, A.N., NEUWIRTH, Z., SUNDSTROM, D.C. and THOMPSON, N.H. (1988) *In vitro* mechanism of inhibition of bacterial cell growth by allicin. *Antimicrob. Agents Chemother.* **32** 1763–1768.

FEREIRA, M.A.S.S. and LUND, B.M. (1996) The effect of nisin on *Listeria monocytogenes* inculture medium and long-life cottage cheese. *Lett. Appl. Microbiol.* **22** 433–438.

GANZLE, M.G. and VOGEL, R.F. (2001) On-line fluorescence determination of pressure mediated outer membrane damage in *Escherichia coli. Syst. Appl. Microbiol.* **24** 477–485.

GARCIA-GRAELLS, C., MASSCHALCK, B. and MICHIELS, C.W. (1999) Inactivation of *Escherichia coli* in milk by high hydrostatic pressure treatment in combination with antimicrobial peptides. *J. Food Prot.* **62** 1248–1254.

GILL, A.O. and HOLLEY, R.A. (2000a) Inhibition of bacterial growth on ham and bologna by lysozyme, nisin and EDTA. *Food Res. Int.* **33** 83–90.

GILL, A.O. and HOLLEY, R.A. (2000b) Surface application of lysozyme, nisin and EDTA to inhibit spoilage and pathogenic bacteria on ham and bologna. *J. Food Prot.* **63** 1338–1346.

GOULD, G.W. (2000) Strategies for food preservation. In: *The Microbiological Safety and*

Quality of Food. B.M. Lund, A.C. Baird-Parker and G.W. Gould (eds), Aspen Publishers Inc. Gaithersburg. pp. 19–35.

HANLIN, M.B., KALCHAYANAND, N., RAY, P. and RAY, B. (1993) Bacteriocins of lactic acid bacteria in combination have greater antibacterial activity. *J. Food Prot.* **56** 252–255.

HARRIS, L.J. , FLEMING, H.P. and KLAENHAMMER, T.R. (1991) Sensitivity and resistance of *Listeria monocytogenes* ATCC19115, Scott A and UAL 500 to nisin. *J. Food Prot.* **54** 836–840.

HAUBEN, K.J. A., WUYTACK, E.Y., SOONTJENS, G.C.F. and MICHIELS K.W. (1996) High pressure transient sensitisation of *Escherichia coli* to lysozyme and nisin by disruption of outer membrane permeability. *J. Food Prot.* **59** 350–355.

HELANDER, I.M., ALAKOMI, H-L., LATVA-KALA, K., MATILLA-SANDHOLM, T., POL, I., SMID, E.J., GORRIS, L.G.M. and VON WRIGHT, A. (1998) Characterization of the action of selected essential oil components on Gram-negative bacteria. *J. Agricultural and Food Chemistry* **46** 3590–3595.

HIRSCH, A., GRINSTED, E., CHAPMAN, H.R. and MATTICK, H.T.R. (1951) A note on the inhibition of an anaerobic sporeformer in Swiss-type cheese by a nisin-producing *Streptococcus. J. Dairy Res.* **18** 205–207.

HOFFMAN, K.L., HAN, I.Y. and DAWSON, P.L. (2001) Antimicrobial effects of corn zein films impregnated with nisin, lauric acid, and EDTA. *J. Food Prot.* **64** 885–889.

IU, J. , MITTAL, G.S. and GRIFFITHS, M.W. (2001) Reduction in levels of *Escherichia coli* O157:H7 in apple cider by pulsed electric fields. *J. Food Prot.* **64** 964–969.

JARVIS, B. (1967) Resistance to nisin and production of nisin-inactivating enzymes by several *Bacillus* species. *J. Gen. Microbiol.* **47** 33–48.

JARVIS, B. (1970) Enzymic reduction of the C-terminal dehydroalanyl-lysine sequence in nisin. *Proc. Biochem. Soc.* **119** 56P.

JARVIS, B. and FARR, J. (1971) Partial purification, specificity and mechanism of action of the nisin inactivating enzyme from *Bacillus cereus. Biochim. Biophys. Acta* **227** 232–243.

JEYAMKONDAN, S., JAYAS, D.S. and HOLLEY, R.A. (1999) Pulsed electric field processing of foods: a review. *J. Food Prot.* **62** 1088–1096.

JUNG, D.S., BODYFELT, F.W. and DAESCHEL, M.A. (1992) Influence of fat and emulsifiers on the efficacy of nisin inhibiting *Listeria monocytogenes* in fluid milk. *J. Dairy Sci.* **75** 387–393.

JUVEN, B.J., KRAMER, J., SCHVED, F. and WEISSLIWISC, H. (1994) Factors that interact with the antibacterial action of thyme essential oil and its active constituents. *J. Applied Bacteriology* **76** 626–631.

KABARA, J. J. (1993) Medium-chain fatty acids and esters. In: *Antimicrobials in Foods,* P.M. Davidson and A.L. Branen (eds) 2nd edn Marcel Dekker Inc., New York pp. 307–342.

KALCHAYANAND, N., HANLIN, M.B. and RAY, B. (1992) Sublethal injury makes Gram negative and resistant Gram positive bacteria sensitive to the bacteriocins, pediocin, AcH and nisin. *Lett. Appl. Microbiol.* **15** 239–243.

KALCHAYANAND, N., SIKES, T., DUNNE, C.P. and RAY, B. (1994) Hydrostatic pressure and electroporation have increased bactericidal efficiency in combination with bacteriocins. *Appl. Environ. Microbiol.* **60** 4174–4177.

KLAENHAMMER, T.R. (1993) Genetics of bacteriocins produced by lactic acid bacteria. *FEMS Microbiol. Rev.* **12** 39–85.

KOMITOPOULOU, E., BOZIARIS, I.S., DAVIES, E.A., DELVES-BROUGHTON, E.A. and ADAMS, M.R.

(1999) *Alicyclobacillus acidoterrestris* and its control by nisin. *Int. J. Food Sci. Technol.* **34** 81–85.

KUSSENDRAGER, K.D. and VAN HOOIJDONK, A.C.M. (2000) Lactoperoxidase: physico-chemical properties, occurrence, mechanism of action and applications. *Brit. J. Nut.* **84** Suppl. 1, S19–S25.

LEE, J. I., LEE, H.-J. and LEE, M.-H. (2002) Synergisitic effect of nisin and heat treatment on the growth of *Escherichia coli* O157:H7. *J. Food Prot.* **65** 408–410.

LEISTNER, L. (2000) Use of combined preservative factors in foods in developing countries. In: *The Microbiological Safety and Quality of Food.* B.M. Lund, A.C. Baird-Parker and G.W. Gould (eds), Aspen Publishers Inc, Gaithersburg. pp. 294–314.

LIANG, Z.W., MITTAL, G.S. and GRIFFITHS, M.W. (2002) Inactivation of *Salmonella typhimurium* in orange juice containing antimicrobial agents by pulsed electric field. *J. Food Prot.* **65** 1081–1087.

MAISNIER-PATIN, S. and RICHARD, J. (1996) Cell wall changes in nisin-resistant variants of *Listeria innocua* grown in the presence of high nisin concentrations. *FEMS Microbiol. Lett.* **140** 29–35.

MANSOUR, M. and MILLIÈRE, J. B. (2001) An inhibitory synergistic effect of a nisin-monolaurin combination on *Bacillus* sp. vegetative cells in milk. *Food Microbiol.* **18** 87–94.

MASSCHALCK, B., VAN HOUDT, R. and MICHIELS, C.W. (2001) High pressure increases bactericidal activity and spectrum of lactoferrin, lactoferricin and nisin. *Int. J. Food Microbiol.* **64** 325–332.

MAZZOTTA, A.S. and MONTVILLE, T.J. (1997) Nisin induces changes in membrane fatty acid composition of *Listeria monocytogenes* nisin resistant strains at 10 °C and 30 °C. *J. Appl. Microbiol.* **82** 32–38.

MING, X. and DAESCHEL, M.A. (1993) Nisin resistance of foodborne bacteria and the specific resistance responses of *Listeria monocytogenes*. *J. Food Prot.* **56** 944–948.

MING, X. and DAESCHEL, M.A. (1995) Correlation of cellular phospholipid content with nisin resistance of *Listeria monocytogenes* Scott A. *J. Food Prot.* **58** 416–420.

MIRON, T., RABINKOV, A., MIRELMAN, D., WILCHEK, M. and WEINER, L. (2000) The mode of action of allicin: its ready permeability through phospholipid membranes may contribute to its biological activity. *Biochim. Biophys. Acta* **1463** 20–30.

MODI, K.D., CHIKINDAS, M.L. and MONTVILLE, T.J. (2000) Sensitivity of nisin-resistant *Listeria monocytogenes* to heat and the synergistic action of heat and nisin. *Lett. Appl. Microbiol.* **30** 249–253.

MONTICELLO, D. (1989) Control of microbial growth with nisin/lysozyme formulations. Eur. Pat. Appl. 89123445.2.

MULDERS J.W., BOERRIGTER I.J., ROLLEMA H.S., SIEZEN R.J. and DE VOS W.M. (1991) Identification and characterization of the lantibiotic nisin Z, a natural nisin variant. *Eur. J. Biochem.* **201**(3) 581–584.

MULET-POWELL, N., LACOSTE-ARMYNOT, A.M., VINAS, M. and SIMEON DE BUOCHBERG, M. (1998) Interactions between pairs of bacteriocins from lactic acid bacteria. *J. Food Prot.* **61** 1210–1212.

MUSTAPHA, A., ARIYAPITIPUN, T. and CLARKE, A.D. (2002) Survival of *Escherichia coli* O157:H7 on vacuum-packaged raw beef treated with polylactic acid, lactic acid and nisin. *J. Food Sci.* **67** 262–267.

NATRAJAN, N. and SHELDON, B.W. (2000a) Efficacy of nisin-coated polymer films to inactivate *Salmonella* Typhimurium on fresh broiler skin. *J. Food Prot.* **63** 1189–

1196.

NATRAJAN, N. and SHELDON, B.W. (2000b) Inhibition of *Salmonella* on poultry skin using protein- and polysaccharide-based films containing a nisin formulation. *J. Food Prot.* **63** 1268–1272.

NATTRESS, F.M., YOST, C.K. and BAKER, L.P. (2001) Evaluation of the ability of lysozyme and nisin to control meat spoilage bacteria. *Int. J. Food Microbiol.* **70** 111–119.

NILSSON, L., HUSS, H. H. and GRAM, L. (1997) Inhibition of *Listeria monocytogenes* on cold-smoked salmon by nisin and carbon dioxide atmosphere. *Int. J. Food Microbiol.* **38** 217–227.

NILSSON, L., CHEN, Y., CHIKINDAS, M.L., HUSS, H. H., GRAM, L. and MONTVILLE, T.J. (2000) Carbon dioxide and nisin act synergistically on *Listeria monocytogenes. Appl. Environ. Microbiol.* **66** 769–774.

NYCHAS, G. (1995) Natural antimicrobials from plants. In: *New Methods of Food Preservation*, G.W. Gould (ed.), Blackie Academic, London, pp. 58–89.

NYKäNEN , A., VESANEN, S. and KALLIO, H. (1998) Synergistic antimicrobial effect of nisin whey permeate and lactic acid on microbes isolated from fish. *Lett. Appl. Microbiol.* **27** 345–348.

PENNA, T.C.V. and MORAES, D.A. (2002) The influence of nisin on the thermal resistance of *Bacillus cereus. J. Food Prot.* **65** 415–418.

PERIAGO, P.M., PALOP, A. and FERNADEZ, P.S. (2001) Combined effect of nisin, carvacrol and thymol on the viability of *Bacillus cereus* heat-treated vegetative cells. *Food Sci. Technol. Int.* **7** 487–492.

PHILLIPS, C.A. and DUGGAN, J. (2001) The effect of EDTA and trisodium phosphate, alone and in combination with nisin, on the growth of *Arcobacter butzleri* in culture. *Food Microbiol.* **18** 547–554.

POL, I.E. and SMID, E.J. (1999) Combined action of nisin and carvacrol on *Bacillus cereus* and *Listeria monocytogenes. Lett. Appl. Microbiol.* **29** 166–170.

POL, I.E., MASTWIJK, H.C., BARTELS, P.V. and SMID, E.J. (2000) Pulsed electric field treatment enhances the bactericidal action of nisin against *Bacillus cereus. Appl. Env. Microbiol.* **66** 428–430.

POL, I.E., MASTWIJK, H.C., SLUMP, R.A., POPA, M.E. and SMID, E.J. (2001a) Influence of food matrix on inactivation of *Bacillus cereus* by combinations of nisin, pulsed electric field treatment, and carvacrol. *J. Food Prot.* **64** 1012–1018.

POL, I.E., VAN ARENDONK, W.G.C., MASTWIJK, H.C., KROMMER, J. , SMID, E.J. and MOEZELAAR, R. (2001b) Sensitivity of germinating spores and carvacrol-adapted vegetative cells and spores of *Bacillus cereus* to nisin and pulsed-electric-field treatment. *Appl. Env. Microbiol.* **67** 1693–1699.

PONCE, E., PLA, R., SENDRA, E., GUAMIS, B. and MOR-MUR, M. (1998) Combined effects of nisin and hydrostatic pressure on destruction of *Listeria innocua* and *Escherichia coli* in liquid whole egg. *Int. J. Food Microbiol.* **43** 15–19.

RAMSEIER, H.R. (1960) The action of nisin on *Clostridium butyricum. Arch. Mikrobiol.* **37** 57–94.

ROBERTS, C.M. and HOOVER, D.G. (1996) Sensitivity of *Bacillus coagulans* spores to combinations of high hydrostatic pressure, heat, acidity and nisin. *J. Appl. Bact.* **81** 363–368.

RODRIGUEZ, E., TOMILLO, J. , NUNEZ, M. and MEDINA, M. (1997) Combined effect of bacteriocin-producing lactic acid bacteria and lactoperoxidase system activation on *Listeria monocytogenes* in refrigerated raw milk. *J. Appl. Microbiol.* **83** 389–395.

SCHILLINGER, U., GEISEN, R. and HOLZAPFEL, W.H. (1996) Potential of antagonisitic microorganisms and bacteriocins for the biological preservation of foods. *Trends Food Sci and Technol.* **7** 158–164.

SCHILLINGER, U., CHUNG, H.S., KEPPLER, K. and HOLZAPFEL, W.H. (1998) Use of bacteriocinogenic lactic acid bacteria to inhibit spontaneous nisin-resistant mutants of *Listeria monocytogenes* Scott A. *J. Appl. Microbiol.* **85** 657–663.

SCHILLINGER, U., BECKER, B., VIGNOLO, G. and HOLZAPFEL, W.H. (2001) Efficacy of nisin in combination with protective cultures against *Listeria monocytogenes* Scott A in tofu. *Int. J. Food Microbiol.* **71** 159–168.

SHEFET, S.M., SHELDON, B.W. and KLAENHAMMER, T.R. (1995) Efficacy of optimised nisin-based treatments to inhibit *Salmonella* Typhimurium and extend shelf-life of broiler carcasses. *J. Food Prot.* **58** 1077–1082.

SHELEF, L.A. (1983) Antimicrobial effects of spices. *J. Food Safety* **6** 29–44.

SILVA, F.V.M. and GIBBS, P. (2001) *Alicyclobacillus acidoterrestris* spores in fruit products and design of pasteurisation processes. *Trends Food Sci. Technol.* **12** 68–74.

SINGH, B., FALAHEE, M.B. and ADAMS, M.R. (2001) Synergistic inhibition of *Listeria monocytogenes* by nisin and garlic extract. *Food Microbiol.* **18**(2) 133–139.

SMID, E.J. and GORRIS, L.G.M. (1999) Natural Antimicrobials for Food Preservation. In: *Handbook of Food Preservation*, M. S. Rahman (ed.), Marcel Dekker Inc., New York, Chapter 9, pp. 285–308.

STANBRIDGE, L.H. and DAVIES, A.R. (1998) The microbiology of chill stored meat. In: *The Microbiology of Meat and Poultry*, A. Davies and R. Board (eds), Blackie Acdemic and Professional, London pp. 174–219.

STEVENS, K.A., SHELDON, B.W., KLAPES, N.A. and KLAENHAMMER, T.R. (1991) Nisin treatment for inactivation of *Salmonella* species and other Gram-negative bacteria. *Appl. Environ. Microbiol.* **57** 3613–3615.

STEVENS, K.A., SHELDON, B.W., KLAPES, N.A. and KLAENHAMMER, T.R. (1992) Effect of treatment conditions on nisin inactivation of Gram-negative bacteria. *J. Food Prot.* **55** 763–766.

TEREBIZNIK, M.R., JAGUS, R.J. , CERRUTTI, P., DE HUERGO, M.S. and PILOSOF, A.M.R. (2000) Combined effect of nisin and pulsed electric fields on the inactivation of *Escherichia coli*. *J. Food Prot.* **63** 741–746.

TEREBIZNIK, M.R., JAGUS, R.J. , CERRUTTI, P., DE HUERGO, M.S. and PILOSOF, A.M.R. (2002) Inactivation of *Escherichia coli* by a combination of nisin, pulsed electric fields, and water activity reduction by sodium chloride. *J. Food Prot.* **65** 1253–1258.

TER STEEG, P.F., HELLEMONS, J. C. and KOK, A.E. (1999) Synergistic actions of nisin, sublethal ultrahigh pressure and reduced temperature on bacteria and yeast. *Appl. Environ. Microbiol.* **65** 4148–4154.

THOMAS, L.V., DAVIES, E.A., DELVES-BROUGHTON, J. and WIMPENNY, J. W.T. (1998) Synergist effect of sucrose fatty acid esters on nisin inhibition of Gram-positive bacteria. *J. Appl. Microbiol.* **85** 1013–1022.

THOMAS, L.V., CLARKSON, M.R. and DELVES-BROUGHTON, J. (2000) Nisin. In: *Natural Food Antimicrobial Systems*, A.S. Naidu (ed.), CRC Press, Boca Raton, pp. 463–524.

TRANTER, H.S. (1994) Lysozyme, ovotransferrin and avidin. In: *Natural Antimicrobial Systems and Food Preservation*, V.M. Dillon and R.G. Board (eds), CAB International, Wallingford, pp. 65–97.

TU, L. and MUSTAPHA, A. (2002) Reduction of *Brochothrix thermosphacta* and *Salmonella* serotype typhimurium on vacuum-packaged fresh beef treated with nisin combined with EDTA. *J. Food Sci.* **67** 302–306.

UECKERT, J. E., TER STEEG, P.F. and COOTE, P.J. (1998) Synergistic antibacterial action of heat in combination with nisin and magainin II amide. *J. Appl. Microbiol.* **85** 487–494.

UKUKU, D.O. and SHELEF, L.A. (1997) Sensitivity of six strains of *Listeria monocytogenes* to nisin. *J. Food Prot.* **60** 867–869.

ULTEE, A., E.P.W. KETS and E.J. SMID (1999) Mechanisms of action of carvacrol on the food borne pathogen *Bacillus cereus. Appl. Environ. Microbiol.* **65** 4606–4610.

ULTEE, A., BENNIK, M.H.J. and MOEZELAAR, R. (2002) The phenolic hydroxyl group of carvacrol is essential for action against the food-borne *Bacillus cereus. Appl. Environ. Microbiol.* **68** 1561–1568.

VERHEUL, A., RUSSELL, N.J., VANT HOFF, R., ROMBOUTS, F.M. and ABEE, T. (1997) Modifications of membrane phospholipid composition in nisin-resistant *Listeria monocytogenes* ScottA. *Appl. Environ. Microbiol.* **63** 3451–3457.

WALKER, J. R.L. (1994) Antimicrobial compounds in food plants. In: *Natural antimicrobial systems and food preservation*, V.M. Dillon and R.G. Board (eds), CAB International, Wallingford, pp. 181–204.

WANDLING, L.R., SHELDON, B.W. and FOEGEDING, P.M. (1999) Nisin in milk sensitizes *Bacillus* spores to heat and prevent recovery of survivors. *J. Food Prot.* **62** 492–498.

WIEDEMANN I., BREUKINK E., VAN KRAAIJ C., KUIPERS O.P., BIERBAUM G., DE KRUIJFF B. and SAHL H.G. (2001) Specific binding of nisin to the peptidoglycan precursor lipid II combines pore formation and inhibition of cell wall biosynthesis for potent antibiotic activity. *J. Biol. Chem.* **276**(3) 1772–1779.

YUSOF, R.M., MORGAN, J. B. and ADAMS, M.R. (1993) Bacteriological safety of a fermented weaning food containing L-lactate and nisin. *J. Food Prot.* **56** 414–417.

YUSTE, J. , MOR-MUR, M., GUAMIS, B. and PLA, R. (2000) Combination of high pressure with nisin or lysozyme to further process mechanically recovered poultry meat. *High Press. Res.* **19** 475–480.

ZAPICO, P., MEDINA, M., GAYA, P. and NUNEZ, M. (1998) Synergistic effect of nisin and the lactoperoxidase system on *Listeria monocytogenes* in skim milk. *Int. J. Food Microbiol.* **40** 35–42.

ZHANG, S.S. and MUSTAPHA, A. (1999) Reduction of *Listeria monocytogenes* and *Escherichia coli* O157:H7 numbers on vacuum packaged fresh beef treated with nisin or nisin combined with EDTA. *J. Food Prot.* **62** 1123–1127.

3

Nisin in the decontamination of animal products

P. L. Dawson, Clemson University, USA and B. W. Sheldon, North Carolina State University, USA

3.1 Introduction

While outbreaks of foodborne disease are associated with many foods, foods of animal origin rank near the top as vehicles of bacterial pathogens. The association of *Campylobacter* and *Salmonella* species with raw foods of animal origin, particularly poultry products, is well established (Bean *et al.*, 1997). Likewise, the link between ground beef contamination with *Escherichia coli* O157:H7 has been firmly established. *Campylobacter* spp. have been recovered from retail cuts of poultry products at rates ranging from 1.8 to 92% (Stern, 1982) whereas the current *Salmonella* incidence rate estimate for poultry carcasses in the USA is approximately 10.9%, down from 20% since the adoption of hazard analysis and critical control point (HACCP) programs in 1998 for very large poultry processing plants (http://www.fsis.usda.gov/OA/news/0095-99.htm). *Salmonella* prevalence in broiler carcasses, market hogs, cows/bulls, steers/heifers, ground beef, and ground turkey in 2000 was 16.4, 14.2, 4.6, 0, 8.7, and 43.4%, respectively (Rose *et al.*, 2002). In various studies, 60% of raw chickens in the UK (Pini and Gilbert, 1988), 23% of retail fresh broilers in the USA (Bailey *et al.*, 1989), 61% of chicken carcasses in Norway (Rorvik and Yndestad, 1991), and 64% of poultry samples in Spain (Franco *et al.*, 1997) were found to contain *Listeria monocytogenes*. Thus, *L. monocytogenes*, *Salmonella* sp., *C. jejuni*, and *E. coli* O157:H7 have been the target of numerous decontamination studies due to their prevalence in foods of animal origin and human pathogenicity. There are many other studies that have addressed the inhibition of other bacterial pathogens and spoilage bacteria using nisin as well as research into extending product shelf-life and these are reviewed in Chapter 2.

3.2 Overview of current meat decontamination practices

Intervention strategies used by the meat and poultry industry to decontaminate carcasses include hot water, steam, high-pressure water, cooking, irradiation and treatment solutions containing organic acids, ozone, chlorinated compounds and trisodium phosphate (TSP). Other disinfectants currently under investigation are hydrogen peroxide, acidified sodium chlorite and electrolyzed water.

The beef carcass is considered to be essentially sterile immediately after removal of the hide. The US Department of Agriculture (USDA) FSIS (Food Safety Inspection Service) regulations require that carcass contamination with feces or ingesta of less than 2.54 cm in size (in the greatest dimension) must be physically removed by either knife trimming or steam vacuuming. Any visual contamination greater than 2.54 cm including lesions, feces, or bruises must be completely removed from the carcass by knife trimming (FSIS, 1996). There are several USDA-approved interventions to reduce surface carcass contamination, including (1) steam vacuuming, (2) pre-evisceration rinses using water followed by an organic acid rinse, (3) chlorinated water washes at 20–50 ppm, (4) 1.5–2.5% organic acid sprays, (5) 8–12% trisodium phosphate sprays at 32–44 °C for not more that 30 seconds, (6) hot water sprays, and (7) steam pasteurization (Dorsa *et al.*, 1997a).

The decontamination of chicken and turkey carcasses differs from beef and swine owing to differences in animal size and the methods used to slaughter these animals. Poultry processing is a rapid and highly automated system involving shackled carcasses moving continuously along a conveyor line. There are numerous points during the process where washing and rinsing occur between and during the unit operations of scalding, feather removal, evisceration, and chilling. The first major carcass washing site is immediately following evisceration at the inside/outside bird washer where chlorinated water (or other antimicrobial) is sprayed on both the outside surface and inside body cavity of each carcass. An additional carcass spray or dip containing TSP or acidified sodium chlorite takes place immediately prior to immersion chilling in water. US poultry processors have used immersion chilling for many years; however, this mode of chilling has recently been approved only under the stipulation that besides lowering carcass temperature, it must also reduce bacterial contamination. Many European poultry processors employ either air or spray chilling systems. In either case, chiller water containing chlorine or other chlorine derivatives will generally lower bacterial populations. Ozone was approved by the USDA in 1984 for use in poultry chill tanks and was approved by the Food and Drug Administration (FDA) in 2001 as an antimicrobial treatment (in solution or as a gas) for meat and poultry. Ozone is a powerful oxidizing agent that is effective against Gram-positive and Gram-negative bacteria, spore-forming bacteria, fungi, viruses, and protozoa (Kim *et al.*, 1999).

3.2.1 Beef products

Steam vacuum systems are effective in reducing fecal associated bacteria and *E. coli* strains from beef (Dorsa *et al.*, 1996a, 1997a) and sheep carcasses (Dorsa *et al.*, 1996b). The surface bacterial population of beef carcasses contaminated with ~6 log CFU per cm^2 of mesophilic aerobic bacteria was reduced by 2.3 to 4.0 log CFU/cm^2 by steam vacuuming. In studies involving higher initial inoculation levels (up to 7.6 log CFU/cm^2), *E. coli* O157:H7 populations were lowered to 2.3 log CFU/cm^2 by steam vacuuming (Dorsa *et al.*, 1996a). Phebus *et al.* (1997) observed a reduction of 3.5 log CFU/cm^2 in *E. coli* O157:H7 on beef tissue surfaces following steam vacuuming (initial inoculum 5.5 log CFU/cm^2). They also reported that steam vacuuming was effective in removing *Salmonella* Typhimurium and *L. monocytogenes* from beef carcass surfaces.

Water washing has been shown to remove approximately 1 log CFU/cm^2 of the microbial flora from beef carcass surfaces (Siragusa, 1995). The addition of chlorine to wash water was reported by several researchers to provide little advantage over water-only washing (Cutter and Siragusa, 1995a; Johnson *et al.*, 1979; Stevenson *et al.*, 1978). The efficacy of organic acids and TSP has been evaluated using a large number of variables including meat type (different cuts, non-cut carcass surfaces), modes of application (dips, sprays of various pressures, times and temperatures), and concentrations (Dorsa *et al.*, 1997a). Some US beef industry processors apply organic acid sprays to hanging carcasses at 1.4 bar (20 lb/in^2) following high pressure water sprays of 8.6–27.6 bar (125–400 lb/in^2). The FDA granted TSP Generally Recognized As Safe (GRAS) status in 1992 for use as a meat antimicrobial treatment. Results of studies simulating a commercial processing set-up to decontaminate beef carcasses indicate that both organic acid sprays and TSP are capable of reducing *E. coli* O157:H7 to below detectable levels (Dorsa *et al.*, 1997b; Hamby *et al.*, 1987). Some studies evaluating antimicrobial sprays and rinses, especially those involving TSP, may have overestimated the impact of this treatment as investigators failed to neutralize the antimicrobial prior to recovering the surviving bacterial population. This is particularly a concern with TSP since it is not easily inactivated by organic matter or enzymes naturally present in the meat.

Steam pasteurization has been applied to the surfaces of beef frankfurters (Cynarowicz-Provost *et al.*, 1994), lamb carcasses (Dorsa *et al.*, 1996b), and beef carcasses (Phebus *et al.*, 1997) with the most effective bacterial reductions reported to be around 3 log CFU/cm^2 for *E. coli* O157:H7, *Salmonella* Typhimurium, and *L. monocytogenes*. The use of condensing steam under sub-atmospheric pressures was shown to reduce *E. coli* O157:H7 in fecal clods on beef hide from 4.2 to 1.2 log CFU/g after 30 s of treatment (McEvoy *et al.*, 2001). Higher reductions in populations were observed on hides having higher initial bacterial loads.

3.2.2 Poultry products

Russell (1998) studied the effectiveness of the most common sanitizers against the major spoilage microorganisms associated with poultry carcasses and found that sodium hypochlorite, lactic acid, TSP, hydrogen peroxide, and Timsen (a combination of a quaternary ammonium compound and urea) eliminated *P. fluorescens*, *P. putida*, and *P. fragi*. The quaternary ammonium product did not consistently inhibit spoilage organisms except for *Schewanella putrefaciens,* which was resistant to all other sanitizers.

The addition of 20 to 50 ppm of chlorine to poultry chiller water was reported to reduce cross-contamination of *Salmonella* spp. between carcasses (James *et al.*, 1992; Morrison *et al.*, 1985; Wabeck *et al.*, 1968). The application of TSP at 10 to 13% (pH > 11.6) has been reported to reduce *E. coli*, *Campylobacter* spp., *Salmonella* spp. and total aerobic bacteria on poultry carcasses by more than 2 log CFU/cm^2 (Salvat, 1996; Somers *et al.*, 1994). Whyte *et al.* (2001) found that 25 ppm chlorine in the final rinse water of chicken carcasses reduced the total carcass bacterial count by 0.5 log CFU/g but did not significantly reduce carcass *E. coli* and Enterbacteriaceae populations recovered from the chicken neck rinse samples. Chicken carcasses dipped in 10% TSP had lower total counts, *E. coli*, and Enterobacteriaceae than non-treated carcasses or those rinsed with water. However, no TSP inactivation step was used in this study and so the efficacy of TSP may have been overestimated.

3.3 The need for alternative decontamination treatments

The drawbacks of steam vacuuming beef carcass surfaces include meat surface discoloration, partial cooking of the raw tissue, variable effectiveness and potential embedding of bacteria in the meat tissue from high-pressure sprays. The drawbacks of organic acids are meat surface discoloration, emergence of acid-resistant pathogens, disposal of acids in wastewater, and off-flavor development in meat. Trisodium phosphate may also cause off flavors if not administered properly. Chlorine, chlorinated compounds, and ozone are reactive and can alter meat quality by promoting oxidative reactions that affect color and flavor. While the negative effects of these treatments have been minimized, the meat industry continues to explore other decontamination methods.

3.4 Factors affecting nisin activity in meat

The effectiveness of nisin on meat surfaces is dependent on the meat system and the type of organisms present (Davies *et al.*, 1999). Organisms normally sensitive to nisin *in vitro* are not necessarily as sensitive to nisin in a meat system. For example, *Cl. botulinum* spore outgrowth was prevented by 50 μg/ml nisin in a trypticase–peptone–yeast–glucose medium but was not prevented in a cooked meat medium at 2.5 times (125 μg/ml) the effective concentration in the culture medium (Scott and Taylor, 1981). Coventry *et al.* (1995) demonstrated that both pediocin and nisin together exerted a strong inhibitory effect against *Lactobacillus curvatus* in a meat model system but, alone, pediocin had no effect

Table 3.1 Effect of intrinsic and extrinsic factors on the inhibitory action of nisin

Function	Component or ingredient	Antagonist	Synergist	No effect	Reference
Major food components	Starches			X	Pirttijarvi *et al.* (2001)
	Fats/lipids	X			Jung *et al.* (1992)
	Proteins	X			DeVuyst and Vandamme (1994)
Binders	Phosphates, sodium chloride		X		Jarvis and Burke (1976); Stevens *et al.* (1992); Davies *et al.* (1999); Pawar *et al.* (2000)
Carriers	Calcium alginate and agar films		X		Cutter and Siragusa (1997); Natrajan and Sheldon (2000a)
Chelators	EDTA, citric acid		X		Stevens *et al.* (1992); Shefet *et al.* (1995)
Emulsifiers	Mono- and diglycerides, glyceride acid esters, phospholipids	X			Henning *et al.* (1986)

Enzymes	Lysozyme		X		Nattress *et al*. (2001); Gill and Holley (2000)
	Nisinase, pancreatin, α-chymotrypsin	X			Jarvis (1967); Jarvis and Farr (1971); Jarvis and Mahoney (1969)
	Trypsin, elastase, carboxypeptidase, pepsin, erepsin			X	Hurst and Hoover (1993)
Preservatives	Organic acids		X		Jarvis and Burke (1976); Avery and Buncic (1997)
Surfactants	Tween 20		X		Stevens *et al*. (1992); Shefet *et al*. (1995)
Whiteners	Sodium bisulfite, titanium dioxide	X			Delves-Broughton *et al*. (1996)
Packaging	Vacuum- packaging		X		Zhang and Mustapha (1999); Cutter and Siragusa (1996a)
	CO_2 packaging		X		Avery and Buncic (1997); Nilsson *et al*. (2000)
	Polymer packaging	X			Natrajan and Sheldon (2000a)

and nisin only a slight inhibitory effect when incorporated in a manufactured cooked meat product. There are numerous factors in meats that affect nisin activity including salt and fat content, pH, the presence of curing agents, and food particle size (Jung *et al.*, 1992). Storage temperature, modified atmosphere packaging, and the presence of other preservatives can also affect nisin efficacy. These external factors can change the bacterial population dynamics, leaving a predominant microflora either more or less resistant to nisin. The state of the meat (raw or cooked) can also affect nisin activity as demonstrated by Murray and Richard (1998) who found that nisin A was more effective in controlling *L. monocytogenes* in raw minced pork than the bacteriocin pediocin ACH due to nisin's greater resistance to meat proteases. In addition, the poor inhibitory activity of nisin in some meat applications may be attributed to binding to meat proteins, uneven mixing, poor absorption into meat, heat sensitivity at neutral pH, or interference with phospholipids (Henning *et al.*, 1986; de Vuyst and Vandamme, 1994; Davies *et al.*, 1999). Masschalck *et al.* (2001) showed that high pressure (155–400 MPa at 20 °C) increased the degree of inactivation by Nisaplin® (100 IU/ml) of *Salmonella* Enteriditis, *P. fluorescens, Staph. aureus*, and *Shigella sonnei*, but not *E. coli* and *Sh. flexneri* in phosphate buffer. Factors affecting nisin activity in animal products are summarized in Table 3.1.

There have been conflicting reports as to the recovery of nisin from meat products, which would be expected to be low due to adsorption to meat proteins and fats. Bell and deLacy (1986) found that nisin recovery was not affected by curing salts or meat particle size but was influenced by fat content. These researchers recovered 26% of nisin activity from beef containing 3% fat and 76% in beef composed of 83% fat. Conversely, Davies *et al.* (1999) recovered the same level of nisin activity from both pork meat and Bologna, regardless of fat content or length of refrigerated storage (0 *vs* 60 days).

3.5 Decontamination using nisin

To allow the reader to make some comparisons between the levels of nisin used in the different studies cited, every attempt has been made to state the form of nisin used, the source, the application method and/or the activity units. One of the major commercial sources of nisin is Nisaplin® which is composed of 2.5% pure nisin with the bulk of the preparation being milk solids and sodium chloride. As mentioned in Chapter 2, Nisaplin® has a standardized activity of 1000 IU/mg.

3.5.1 Raw meat products

Combinations of nisin with organic acids have been tested in several applications. Nisin (10 μg/g) alone and in combination with sorbic acid (100 ppm) and polyphosphate (2.5%, wt/vol) lengthened the lag phase and population doubling time for all groups of bacteria (*Microbacterium thermosphactum* (now renamed *Brochothrix thermosphacta*), coliforms, lactic acid bacteria, and pseudomonads) in raw pork sausages (Jarvis and Burke, 1976). Avery and Buncic (1997) used a combination of sorbate (0.2%, wt/vol) and nisin (40 IU/ml) to reduce *L. monocytogenes* counts by 1 log CFU/g on vacuum-packaged beef and by slightly less than 1 log CFU/g on CO_2-packed beef. In a subsequent study, nisin (200 IU/ml), polylactic acid (2%), lactic acid (2%), and combinations of nisin with each acid were used as dipping solutions for fresh beef that was then vacuum-packaged and stored at 4 ºC for up to 56 days (Ariyapitipun *et al.*, 1999). Both acid dips alone or in combination with nisin were equally effective in reducing psychrotrophic aerobes, Enterobacteriaceae, and *Pseudomonas* populations by 0.7 to 2.6 log CFU/g compared with untreated meat, water-dipped meat, or meat dipped in nisin solutions alone. In a subsequent study, Ariyapitipun *et al.* (2000) dipped *L. monocytogenes*-inoculated raw beef cubes in solutions of 2% lactic acid, 2%b polylactic acid, 400 IU/ml nisin, or combinations of the acids with nisin then stored the vacuum-packed samples at 4 ºC for 42 days. These researchers observed an initial reduction from 5 log CFU/g to 3–4 log CFU/g for all treatments. After 42 days, there were 4 log CFU/g *L. monocytogenes* cells remaining in controls while the lactic acid and nisin + lactic acid treatments had less than 1.5 log CFU/g remaining. Cubes dipped in the solution containing nisin only had 2.5 log CFU/g *L. monocytogenes* after 42 days. In an effort to find a replacement for sulfur dioxide, Scannell *et al.* (1997) found that the combination of sodium lactate and nisin inhibited *Staph. aureus* and *Salmonella* spp. in fresh pork sausage. Scannell *et al.* (2000a) later showed that a combination of 2% sodium lactate or 2% sodium citrate enhanced nisin (500 IU/g) inhibition of *L. innocua* by 1 log cfu/g but did not enhance inhibition of *Salmonella* Kentucky in fresh pork sausage.

Zhang and Mustapha (1999) reduced *L. monocytogenes* populations on vacuum-packaged beef cubes by dipping the cubes in solutions of either 5000 IU/ml nisin (2 log CFU/cm^2) or 5000 IU/ml nisin supplemented with 20 mM ethylenediaminetetraacetic acid (EDTA) (0.99 log CFU/cm^2). Reductions in *E. coli* O157:H7 surface populations were less than those achieved for *L. monocytogenes* for both 5000 IU/ml nisin (1 log CFU/cm^2) or 5000 IU/ml nisin plus 20 mM EDTA (0.8 log CFU/cm^2). A combination of 5000 IU/ml nisin-spray and packaging in a modified atmosphere or vacuum inhibited growth of *L. innocua* on beef by 1.5 to 2 log CFU/g for 14 days and 2 to 2.5 log CFU/g for 21 days compared with water rinsed or untreated meat (Cutter and Siragusa, 1996a). Nisin immobilized in calcium alginate gels was effective in decontaminating fresh pork (Fang and Lin, 1995) and beef carcasses of *Br. thermosphacta* by 2.4 log CFU/g over 7 days using 100 μg Nisaplin®/ml alginate (Cutter and Siragusa, 1996b). Raw beef was treated in a pilot-scale carcass washer using 5000 IU/ml nisin rinses (Cutter and Siragusa, 1994) and resulted in a 1.8 to 3.5 (Day 0) and 2.0 to 3.6 log CFU/g (Day 1) reduction in *Br. thermosphacta, L. monocytogenes*, and *Carnobacterium divergens* populations. Cutter and Siragusa (1998) added 10 μg Nisaplin®/ml to a commercial meat binder (Fibrimex) to inhibit *Br.*

thermosphacta on pre- and post-rigor lean beef. The fibrimex–nisin mixture maintained the lean beef tissue below detection levels of *Br. thermosphacta* for seven days while untreated samples increased from 4.9 to 8.0 log CFU/g.

Beef carcasses submerged in a solution containing 50 μg Nisalpin®/ml (combined with 50mM EDTA, 100 mM lactate and 500 mM hexametaphosphate) did not reduce the populations of *E. coli* O157:H7 or *Salmonella* Typhimurium on lean beef muscle (Cutter and Siragusa, 1995b). However, Shefet *et al.* (1995) determined the optimum treatment combination of nisin, chelating agents (EDTA, citric acid), pH, and surfactant (Tween 20) to inhibit *Salmonella* Typhimurium on fresh drumstick skin and to extend the refrigerated shelf-life of chicken drumsticks. Of nine optimized formulations compared, four formulations containing 100 μg purified nisin/ml plus the other components yielded significant bactericidal activity (3.12–4.86 log reduction). They also demonstrated that nisin concentration could be reduced to 25 μg/ml without a significant loss in inhibitory activity against the *Salmonella* Typhimurium test strain. Moreover, the refrigerated shelf-life of the treated broiler drumsticks was extended from 1.5 to 3 days following immersion for 30 minutes in the optimized nisin formulation.

Raw beef soaked in 10^4 IU/ml solution (1 g Nisaplin® in 100 ml 0.02 N HCl, 0.75% NaCl, pH 3.0) of nisin delayed the growth of *L. monocytogenes* by one day at room temperature and two weeks at 5 °C and prevented the growth of *Staph. aureus* for up to two weeks at 5 °C (Chung *et al.*, 1989). The Nisaplin® solution had no inhibitory effect on *Serratia marcescens*, *Salmonella* Typhimurium, and *P. aeruginosa* on inoculated meat surfaces. In addition, the extractable nisin activity rapidly decreased when meat was incubated at ambient temperatures compared with 5 °C.

Nisaplin® (2.5 μg/l) reduced *L. monocytogenes* counts in scalding water by 1 log CFU/ml (from 5 log CFU/ml initial load), while the combination of heating (52 °C for 3 min) and Nisaplin® (2.5 μg/l) resulted in a 2 log CFU/ml reduction (Mahadeo and Tahini, 1994). The heat/nisin combination resulted in no detectable *L. monocytogenes* cells in the scalding water after 48 h. Similar initial population reductions were found when the same nisin/heat treatments were applied to $3 \times 2\,cm^2$ inoculated turkey skin surfaces. However, 3.5 log CFU/ml of *L. monocytogenes* was still viable after 48 h.

In another study, chicken drummettes were soaked for 30 min in nisin solutions containing either 25, 50, or 100 μg Nisaplin®/ml and 20 or 50 mM EDTA. These treatments resulted in a 2.4–2.8 log CFU reduction in total aerobic counts per ml of drummette rinse water after 21 days of refrigerated storage at 4 °C (Cosby *et al.*, 1999). Dykes and Moorhead (2002) evaluated Nisaplin® at 5000 IU/ml with and without a listeriophage (3000 CFU/ml) against *L. monocytogenes* in broth, buffer, and raw beef cubes. Nisin alone reduced the *L. monocytogenes* population from 9 to 1 log CFU/ml over 28 days of exposure, while nisin and listeriophage reduced the population from 9 to 2.5 log CFU/ml. Vacuum-packaged beef cubes dipped in the nisin or nisin plus listerophage solutions reduced the *L. monocytogenes* populations by 1 and 1.5 log CFU/ml, respectively, over 28 days of refrigerated storage.

Three bacteriocins (lactocin, enterocin, and nisin) produced by lactic acid bacteria were more inhibitory against *L. monocytogenes* and *L. innocua* when paired together than when tested individually (Vignolo *et al.*, 2000). Nisin (2000 IU/ml), lactocin 705 (17 000 AU/ml), and enterocin CRL 35 (17 000 AU/ml) reduced the *L. monocytogenes* populations in lean beef slurry by 4 log CFU/g after 10 and 24 h of exposure. While two-way bacteriocin combinations yielded reductions of 5 to 6 log CFU/g, the three-way combination resulted in no detectable cells (from an initial 8 log CFU/g inoculum) after 3, 10 and 24 h of exposure. The bacteriocin combinations were effective both in meat broth and meat systems. The anti-listerial activity of Nisaplin® at 400 or 800 IU/g in combination with 2% sodium chloride was evaluated in raw buffalo meat mince stored at 4 or 37 °C (Pawar *et al.*, 2000). Compared with the controls, the degree of nisin inhibition increased with increasing nisin concentration and with lower storage temperature. Moreover, the addition of 2% sodium chloride increased the efficacy of nisin at both storage temperatures.

Calcium alginate containing purified nisin (100 μg/ml) was more effective in inhibiting *Br. thermosphacta* (by 2 log CFU/g at day 1) than nisin alone when added to ground beef (Cutter and Siragusa, 1997). However, the calcium alginate/nisin mixture did not prevent outgrowth of the organism (Day 7, 6.0 log CFU/g; Day 14, >7 log CFU/g). Nattress *et al.* (2001) evaluated the ability of lysozyme and nisin to inhibit *Br. thermosphacta* and *Carnobacterium* sp. inoculated into culture media, pork juices, and onto lean pork and fat tissue. Lysozyme and nisin inhibited *Br. thermosphacta* to varying degrees in test media, meat juice, and tissue. Nisin (1000 IU/mg) and lysozyme (22 800 Shugar units/mg) and mixtures of nisin and lysozyme at 3:1, 1:1, and 1:3 ratios were tested at concentrations of 250, 500, and 1000 μg/ml of APT broth and at 125, 250, 500, and 1000 μg/ml of pork juice medium. The same preparation of nisin, lysozyme and their combinations at 1:3 and 1:1 nisin:lysozyme were used to coat lean and fat pork cores at a concentration of 65, 130, and 260 $\mu g/cm^3$. The Shugar unit is a measure of lysozyme activity based on the amount of enzyme in 1 ml of bacterial cell suspension that causes a 0.001 decrease in absorbance per minute measured at 450 nm. Nisin, lysozyme, and their mixtures inhibited *Br. thermosphacta* in broth at 250 μg/ml for 10 days at 2 °C while the 130 $\mu g/cm^3$ and greater concentrations of either treatment prevented growth on lean and fat cores for 6 weeks at 2 °C. Nisin and lysozyme together were also effective in reducing *Carnobacterium* sp. by 3 log CFU/g compared with untreated samples (Nattress *et al.*, 2001).

A nisin-producing lactic acid bacterial strain was inoculated into raw beef mince then stored five-to-seven days prior to being processed into frankfurters (Wang *et al.*, 1986). This approach resulted in an extension of the refrigerated shelf-life of the vacuum-packaged frankfurters compared with product produced with beef mince not inoculated with the nisin producer.

Pediocin (1400 AU/g), nisin (1400 AU/g), and Nisaplin® (500 IU/g) applied individually to raw vacuum-packaged beef reduced *Leuconostoc* spp. by 1–2 log CFU/g and kept bacterial populations below 6 log CFU/g for eight weeks at

33 °C (Rozbeth *et al.*, 1993). In another study, mechanically deboned poultry meat was treated with a combination of high pressure (350 MPa), nisin (100 μg/g), and 1% glucono-delta lactone as a means of extending shelf-life. This combination treatment reduced the mesophilic bacteria populations by 3.4 to 5.4 log CFU/g and held the psychrotrophic population below detection levels for 30 days at 2 °C (Yuste *et al.*, 1998). Examples of multifactorial nisin applications in raw meat and meat products are shown in Table 3.2.

3.5.2 Cooked meat products

Nisin may have a greater application for cooked or 'further-processed' products than raw products for several reasons. Firstly, many of these products are produced from minced meat, allowing for the homogeneous mixing of nisin throughout the product. Secondly, since most of these products are fully cooked, the primary location of bacterial recontamination and growth is limited more to the food surface, thus allowing nisin to be applied to the surface. Thirdly, many meat products are vacuum-packaged, creating a microenvironment that favors the growth of Gram-positive anaerobic or microaerophilic bacteria such as lactic acid bacteria that are more sensitive to nisin than Gram-negative bacteria, which grow more readily in an aerobic environment. Fourthly, proteolytic enzymes are inactivated in cooked products, thereby prolonging the activity of nisin.

Stankiewicz-Berger (1969) observed a protective action of 100 IU Nisaplin®/g against the greening of frankfurter slants by inhibiting *Lac. virdescens* and *Lactobacillus cellobiosus* populations by 5 and 2 log CFU/g, respectively. Similarly, frankfurter slants pasteurized at 72 °C for 30 minutes and treated with 100 IU Nisaplin®/g yielded population reductions of 7 and 4 log CFU/g, respectively, for *Lac. viridescens* and *Lac. cellobiosus*. Adding 200 IU Nisaplin®/g to frankfurters protected them against greening throughout a five-day period at 22 °C, while frankfurters with no nisin turned green in four days (Stankiewicz-Berger, 1969). Nisin (12.5 μg/g) combined with sorbic acid (0.125%) and monolaurin (0.25–0.5%) prevented growth of the spoilage microorganism *Br. licheniformis* on cured meats (Bell and deLacy, 1987). In related studies, 12.5 μg nisin/g prevented the outgrowth of *Br. licheniformis* spores in inoculated cooked luncheon meat stored at 20 °C for 10 days (Bell and deLacy, 1986).

Davies *et al.* (1999) demonstrated that the type of phosphate (added at 3 g/kg meat) used in minimally heated, vacuum-packed meats affected the efficacy of nisin against lactic acid bacteria (LAB). Meat products inoculated with LAB and treated with 25 μg/g Nisaplin® were unspoiled for 50 days at 8 °C while products without nisin spoiled after 7 days (>10^8 LAB/g). Nisin activity against LAB in these products decreased with increasing fat content. A longer preservation of vacuum-packaged sausage was achieved using 6.25–25 μg/g of nisin along with diphosphate as an emulsifier compared to orthophosaphate. Yang and Ray (1994) postulated that nisin could be added to control *Leuc. carnosum* and *Leuc.*

Table 3.2 Examples of multifactorial nisin applications in raw meat and meat products

Meat type and storage conditions	Nisin concentration* and co-agents	Bacterial target(s)	Effectiveness	Reference
Beef carcass after 21 days	5000 IU/ml spray	*L. innocua*	2.5 log CFU/g reduction	Cutter and Siragusa (1996a)
Beef carcasses	100 μg/ml Nisaplin® in alginate gel	*Br. thermosphacta*	2.4 log CFU/g reduction	Cutter and Siragusa (1996b)
Beef, lean tissue refrigerated for 7 days	10 μg/ml Nisaplin® in Fibrimex	*Br. thermosphacta*	4–8 log CFU/g reduction	Cutter and Siragusa (1998)
Beef, nisin solution applied to lean muscle	50 μg/ml Nisaplin®, 50 mM EDTA, 100mM lactate	*E. coli* O157:H7 and *Salmonella* Typhimurium	No effect	Cutter and Siragusa (1995b)
Beef, vacuum-packaged, at 4 °C for 56 days	200 IU/ml + 2% polylactic acid, 2% lactic acid	*Enterbacteriaceae*, *Pseudomonas*, psychrotrophic aerobes	Up to 2.6 log CFU/g reduction (acids alone or with nisin)	Ariyapitipun *et al.* (1999)
Beef, vacuum-packaged	40 IU/ml + 0.2% sorbate	*L. monocytogenes*	1 log CFU/g reduction	Avery and Buncic (1997)
Beef, raw cubes dipped then vacuum-packaged, stored at 4 °C for 28 days	5000 IU/ml nisin with or without 3000 PFU/ml listeriophage solution	*L. monocytogenes*	1 to 1.5 log reduction in rinse water	Dykes and Moorhead (2002)

Table 3.2 Continued

Meat type and storage conditions	Nisin concentration* and co-agents	Bacterial target(s)	Effectiveness	Reference
Chicken (broiler), nisin solution applied to drumsticks	25 μg/ml Nisaplin®, EDTA, citric acid, Tween 20	*Salmonella* Typhimurium; spoilage microorganisms	2 log CFU/g reduction; 3-day shelf-life extension	Shefet *et al.* (1995)
Chicken drummettes stored at 4 °C for 21 days	25–100 μg/ml Nisaplin® + 20 or 50 mM EDTA solution for 30 m	Total aerobic counts	2.4 to 2.8 log reduction in rinse water	Cosby *et al.* (1999)
Pork, lean and fat cores, coated with nisin, stored at 2 °C	1000 IU/mg nisin + lysozyme (22 800 Shugar units)	*Br. thermosphacta*	Nisin/lysozyme mixtures prevented growth for 6 weeks	Nattress *et al.* (2001)
Sausage, vacuum-packaged	6.25–25 μg/g Nisaplin® + diphosphate	Lactic acid bacteria	Inhibited outgrowth over 50 days *vs.* 7 days for controls	Davies *et al.* (1999)
Sausages, raw pork	10 μg/g + 100 ppm sorbic acid + 0.5% polyphosphate	*Br. thermosphacta,* coliforms, pseudomonads, and lactic acid bacteria	Delayed growth	Jarvis and Burke (1976)
Cured meats, luncheon meats at 20 °C for 10 days	12.5 μg/g Nisaplin®, 0.125% sorbic acid, 0.25–0.5% monolaurin	*Br. licheniformis*	Prevented growth on cured meats and outgrowth of spores on luncheon meats	Bell and deLacy (1986, 1987)

Ham and Bologna, vacuum-packaged, stored at 8 °C	500 mg/kg lysozyme:nisin (1:3) + 500 mg/kg EDTA	*L. monocytogenes*, *Br. thermosphacta*, *Leuc. mesenteriodes*, *E. coli* O157:H7, *Salmonella* Typhimurium	*Br. thermosphacta* and *L. mesenteriodes* inhibited in ham and Bologna, *L. monocytogenes* in Bologna, and *E. coli* and *S.* Typhimurium in ham	Gill and Holley (2000)
Buffalo mince, raw, stored at 4 °C or 37 °C	400 or 800 IU/ml Nisaplin® with and without 2% NaCl	*L. monocytogenes*	Nisin more effective at 4 °C and with NaCl	Pawar *et al*. (2000)
Ground meat	100 μg/ml Nisaplin® in calcium alginate	*Br. thermosphacta*	Nisin in calcium alginate more effective than nisin alone, 2 log CFU/g reduction on Day 1	Cutter and Siragusa (1997)
Sous vide pork and ground beef at 4 and 8 °C	250 or 500 IU/g Nisaplin®	*C. botulinum* spore outgrowth, total and lactic acid bacteria	Did not prevent *C. botulinum* spore outgrowth but inhibited total and lactic acid bacteria	Lindstrom *et al*. (2001)
Turkey skin in scalding water	2.5 μg/L Nisaplin® + 52 °C for 3 min	*L. monocytogenes*	No cells in water after 48 h, 3.5 log CFU/g on skin after 48 h	Mahadeo and Tahini (1994)

* 1 mg Nisaplin® is equivalent to 1000 IU (see also Chapter 2, Section 2.3).

mesentenoides, the primary spoilage organisms in low-heat processed, vacuum-packaged meats. Gill and Holley (2000) prepared a cooked ham and Bologna product with 500 mg/kg of a lysozyme : nisin (1 : 3) mixture and 500 mg/kg EDTA. Following cooking, products were inoculated with 10^4 CFU/cm^2 of one of nine different bacteria. Vacuum-packaged and refrigerated (8 °C) ham and Bologna treated with nisin/lysozyme/EDTA were not different in populations of *Br. thermosphacta* compared with controls. *Listeria monocytogenes* was reduced by 1 log CFU/g for two weeks at 8 °C but after four weeks the counts did not differ from the controls at 8 log CFU/cm^2. The nisin/lysozyme/EDTA treatment reduced *E. coli* O157:H7 in Bologna stored at 8 °C to 3.5 from 6.5 log CFU/g at two weeks; however, by four weeks the counts exceeded those in the controls. The nisin/lysozyme EDTA treatment reduced *Salmonella* Typhimurium from 4 to 2.5 CFU/cm^2 while the untreated samples grew to 7 log CFU/cm^2 by four weeks in ham and Bologna. In a similar study, surface populations of *L. monocytogenes* were reduced by over 2 log CFU/g in a cooked pork product using a surface treatment of 250 μg/ml nisin with or without modified atmosphere (80–100% CO_2) packaging (Fang and Lin, 1994) and in liver paté using a combination of nisin, lysozyme and citrate (Ter Steeg, 1993). Aymerich *et al.* (2002) evaluated nisin and other bacteriocins (sakacin and enterocins) and bacteriocin-producing cultures on cooked pork. Nisin (128 AU/g) inhibited *Leuc. carnosum* but not *Lac. sakei*. Moreover, nisin and sakacin (9.5 AU/g) withstood pasteurization temperatures, while interaction with the meat matrix did not significantly alter inhibitory activity. The enterocins and the bacteriocin-producing strains did not retain activity after pasteurization. In contrast, the addition of 100 mg/kg nisin (1000 IU/mg) to a ham mince did not inhibit aerobic, anaerobic, lactic acid or *Leuconostoc* spp. in dry-cured vacuum-packaged Chinese sausage stored at 20 °C (Wang, 2001). Moreover, Lindstrom *et al.* (2001) found that 250 IU/g or 500 IU/g Nisaplin® did not prevent the outgrowth of non-proteolytic *C. botulinum* spores in minimally processed *sous vide* pork cubes and ground beef stored at 4 and 8 °C. Five hundred IU/g Nisaplin® reduced the growth of lactic acid bacteria (from 5.5 to 2.2 $\log_{10}$ CFU/g) and total aerobic bacteria (from 5.6 to 1.3 $\log_{10}$ CFU/g). Both meat types showed botulinal growth and all beef treated with nisin and then stored at 8 °C became toxic within 21 to 28 days.

3.5.3 Seafood products

Eviscerated tilapia soaked in a nisin-brine solution prior to dehydration were shelf stable for three months at room temperature (Zaki *et al.*, 1976). Moreover, tilapia dipped in a nisin solution after gutting had an extended shelf-life of nine more days compared with fish not treated with nisin (El-Bedawey *et al.*, 1985).

Refrigerated seafood products that are stored for extended periods of time are susceptible to *L. monocytogenes* and *C. botulinum* outgrowth. Moreover, *C. botulinum* growth is magnified in vacuum or modified atmosphere packaging. In 1989, the FDA established a zero tolerance for *L. monocytogenes* in ready-to-eat seafood. High-value products such as cold smoked fish are of particular concern for contributing to listeriosis outbreaks since these products are often shipped through the mail and are usually not heated prior to being consumed. Duffes *et al.* (1999) found that semi-purified Nisaplin® reduced *L. monocytogenes* by 2 log CFU/g in cold-smoked, vacuum-packed salmon for 21 days at 4 °C and by 0.5 log CFU/g at 8 °C. When cold-smoked salmon slices were inoculated with 10^4 CFU/g *L. monocytogenes* and treated with nisin-producing *L. lactis* subsp. *lactis*, *L. monocytogenes* declined by only 0.5 log CFU/g and then increased at a slightly slower rate than in the untreated control at 10 °C (Wessels and Huss, 1996).

The combination of 0.25% lactic acid, 3% nisin whey permeate, and 2% sodium chloride inhibited the total aerobic microflora in minced rainbow trout (Nykanen *et al.*, 1998). In a follow-up study, Nykanen *et al.* (2000) injected 12 ml/200 g of a nisin (360 IU/g), sodium lactate (36 g/kg), or nisin combined with sodium lactate solution into pouches containing cold-smoked rainbow trout inoculated with *L. monocytogenes* that was subsequently stored at 8 °C for 17 days. The nisin, sodium lactate, and their combination treatments resulted in *L. monocytogenes* counts of 5.0, 2.9, and 1.9 log CFU/g after 17 days, respectively, compared with 6.0 log CFU/g for the untreated controls. Moreover, combining nisin with a heat treatment enhanced the killing effect against *L. monocytogenes* in cold-packaged canned lobster (Budu-Amoaka *et al.*, 1999). The combination of Nisaplin® (25 mg/kg) with retort processing at 60 ° C for 5 minutes or 65 ° C for 2 min increased the inhibition of *L. monocytogenes* in canned lobster by 3 and 5 log CFU/g, respectively compared with 1 to 3 log CFU/g with heat or nisin alone (Budu-Amoaka *et al.*, 1999). Degnan *et al.* (1994) evaluated the lactic acid bacteria fermentation products from Alta 2341 (2000 AU/ml), Enterocin 1083 (5000 AU/ml), and Nisaplin® (10 000 AU/ml) using 200 ml rinses per 200 g of blue crab meat. From an initial inoculum level of 10^5 log CFU/g, *L. monocytogenes* populations were 5, 4, 3.9, and 3 log CFU/g after six days at 4 °C for the controls, Alta, Enteriocin, and Nisaplin® treatments, respectively. Similar reductions of 1–2 log CFU/g were achieved by Nilsson *et al.* (1997) who treated cold-smoked salmon with a mixture of 500 or 1000 IU/g nisin solutions followed by packaging in a 70% carbon dioxide/30% nitrogen modified atmosphere. Following 45 days of storage, the final *L. monocytogenes* populations were 3, 5, and 7 log CFU/g, respectively, for the 1000, 5000, and 0 IU nisin/g treatments. In a subsequent study, Nilsson *et al.* (2000) found a synergistic effect between Nispalin (2.5 μg/ml) and 100% CO_2 atmosphere in controlling wild-type *L. monocytogenes* in cold-smoked salmon. The carbon dioxide atmosphere slowed the growth of both the nisin-resistant and wild-type strains of *L. monocytogenes*; however, the wild-type strain populations were reduced an additional 2 log CFU/g by the nisin plus CO_2 control.

The outgrowth of *C. botulinum* spores in modified atmosphere packaged seafood was prevented by spraying Nisaplin® solutions containing 8000 to 32 000 IU/ml (250 to 1000 IU/g) on fish prior to packaging (Taylor *et al.*, 1990). Nisaplin® treatment delayed the outgrowth of type E *C. botulinum* spores in cod

(1 day), herring (1–2 days), and smoked mackerel fillets (2-4 days) stored at both 10 °C and 26 °C. Moreover, the addition of 333 or 675 IU/g nisin extended the safe shelf-life of cod stored at 10 °C to 14 days compared with 9 days without the addition of nisin. Nisin sprays at 173 and 409 IU/g extended the safe shelf-life of smoked mackerel stored at 10 °C to 12 and 13 days, respectively, compared to 7 days when nisin was not used. Examples of applications of nisin in seafood products are summarized in Table 3.3.

3.5.4 Nisin in packaging materials

Ming *et al.* (1997) reported that casings containing nisin (5000 IU/g) and pediocin (640 000 AU/g) were ineffective in inhibiting *L. monocytogenes* on the surfaces of beef and turkey. However, casings sprayed with 9.3 $\mu g/cm^2$ of a pediocin solution completely inhibited *L. monocytogenes* growth for 12 weeks of refrigerated storage on turkey breast. Kassaify (1998) dipped collagen and natural meat casings into solutions containing various levels of Nisaplin® and found good retention of nisin activity. Natrajan and Sheldon (2000a) evaluated the inhibitory activity of three polymer films having varying hydrophobicities (polyvinyl chloride, linear low-density polyethylene (LLDPE), and nylon) and treated with a nisin formulation including 100 μg/ml nisin, and varying concentrations of EDTA (0., 5.0, 7.5 mM), citric acid (3.1, 3.0, 3.0%), and Tween 80 (0, 0.01, 0.5%) against *Salmonella* Typhimurium inoculated on broiler drumstick skin. The PVC-treated films were more effective than the other two films in inhibiting *Salmonella* Typhimurium (i.e. 1.6 log CFU/g reduction after 24–96 min at 4 °C). The two critical factors impacting the effectiveness of nisin were the presence of EDTA and the wetness of the chicken skin surface. In the same report, chicken drumsticks dipped in an aqueous nisin (50 μg/ml) solution for 3 min and then packaged with a nisin-treated PVC overwrap film in a foam tray containing an absorbent pad treated with 2 ml of a 50–150 μg/ml nisin solution extended the refrigerated shelf-life between one and two days based on either mesophilic or psychrotrophic bacterial counts. In a subsequent study (Natrajan and Sheldon, 2000b), chicken skin samples were coated with calcium alginate containing either 0, 200, 500, 600, or 1000 μg/ml nisin with 5.0 mM EDTA, 3.0% citric acid, and 0.5% Tween 80 and then stored at 4 °C. The level of kill was proportional to increasing exposure time and produced a 1.8 to 4.6 log CFU/g reduction in *Salmonella* Typhimurium populations after 96 hours of exposure. Cutter *et al.* (2001) incorporated 0.1% Nisaplin® and EDTA into polyethylene and polyethylene oxide films as a means for reducing *Br. thermosphacta* populations inoculated on vacuum-packaged beef. The inhibition achieved was from 1.7 to 3.5 log CFU/g over 21 days of refrigerated storage.

McCormick (2001) found that wheat gluten films containing 5% Nisaplin® reduced *L. monocytogenes* on inoculated in-package heat pasteurized Bologna slices from 2.5 log CFU/g to below the detection limit for eight weeks of refrigerated storage while counts in the controls remained at 3 log CFU/g (Fig. 3.1). Coma *et al.* (2001) tested the inhibitory properties of hydroxypropyl

Table 3.3 Examples of multifactorial nisin applications in seafood

Food type and storage conditions	Nisin concentration* and co-agents	Bacterial target(s)	Effectiveness	Reference
Cod, herring, smoked mackerel	250–1000 IU Nisaplin®/ ml spray	*C. botulinum* spore outgrowth	Delayed outgrowth 1 day (cod), 1–2 days (herring), and 2–4 days (smoked mackerel)	Taylor *et al.* (1990)
Crab, cooked meat stored for 6 days	10 000 AU/ml Nisaplin®/ml in 200 ml rinse of 200 g meat	*L. monocytogenes*	3 log CFU/g reduction from an initial 5 log CFU/g inoculation	Degnan *et al.* (1994)
Dried and fresh fish	Nisin or nisin–brine solutions	Total plate counts	3 month shelf-life for dried fish and 9 day shelf-life extension for fresh fish	Zaki *et al.* (1976); El-Bedawey *et al.* (1985)
Lobster, canned	25 mg/kg Nisaplin® with retorting at 65°C for 2 min	*L. monocytogenes*	3 log CFU/g reduction	Budu-Amoako *et al.* (1999)
Salmon slices stored at 10°C	300-fold rinse with *Lactococcus lactis* (nisin-producer)	*L. monocytogenes*	0.5 log CFU/g reduction	Wessells and Huss (1996)
Salmon, cold-smoked, stored at 8°C	500 or 1000 IU/g nisin solutions with 70% CO_2/30%N_2 MAP	*L. monocytogenes*	4 and 2 log CFU/g reductions for 100 and 500 IU/g after 45 days	Nykanen *et al.* (1998)
Salmon, cold-smoked, vacuum-packaged	Semi-purified nisin	*L. monocytogenes*	1 log reduction after 2 days at 4 or 8°C	Duffes *et al.* (1999)
Salmon, cold-smoked	2.5 μg Nisaplin®/ml with 100% CO_2	*L. monocytogenes*	2 log CFU/g reduction	Nilsson *et al.* (2000)
Salmon, cold-smoked, stored at 8°C	360 IU/g nisin with 36 g/kg sodium lactate	*L. monocytogenes*	4 log CFU/g reduction after 17 days	Nykanen *et al.* (2000)

* 1 mg Nisaplin® is equivalent to 1000 IU (see also Chapter 2, Section 2.3).

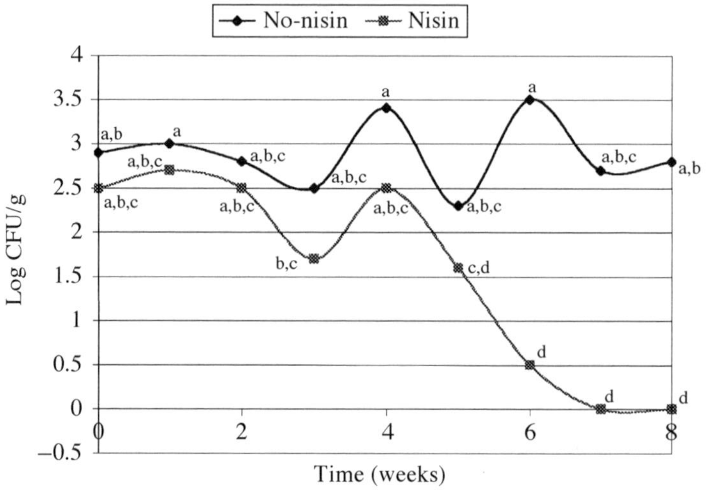

Fig. 3.1 Survival of *L. monocytogenes* in pasteurized, packaged turkey Bologna with nisin-impregnated or no-nisin packaging film. Data points represent means of three replications with duplicate plates. Data labels with different letters indicate significant differences between data points (McCormick, 2001).

methylcellulose films supplemented with nisin and up to 50% added fatty acids (stearic acid, oleic acid, methyl palmitate, or methyl stearate) against *L. innocua* and *Staph. aureus* and *Micrococcus luteus* in laboratory media. The presence of the fatty acids interfered with the antimicrobial activity of nisin. Further details on the uses of nisin in edible films and coatings are given in Chapter 12.

3.5.5 Nisin in absorbent tray pads

Absorbent tray pads containing bacterial inhibitors for controlling pathogens and spoilage microorganisms in the drip loss from fresh poultry and meat products represent a relatively recent development (Table 3.4). In a study by Sheldon *et al.* (1996), a formulation composed of 50 μg/ml nisin, 5 mM EDTA, 3% citric acid, and 5% Tween 20 was introduced into three commercial absorbent tray pads, inoculated with *Salmonella* Typhimurium or an exudate from a tray pad taken from a commercial pack of cut-up chicken, and then stored at 4 °C for up to 168 h. Reductions in count of mesophilic bacteria and *Salmonella* Typhimurium in the pads averaged 6.3 and 3.1 log CFU/g, respectively, after 168 h. Moreover, no detectable *Salmonella* Typhimurium were recovered from the nisin-treated pads. In related studies, Sheldon (2001) introduced several bactericidal treatments including nisin-containing formulations or sodium lactate into cellulose gum tray pads and subsequently challenged the pads with a fluid exudate from a pack of chicken pieces. The range of bacterial inhibition varied from 1.0 log CFU/g at 0.5 h to 4.2 log CFU/g after 168 h. Moreover, when nisin-treated pads were introduced into packs of skinless, boneless chicken

Table 3.4 Examples of studies on nisin applications in non-edible packaging and absorbent packaging pads

Food type and storage conditions	Nisin concentration* and co-agents	Bacterial target(s)	Effectiveness	Reference
Beef, nisin added to films before vacuum-packaging, stored at chill temperatures	Nisin and EDTA in PE and PE-oxide films	*Br. thermosphacta*	Reduced by 1.7 to 3.5 log CFU/g over 21 days	Cutter *et al.* (2001)
Chicken drumsticks packaged with PVC, LLDPE, and nylon films	100 μg nisin/ml film with 0–7.5 mM EDTA, 3.0–3.1% citric acid, 0–0.5% Tween 80	*Salmonella* Typhimurium	PVC films most effective, 0.4 to 2.1 log CFU/g reduction	Natrajan and Sheldon (2000a)
Chicken drumsticks, calcium alginate or agar applied to surface, stored at 4 °C	0–1000 μg nisin/ ml calcium alginate or agar with 5.0 mM EDTA, 3.0% citric acid, 0.5% Tween 80	*Salmonella* Typhimurium	1.8 to 4.6 log CFU/g reduction after 96 h	Natrajan and Sheldon (2000b)
Chicken pieces, commercial absorbent tray pads and pads inoculated with *S.* Typhimurium, stored at 4 °C for 168 h	50 μg/ml nisin, 5mM EDTA, 3% citric acid, 5% Tween 20 added to pads	*Salmonella* Typhimurium and mesophilic bacteria	6.3 and 3.1 log CFU/g reductions in mesophilic bacteria and *S.* Typhimurium, respectively	Sheldon *et al.* (1996)

Exudate from chicken pieces stored at 4 °C	50 μg/ml nisin, 5mM EDTA, 3% citric acid, 5% Tween 20 added to pads with sodium lactate	Mesophilic and psychrotrophic bacteria	4.2 log CFU/g reduction after 168 h and 4 day shelf-life extension	Sheldon (2001)
Meat casing, stored for 12 weeks under refrigeration	10% solutions of 5000 IU nisin /ml dialysed onto meat casing	*L. monocytogenes*	No effect	Ming *et al.* (1997)
Ham, nisin-dipped, then stored on pads and chilled	10 000 or 163 480 AU/ml Nisaplin® soaked then dried, cellulose pads and PE/nylon films	*L. innocua Staph. aureus*	*L. innocua* reduced by > 1 log and *Staph. aureus* by 2.8 log CFU/g	Scannell *et al.* (2000b)

* 1 mg Nisaplin is equivalent to 1000 IU (see also Chapter 2, Section 2.3).

breast fillets under a modified atmosphere, the shelf-life was extended from two to six days. Similarly, Scannell *et al.* (2000b) soaked absorbent cellulose pad inserts and polyethylene/polyamide packaging in Nispalin® solutions containing 10 240 and 163 480 AU/ml, respectively. After drying, the pads and polymer films were used to package sliced cheese and ham. Nisin-adsorbed cellulosic pads reduced *L. innocua* by more than 2 log CFU/g in both ham and cheese and by 2.8 log CFU/g for *Staph. aureus* in ham. When the foods were packaged under a modified atmosphere, lactic acid bacteria were reduced from 4 log CFU/g to below detection after 24 days under refrigeration. These findings clearly demonstrated the positive impact of using nisin-treated absorbent tray pads to extend product shelf-life and inhibit spoilage and pathogenic microflora in both raw and ready-to-eat proteinaceous foods.

3.5.6 Nisin in egg products

Addition of nisin to liquid egg products represents another novel approach to enhancing microbiological safety and shelf-life during refrigerated storage. Examples of some of these studies are given in Table 3.5. In addition, nisin may have potential as a thermal processing adjunct that would reduce the thermal resistance of vegetative bacteria and/or spores in food systems. Several investigators have suggested that the use of nisin or other bacteriocins could permit a reduction in the severity of thermal processes used for a variety of foods (Pflug and Holcomb, 1991). However, the varying susceptibility of microorganisms (even within a genus or species) to nisin and the diversity of the microbial flora which may contaminate foods dictate that a conservative approach must be taken to ensure consumer safety.

Further studies evaluating lower, more cost-effective levels of nisin as a thermal processing adjunct for liquid egg products would be worthy of investigation. Of particular interest would be the effect of such thermal processes on important egg-associated pathogens (e.g. *Salmonella* sp., *L. monocytogenes, B. cereus*). The influence of product pH on the solubility of nisin and its stability during heat processing and subsequent storage is an important consideration (Pflug and Holcomb, 1991). Schuman (1996) evaluated the effect of sublethal concentrations of nisin on the thermal resistance of *L. monocytogenes* in raw liquid whole egg. Dilute solutions of purified nisin in 0.02 N HCl were added to each sample of inoculated LWE (pH 6.5) to yield nisin concentrations of 0, 20, 40, or 60 IU/ml (marginal minimum inhibitory concentration for *L. monocytogenes*). Following an 18–20 h or 30 min incubation period at 4 °C, the D_{60}-value (D value at 60 °C) was determined. The presence of nisin reduced the D_{60}-value of *L. monocytogenes* by around 30%. The heat-sensitizing effect of nisin could conceivably be attributed to structural destabilization of the cell wall (i.e. sub-lethal injury). Alternatively, heat-induced injury may have made a proportion of the cells more susceptible to the biocidal activity of nisin.

In 1992 Delves-Broughton *et al.* evaluated the addition of nisin at 200 IU/ml (5 mg/l) as a means of extending the refrigerated shelf-life of conventionally

Table 3.5 Examples of multifactorial nisin applications in liquid whole egg (LWE) and albumen

Nisin concentration* (IU/g)	Bacterial target(s)	Treatment conditions	Effectiveness	Reference
560	*C. sporogenes*	LWE at 120.1 °C in phosphate buffer	Reduced *D*-values by 55 to 57%	Lewis *et al.* (1954)
8000	*Staph. aureus*	Albumen and LWE at 55 or 57 °C for 1.5 min	Enhanced destruction of *Staph. aureus* at 57 °C only	Niewiarowicz *et al.* (1980)
8000	*Staph. aureus*	LWE at 56 °C for 2 min	3.8 log CFU/g reduction with nisin and phosphate 2.8 log CFU/g reduction with nisin alone 0.4 log CFU/g reduction with heat alone	Niewiarowicz *et al.* (1980)
200	*B. cereus*	LWE pasteurized at 64.4 °C for 2.5 min	Extended shelf life (prevented *B. cereus* growth) from 6–11 days to 17–20 days at 6 °C	Delves-Broughton (1992)
100 to 250	Total aerobic plate count	Ultrapasteurized, aseptically packaged LWE	Greater than 12 week shelf- life (APC < 4 log CFU/g) when stored at 4.4 °C	Samimi (1992)
20 to 60	*L. monocytogenes*	LWE + nisin incubated at 4 °C for 18 or 0.5 h	D_{60} values reduced by 27–32%	Schuman (1996)
1000	*L. monocytogenes*	Ultrapasteurized LWE, pH adjusted to 7.5 or 6.6	Nisin delayed growth (pH 7.5) and prevented growth (pH 6.6) for 8 and 12 weeks at 4 and 10 °C, respectively	Schuman (1996) Sheldon and Schuman (1996)

* 1 mg Nisaplin® is equivalent to 1000 IU (see also Chapter 2, Section 2.3)

pasteurized LWE. Nisin was added to commercially processed raw LWE before pasteurization according to UK requirements (i.e. 64.4 °C for 2.5 min). In two trials, nisin-free controls had a shelf-life of 6 to 11 days at 6 °C. In contrast, pasteurized LWE samples containing nisin had a shelf-life of 17 to 20 days at 6 °C. Nisin-supplemented samples were free of detectable *B. cereus*, eventually spoiling due to the growth of *Pseudomonas* species. Control samples underwent an acidic-type spoilage due to the growth of *B. cereus*.

Nisin (at 100 to 250 IU/gram) was also evaluated as a biopreservative in commercially processed ultrapasteurized and aseptically packaged LWE containing 0.15% citric acid (Samimi, 1992). The presence of 250 IU/g of nisin (added before ultrapasteurization) yielded a shelf-life of $\geq$12 weeks at 4.4 °C. Based on a conservative criterion for the end of shelf-life (i.e. an aerobic plate count (APC) $>10^4$ CFU/g, regardless of whether sensory defects were detectable), nisin-supplemented ultrapasteurized LWE had a storage life of 3 days and 14 days under simulated abusive storage at 21.1 and 10 °C, respectively. Because of the small microbial populations present in the raw LWE and the severity of the thermal process, a nisin level as low as 100 IU/g was effective in delaying spoilage during in-plant shelf-life studies (Samimi, 1992).

3.6 Future prospects

Future applications of nisin in animal products will probably have two foci. First, since nisin reduces the thermal resistance of bacteria, it may be used in processed meat products to reduce the severity of the heat treatment, thereby improving the sensory quality of the product. As more and different technologies such as irradiation, high pressure, high-intensity light, and microwave processing become commercialized, the combination of nisin with these technologies would be a logical way to increase their efficacy. Nisin may have a second prospective application in packaging to inhibit bacterial growth during distribution. Processed meats and eggs have an extended shelf-life that is often limited by the outgrowth of psychrotrophic bacteria. Nisin-impregnated packaging that releases antimicrobial activity continuously during distribution and storage may prevent the outgrowth of bacteria for weeks and months after processing.

3.7 References

ARIYAPITIPUN, T., MUSTAPHA, A. and CLARKE, A.D. (1999) Microbial shelf life determination of vacuum-packaged fresh beef treated with polylactic acid, lactic acid, and nisin solutions. *J. Food Prot.* **62**:913–920.

ARIYAPITIPUN, T., MUSTAPHA, A. and CLARKE, A.D. (2000) Survival of *Listeria monocytogenes* Scott A on vacuum-packaged raw beef treated with polylactic acid, lactic acid and nisin. *J. Food Prot.* **63**:131–136.

AVERY, S.M. and BUNCIC, S. (1997) Antilisterial effects of a sorbate-nisin combination in vitro and on packaged beef at refrigerated temperature. *J. Food Prot.* **60**:1075–1080.

AYMERICH, M.T., GARRIGA, M., COSTA, S., MONFORT, J.M. and HUGAS, M. (2002) Prevention of ropiness in cooked pork by bacteriocinogenic cultures. *Int. Dairy J.* **12**:239–246.

BAILEY, J.S., FLETCHER, D.L. and COX, N.A. (1989) Recovery and serotype distribution of *Listeria monocytogenes* from broiler chickens in the southeastern United States. *J. Food Prot.* **52**:148–150.

BEAN, N.H., GOULDING, J.S., DANIELS, M.T. and ANGULO, F.J. (1997) Surveillance for foodborne disease outbreaks United States, 1988–1991. *J. Food Prot.* **60**:1265–1286.

BELL, R.G. and DELACY, K.M. (1986) Factors influencing the determination of nisin in meat products. *Food Technol.* **21**:1–7.

BELL, R.G. and DELACY, K.M. (1987) The efficacy of nisin, sorbic acid, and monolaurin as preservatives in pasteurized cured meat products. *Food Microbiol.* **4**:277–283.

BUDO-AMOAKO, E., ABLETT, R.F., HARRIS, J. and DELVES-BROUGHTON, J. (1999) Combined effect of nisin and moderate heat on destruction of *Listeria monocytogenes* in cold-pack lobster meat. *J. Food Prot.* **62**:46–50.

CHUNG, K., DICKSON, J.S. and CROUSE, J.D. (1989) Effects of nisin on growth of bacteria attached to meat. *Appl. Environ. Microbiol.* **55**:1329–1333.

COMA, V. SEBTI, I., PARDON, P., DESCHAMPS, A. and PICHAVANT, F.H. (2001) Antimicrobial edible packaging based on cellulose esters, fatty acids, and nisin incorporation to inhibit *Listeria innocua* and *Staphylococcus aureus*. *J. Food Prot.* **64**:470–475.

COSBY, D.E., HARRISON, M.A., TOLEDO, R.T. and CRAVEN, S.E. (1999) Vacuum or modified atmosphere packaging and EDTA-nisin treatment to increase poultry product shelf life. *J. Appl. Poult. Res.* **8**:185–190.

COVENTRY, M.J., MUIRHED, K. and HICKEY, M.W. (1995) Partial characterization of pedoicin PO2 and comparison with nisin for biopreservation of meat products. *Int. J. Food Microbiol.* **26**:133–145.

CUTTER, C.N. and SIRAGUSA, G.R. (1994) Decontamination of beef carcass tissue with nisin using a pilot scale carcass washer. *Food Microbiol.* **11**:481–489.

CUTTER, C.N. and SIRAGUSA, G.R. (1995a) Application of chlorine to reduce population of *Escherichia coli* on beef. *J. Food Safety.* **15**:67–75.

CUTTER, C.N. and SIRAGUSA, G.R. (1995b) Treatments with nisin and chelators to reduce *Salmonella* and *Escherichia coli* on beef. *J. Food Prot.* **58**:1028–130.

CUTTER, C.N. and SIRAGUSA, G.R. (1996a) Reductions of *Listeria monocytogenes* and *Brochothrix thermosphacta* on beef following nisin spray treatments and vacuum packaging. *Food Microbiol.* **13**:23–33.

CUTTER, C.N. and SIRAGUSA, G.R. (1996b) Reduction of *Brochothrix thermosphacta* on beef surfaces following immobilization of nisin in calcium alginate gels. *Letts. Appl. Microbiol.* **23**:9–12.

CUTTER, C.N. and SIRAGUSA, G.R. (1997) Growth of *Brochothrix thermosphacta* in ground beef following treatments with nisin in calcium alginate gels. *Food Microbiol.* **14**:425–430.

CUTTER, C.N. and SIRAGUSA, G.R. (1998) Incorporation of nisin into a meat binding system to inhibit bacteria on beef surfaces. *Letts. Appl. Microbiol.* **27**:19–23.

CUTTER, C.N., WILLETT, J.L. and SIRAGUSA, G.R. (2001) Improved antimicrobial activity of nisin-incorporated polymer films by formulation change and addition of food grade chelator. *Letts. Appl. Microbiol.* **33**:325–328.

CYNAROWITZ-PROVOST, M., WHITING, R.C. and CRAIG (JR), J.C. (1994) Steam surface pasteurization of beef framkfurters. *J. Food Sci.* **59**:1–5.

DAVIES, E.A., MILNE, C.F., BEVIS, H.E., POTTER, R.W., HARRIS, J.M., WILLIAMS, G.C., THOMAS, L.V. and DELVES-BROUGHTON, J. (1999) Effective use of nisin to control lactic acid bacterial spoilage in vacuum-packed bologna-style sausage. *J. Food Prot.* **62**:1004–1010.

DE VUYST, L. and VANDAMME, E.J. (1994) Nisin, a lantibiotic produced by *Lactococcus lactis* subsp. *lactis*: properties, biosynthesis, fermentation, and applications. In *Bacteriocins of Lactic acid Bacteria: Microbiology, Genetics, and Applications*. L. de Vuyst and E.J. Vandamme (eds), pp. 151–221. Elsevier Applied Sciences. London.

DEGNAN, A.J., KASPAR, C.W., OTWELL, W.S., TAMPLIN, M.L. and LUCHANSKY, J.B. (1994) Evaluation lactic acid bacterium fermentation products and food-grade chemicals to control *Listeria monocytogenes* in blue crab (*Callinectes sapidus*) meat. *Appl. Environ. Microbiol.* **60**:3198–3203.

DELVES-BROUGHTON, J., WILLIAMS, G.C. and WILKINSON, S. (1992) The use of the bacteriocin, nisin, as a preservative in pasteurized liquid whole egg. *Letts. Appl. Microbiol.* **15**:133–136.

DELVES-BROUGHTON, J., BLACKBURN, P., EVANS, R.J. and HUGENHOLTZ, J. (1996) Applications of the bacterocins nisin. *Antonio von Leeuwenhoek* **69**:193–202.

DORSA, W.J., CUTTER, C.N. and SIRAGUSA, G.R. (1996a) Effectiveness of a steam-vacuum sanitizer for reducing *Escherichia coli* O157:H7 inoculated beef carcasses surface tissue. *Letts. Appl. Microbiol.* **23**:61–63.

DORSA, W.J., CUTTER, C.N., SIRAGUSA, G.R. and KOOHMARAIE, M. (1996b) Microbial decontamination of beef and sheep carcasses by steam, hot water spray washes, and a steam vacuum sanitizer. *J. Food Prot.* **59**:127–135.

DORSA, W.J., CUTTER, C.N. and SIRAGUSA, G.R. (1997a) Effects of acetic acid, lactic acid, and trisodium phosphate on the microflora of refrigerated beef carcass surface tissue inoculated with *Escherichia coli* O157:H7, *Listeria innocua*, and *Clostridium sporogenes*. *J. Food Prot.* **60**:619–624.

DORSA, W.J., CUTTER, C.N. and SIRAGUSA, G.R. (1997b) Effects of steam-vacuuming and hot water spray wash on the microflora of refrigerated beef carcass tissue inoculated with *Escherchia coli* O157:H7, *Listeria innocua*, and *Clostridium sporogenes*. *J. Food Prot.* **60**: 114–119.

DUFFES, F. CORRE, C., LEROI, F. DOUSSET, X. and BOYAVOL, P. (1999) Inhibition of *Listeria monocytogenes* by in situ produced and semipurified bacteriocins of *Carnobacterium* spp. on vacuum-packed, refrigerated cold-smoked salmon. *J. Food Prot.* **62**:1394–1403.

DYKES, G.A. and MOORHEAD, S.M. (2002) Combined antimicrobial effect of nisin and a listerophage against *Listeria monocytogenes* in broth but not in buffer or on raw beef. *Int. J. Food Microbiol.* **73**:71–81.

EL-BEDAWEY, A.E., EL-SHERBINY, A.M., ZAKI, M.S. and KHALIL, A.H. (1985) The effect of certain antibiotics on the keeping quality of bolti fish (*Tilapia nilotica*). *Nahr.* **29**:665–670.

FANG, T.J. and LIN, C.C. (1994) Growth of *Listeria monocytogenes* and *Pseudomonas fragi* in cooked pork in a modified atmosphere packaging/nisin combination system. 1994. *J. Food Prot.* **57**:479–485.

FANG, T.J. and LIN, C.C. (1995) Inhibition of *Listeria monocytogenes* on pork tissue by immobilized nisin. *J. Food Drug Analysis.* **3**:269–274.

FRANCO, C.M., QUINTO, E.J., FENTE, C.A., MENENDEZ, S., VAZQUEZ, B., DOMINQUEZ, L. and CEPEDA, A. (1997) Detection of *Listeria monocytogenes* on selective media without subculturing in food. *Arch. Lebesmittelhyg.* **49**:54–56.

FSIS DIRECTIVE # 63501. (1996) Food Safety and inspection Service. US Department of Agriculture, Washington, DC.

GILL, A.O. and HOLLEY, R.A. (2000) Inhibition of bacterial growth on ham and bologna by lysozyme, nisin, and EDTA. *Food Res. Int.* **33**:83–90.

HAMBY, P.L. SAVELL, J.W., ACUFF, G.R., VANDERZANT, C. and CROSS, H.R. (1987) Spray-chilling and carcass decontamination systems using lactic acid and acetic acid. *Meat Sci.* **21**:1–14.

HENNING, S., METZ, R. and HAMMES, W.P. (1986) Studies on the mode of action of nisin. *Int. J. Food Microbiol.* **3**:121–134.

HURST, A. and HOOVER, D.G. (1993) Nisin. In *Antimicrobials in Foods*, 2nd edition P.M. Davidson and A.L. Bronen (eds). Marcel Dekker, New York.

JAMES, W.O., BREWER, R.L., PRUCHA, J.C., WILLIAMS, W.O. and PARHAM, D.R. (1992) Effects of chlorination of chill water on the bacteriologic profile of raw chicken carcasses and giblets. *J. Am. Vet. Med. Assoc.* **200**:60–63.

JARVIS, B. (1967) Resistance to nisin and production of nisin-inactivating enzymes by several *Bacillus* species. *J. Gen. Microbiol.* **47**:33–48.

JARVIS, B. and BURKE, C.S. (1976) Practical and legislative aspects of the chemical preservatives of food. In *Inhibition and Inactivation of Vegetative Microbes.* F.A. Skinner and W.B. Hugo (eds). pp. 345–367. Academic Press, New York.

JARVIS, B. and FARR, J. (1971) Partial purification, specificity, and mechanism of action of the nisin-inactivating enzyme from *Bacillus cereus. Biochim. Biophys. Acto* **227**:232–240.

JARVIS, B. and MAHONEY, R.R. (1969) Inactivation of nisin by alpha-chymotrypsin. *J. Dairy Sci.* **52**:1148–1150.

JOHNSON, M.J., TITUS, T.C., MCCASKILL, L.H. and ACTON, J.C. (1979) Bacterial counts on surfaces of carcasses and in ground beef from carcasses sprayed or not sprayed with hypochlorous acid. *J. Food Sci.* **44**:169–173.

JUNG, D.S., BODYFELT, F.W. and DAESCHEL, M.A. (1992) Influence of fat and emulsifier on the efficacy of nisin in inhibiting *Listeria monocytogenes* in fluid milk. *J. Dairy Sci.* **75**:387–393.

KASSAIFY, Z.G. (1998) The potential use of nisin on fresh sausages and onto sausage casings. M.Sc. Thesis. Oxford University, Oxford.

KIM, J.G. YOUSEF, A.E. and DAVIES, S. (1999) Application of ozone for enhancing the microbiological safety and quality of food: a review. *J. Food Prot.* **62**:1071–1087.

LEWIS, J.C., MICHENER, H.D., STUMBO, C.R. and TITUS, D.S. (1954) Additives accelerating death of spores by moist heat. *J. Agric. Food Chem.* **2**:298–302.

LINDSTROM, M., MOKKILA, M., SKYTTA, E., HYYTIA-TREES, E., LAHTEENMAKI, L., HIELM, S., AHVENAINEN, R. and KORKEALA, H. (2001) Inhibition of growth of nonproteolytic *Clostridium botulinum* type B in sous vide cooked meat products is achieved by using thermal processing but not nisin. *J. Food Prot.* **64**:838–844.

MAHADEO, M. and TAHINI, S.R. (1994) The potential use of nisin to control *Listeria monocytogenes* in poultry. *Letts. Appl. Microbiol.* **18**:323–326.

MASSCHALCK, B., VAN HOUDT, R. and MICHIELS, C.W. (2001) High pressure increases bacteriocidal activity and spectrum of lactoferrin, lactoferricin, and nisin. *Int. J. Food Microbiol.* **64**:325–332.

MCCORMICK, K. (2001) In-package Pasteurization Combined with Nisin-impregnated Packaging Film to Inhibit *Listeria monocytogenes* and *Salmonella* Typhimurium on ready-to-eat meat. MS Thesis, Clemson University, Clemson, SC.

MCEVOY, J.M. DOHERTY, A.M., SHERIDAN, J.J., BLAIR, I.S. and MCDOWELL, D.A. (2001) Use of steam condensing at subatmospheric pressures to reduce *Escherichia coli* O157:H7 numbers on bovine hide. *J. Food Prot.* **64**(11):1655–1660.

MING, X., WEBER, G.H., AYERS, J.W. and SANDINE, W.E. (1997) Bacteriocins applied to food packaging materials to inhibit *Listeria monocytogenes* on meats. *J. Food Sci.* **62**:413–415.

MORRISON, G.J. and FLEET, G.H. (1985) Reduction of *Salmonella* on chicken carcasses by immersion treatments. *J. Food Prot.* **48**:939–943.

MURRAY, M. and RICHARD, J.A. (1998) Comparative study on the antilisterial activity of nisin A and pediocin ACH in fresh ground pork stored aerobically at 5 °C. 1998. *J. Food Prot.* **60**:1534–1540.

NATRAJAN, N. and SHELDON, B.W. (2000a) Efficacy of nisin-coated polymer films to inactivate *Salmonella* Typhimurium on fresh broiler skin. *J. Food Prot.* **63**:1189–1196.

NATRAJAN, N. and SHELDON, B.W. (2000b) Inhibition of *Salmonella* on fresh poultry skin using protein- and polysaccharide-based films containing nisin formulations. *J. Food Prot.* **63**:1268–1272.

NATTRESS, F.M., YOST, C.K. and BAKER, L.D. (2001) Evaluation of the ability of lysozyme and nisin to control meat spoilage bacteria. *Int. J. Food Microbiol.* **70**:111–119.

NIEWIAROWICZ, A., STAWICKI, S., KUJAWSKA-BIERNAT, B. and LASKOWSKA, D. (1980) The influence of nisin and sodium polyphosphate in the lowering of the heat resistance of bacteria in egg products. *Medycyna Weterynaryjna* **36**:303–306 (in Polish).

NILSSON, L., HUSS, H.H. and GRAM, L. (1997) Inhibition of *Listeria monocytogenes* on cold-smoked salmon by nisin and carbon dioxide atmosphere. *Int. J. Food Microbiol.* **38**:217–227.

NILSSON, L., CHEN, Y., CHIKINDAS, M.L., HUSS, H.H., GRAM, L. and MONTVILLE, T.J. (2000) Carbon dioxide and nisin act synergistically on *Listeria monocytogenes*. *Appl. Environ. Microbiol.* **66**:769–774.

NYKANEN, A., LAPVETELAINEN, N., HIETANEN, R.M. and KALLIO, H. (1998) Acceptability of lactic acid and nisin to improve the microbiological quality of cold-smoked rainbow trout. *Z. Lebensm. Unters. Forsch.* **208**:116–120.

NYKANEN, A., WECKMAN, K. and LAPVETALAINEN, A. (2000) Synergistic inhibition of *Listeria monocytogenes* on cold-smoked rainbow trout by nisin and sodium lactate. *Int. J. Food Microbiol.* **61**:63–72.

PAWAR, D.D., MALIK, S.V.S., BHILEGAONKAR, K.N. and BARBUDDHE, S.B. (2000) Effect of nisin and its combination with sodium chloride on the survival of *Listeria monocytogenes* added to raw buffalo meat mince. *Meat Sci.* **56**:215–219.

PFLUG, I.J. and HOLCOMB, R.G. (1991) Principles of the thermal destruction of microorganisms, In *Disinfection, sterilization, and preservation*, S.S. Block (ed.), Ch. 6. Lea and Febiger, Philadelphia, PA.

PHEBUS, R.K., NUTSCH, A.L. SCHAFER, D.E., WILSON, R.C., RIEMANN, M.J., LEISING, J.D., KASTNER, C.L., WOLF, J.R. and PRASAI, R.K. (1997) Comparison of steam pasteurization and other methods for reduction of pathogens on surfaces of freshly slaughtered beef. *J. Food Prot.* **60**(5):476–484.

PINI, P.N. and GILBERT, R.J. (1988). The occurrence in the UK of *Listeria* species in raw chickens and soft cheeses. *Int. J. Food Microbiol.* **6**:317–326.

PIRTTIJARVI, T.S.M., WAHLSTROM, G., RAINEY, F.A., SARIS, P.E.J. and SALANEN, M.S. (2001) Inhibition of bacilli in industrial starches by nisin. *J. Ind. Microbiol. Biotechnol.* **26**:107–114.

RORVIK, L.M. and YNDESTAD, M. (1991) *Listeria monocytogenes* in foods in Norway. *Int. J. Food Microbiol.* **13**:97–104.

ROSE, B.E., HILL, W.E., UMHOLTZ, R., RANSOM, G.M. and JAMES, W.O. (2002) Testing for *Salmonella* in raw meat and poultry products collected at federally inspected establishments in the United States, 1998 through 2000. *J. Food Prot.* **65**:937–947.

ROZBETH, M., KALCHAYANAND, N., FIELD, R.A., JOHNSON, M.C. and RAY, B. (1993) The influence of biopreservatives on the level of refrigerated vacuum-packaged beef. *J. Food Safety.* **13**:99–111.

RUSSELL, S.M. (1998) Chemical sanitizing agents and spoilage bacteria on fresh broiler carcasses. *J. Appl. Poult. Res.* **7**:273–280.

SALVAT, G. (1996) Effects of phosphate treatment on the microbiological flora of poultry carcasses. In *COST Action 97, Status and Prospects of Decontamination and Preservation of Poultry and Egg Products*. Ploufragan, France.

SAMIMI, M.H. (1992) Personal communication. M.G. Waldbaum Co., Gaylord, MN.

SCANNELL, A.G.M., HILL, C., BUCKLEY, D.J. and ARENDT, E.K. (1997) Determination of the influence of organic acid and nisin on shelf life and microbiological safety aspects of fresh pork sausage. *J. Appl. Microbiol.* **83**:407–412.

SCANNELL, A.G.M., ROSS, R.P., HILL, C. and ARENDT. E.K. (2000a) An effective lacticin biopreservatives in fresh pork sausage. *J. Food Prot.* **63**:370–375.

SCANNELL, A.G.M., HILL, C., ROSS, R.P., MARX, S., HARTMEIER, W. and ARENDT, K. (2000b) Development of bioactive food packaging materials using immobilized bacteriocins Lacticin 3147 and Nisaplin®. *Int. J. Food Microbiol.* **60**:241–249.

SCHUMAN, J.D., (1996) Thermal and biological inactivation of bacterial pathogens in liquid egg, Doctoral Dissertation, North Carolina State University.

SCOTT, V.N. and TAYLOR, S.L. (1981) Temperature, pH, and spore load effects on the ability to prevent the outgrowth of *Clostridium botulinum* spores. *J. Food Sci.* **46**:121–126.

SHEFET, S.M., SHELDON, B.W. and KLAENHAMMER, T.R. (1995) Efficacy of optimized nisin-based treatments to inhibit Salmonella Typhimurium and extend shelf life of broiler carcasses. *J. Food Prot.* **58**:1007–1082.

SHELDON, B.W. (2001) Development of an inhibitory absorbent cellulose gum tray pad for reducing spoilage microorganisms and the risk of cross-contamination. *Poultry Sci.* **80**(supplement 1):17.

SHELDON, B.W. and SCHUMAN, J.D. (1996) Thermal and biological treatments to control psychrotrophic pathogens. *Poultry Sci.* **75**:1126–1132.

SHELDON, B.W., HALE, S.A. and BEARD, B.M. (1996) Efficacy of incorporating nisin-based formulations into absorbent meat tray pads to control pathogenic and spoilage microorganisms. *Poultry Sci.* **75**(supplement 1):97.

SIRAGUSA, G.R. (1995) The effectiveness of carcass decontamination systems for controlling the presence of pathogens on the surfaces of meat animal carcasses. *J. Food Sci.* **15**:227–238.

SOMERS, E.B., SCHOENI, J.L. and WONG, A.C. (1994) Effect of trisodium phosphate on biofilm and planktonic cells of *Campylobacter jejuni*, *Escherichia coli* O157:H7, *Listeria monocytogenes*, and *Salmonella typhimurium*. *Int. J. Food Microbiol.* **22**:269–276.

STANKIEWICZ-BERGER, H. (1969) Effect of nisin on the *Lactobacilli* that cause greening of cured meat precuts. *Acta Microbiol. Polon. Ser. B* **13**:117–120.

STERN, N.J. (1982) Selectivity and sensitivity of three media for the recovery of inoculated *Campylobacter fetus* spp. *Jejuni* from ground beef. *J. Food Safety* **4**:159–175.

STEVENS, K.A., SHELDON, B.W., KLAPES, N.A. and KLAENHAMMER, T.R. (1992) Effect of treatment conditions on nisin inactivation of Gram-negative bacteria. *J. Food Prot.* **55**:763–766.

STEVENSON, K.E., MERKEL, R.A. and LEE, H.C. (1978) Effects of chilling rate, carcass fatness, and chlorine spray on microbiological quality and case life of beef. *J. Food Sci.* **43**:849–852.

TAYLOR, L.Y., CANN, D.D. and WELCH, B.J. (1990) Antibotulinal properties of nisin in fresh fish packaged in an atmosphere of carbon dioxide. *J. Food Prot.* **53**:953–957.

TER STEEG, P.F. (1993) Interacties tussen Nisine, Lysozym en Citrat in Bioconserving. *De Ware Chemicus.* **23**:183–190.

VIGNOLO, G., PALACIOS, J., FARIAS, M.E., SESMA, F., SCHILLINGER, U., HOLZAPFEL, W. and OLIVER, G. (2000) Combined effect of bacteriocins on the survival of various *Listeria* species in broth and meat system. *Curr. Microbiol.* **41**:410–416.

WABECK, C.J., SCHURALL, D.V., EVANCHO, G.M., HECK, J.G. and ROGERS, A.B. (1968) *Salmonella* and total count reduction in poultry treated with sodium hypochlorite solutions. *Poultry Sci.* **47**:1090–1094.

WANG, F-S. (2001) Effects of three preservative agents on the shelf life of vacuum packaged Chinese-style sausage stored at 20 °C. *Meat Sci.* **56**:67–71.

WANG, S.Y., DOCKERTY, T.R., LEDFORD, R.A. and STOUFFER, J.R. (1986) Shelf life extension of vacuum-packaged frankfurters made from beef inoculated with *Streptococcus lactis*. *J. Food Prot.* **49**:130–134.

WESSELLS, S. and HUSS, H.H. (1996) Suitability of *Lactococcus lactis* subsp. *lactis* ATCC 11454 as a protective culture for lightly preserved fish products. *Food Microbiol.* **13**:323–332.

WHYTE, P., COLLINS, J.D., MCGILL, K., MONAHAN, C. and O'MAHONY, H. (2001) Quantitative investigation of the effects of chemical decontamination procedures on the microbiological status of broiler carcasses during processing. *J. Food Prot.* **64**:179–183.

YANG, R. and RAY, B. (1994) Prevalence and biological control of bacteriocin-producing psychrotrophic leuconostocs associated with spoilage of vacuum-packaged meats. *J. Food Prot.* **57**:209–217.

YUSTE, J., MOR-MOR, M. CAPELLAS, M., GUAMIS, B. and PLA, R. (1998) Microbiological quality of mechanically recovered poultry meat treated with high hydrostatic pressure and nisin. *Food Microbiol.* **15**:407–414.

ZAKI, M.S., EL MANSY, A.H., HASSAN, Y.M. and RAHMA, E.H.A. (1976) Effect of nisin in saturated brine and storage on the quality of dried bolti fish (*Tilapia nilotica*). *Die Nahrung* **20**:691–697.

ZHANG, S. and MUSTAPHA, A. (1999) Reduction of *Listeria monocytogenes* and Escherichia coli 0157:H7 numbers on vacuum-packaged fresh beef treated with nisin or nisin combined with EDTA. *J. Food Prot.* **62**:1123–1127.

4

Bacteriocins other than nisin: the pediocin-like cystibiotics of lactic acid bacteria

B. Ray and K. W. Miller, University of Wyoming, USA

4.1 Introduction: the lactic acid bacteria (LAB)

The group of bacteria known as the lactic acid bacteria (LAB) currently include species in the genera *Lactococcus, Streptococcus* (one species only), *Enterococcus, Pediococcus, Tetragenococcus, Aerococcus, Alloiococcus, Oenococcus, Vagococcus, Lactosphera, Leuconostoc, Weissella, Lactobacillus, Dolosigranulum, Globicatella*, and *Carnobacterium* (Axelson, 1998; Ray, 2000). Species from the genus *Bifidobacterium* are not considered in the LAB group but they share many of the characteristics of LAB. Several of these genera have been added fairly recently. Phylogenetic analysis by 16S rRNA sequencing has aided the creation of new genera and the re-classification of old species. Given the heterogeneity in morphological, physiological, and biochemical properties of species belonging to *Lactobacillus, Leuconostoc*, and *Enterococcus*, it can be assumed that additional new genera will be created in the future (Sneath, 1986; Yang and Ray, 1994a).

Originally, the term 'lactic acid bacteria' was coined by individuals working on dairy fermentations to indicate desirable species/strains that produce large amounts of lactic acid from the metabolism of lactose. These bacteria are also designated by the generic term 'starter cultures' as they are used to initiate fermentations. Over time, both terms have been used interchangeably for bacteria associated with production of dairy, meat, vegetable, and other fermented food products. Since humans have been consuming the live cells and metabolites of these bacteria for a long time without adverse or hazardous effects on health, starter culture bacteria are considered as safe, food grade, and even some as beneficial for the well-being of consumers (Ray and Daeschel, 1992; Salminen and von Wright, 1998). Presently, it is recognized that only a few species of *Lactococcus, Lactobacillus, Leuconostoc*, and *Pediococcus* are associated with food fermentation, and several species of *Lactobacillus* and *Bifidobacterium* have a beneficial effect on the normal health of the human gastrointestinal tract. While these species can be considered as food grade and safe, the food grade status of other species in these genera that are not used in food fermentations is unclear. Some species/strains of *Enterococcus* are pathogenic, isolation of *Lactobacillus* and *Pediococcus* from human infections has been reported (Ray, 2000) and there is a lack of real health hazard information for many of these species. Therefore, it is not unjustifiable to question whether the two terms 'lactic acid bacteria' and 'food grade bacteria' can be used synonymously. Until this is clarified, it is probably better to consider only starter culture lactic acid bacteria as food grade and safe, especially when live cells are consumed in foods.

4.2 Bacteriocin-producing lactic acid bacteria

One of the most important characteristics of LAB is their ability to produce diverse metabolites with antimicrobial properties. Some of these agents have been characterized, but many remain to be identified. Depending upon the type of species/strain and the nutritional, physical, and chemical environments of growth, LAB have been shown to produce lactic acid, acetic acid, ethanol, diacetyl, CO_2 (as carbonic acid), H_2O_2, reuterine, derivatives of lactic acid (hydroxy lactic acid), and small peptides (Ray and Daeschel, 1992). These antimicrobials can either inhibit or kill target microorganisms such as molds, yeasts, vegetative bacteria, bacterial spores, and even viruses. The spectrum of antimicrobial action varies depending upon the specific metabolite being considered.

In recent years, there has been an explosion of research activity in the area of antibacterial peptides produced by many strains of LAB. These ribosomally synthesized, heat-resistant, cationic, and relatively small peptides are known collectively as bacteriocins of the LAB (note that bacteriocins are also produced by many other Gram-positive and Gram-negative bacteria). Many reports have been published describing the isolation of bacteriocin-producing LAB strains (Ray and Daeschel, 1992; Moll *et al.*, 1999; Nes and Holo, 2002). Several studies have shown that different strains of a species or even different species can produce the same bacteriocin (Ray and Miller, 2000). Because in many cases the amino acid sequences of the bacteriocins and/or the nucleotide sequences of the structural genes have not been determined, it is possible that some of the newly isolated bacteriocin-producing strains actually produce peptides that have been previously identified in other bacteria. In addition, not all researchers have used standard, reliable methods to confirm that the activity observed is due to a bacteriocin. For this reason, it is possible that some studies might have been conducted with non-producing strains.

4.3 Class II bacteriocins and cystibiotics of lactic acid bacteria

Bacteriocins produced by different species/strains of LAB differ greatly in physical and chemical characteristics. Broadly, bacteriocins are grouped based on the presence or absence of the unusual amino acids, lanthionine and β-lanthionine (Jack *et al.*, 1995). Nisin A and Z, lacticin 481, lactocin S, and lacticin 3147 are examples of lanthionine-containing or class I bacteriocins of LAB. These peptides are also designated as lantibiotics. Nisin is discussed in Chapters 2 and 3 in this book. In this chapter, we focus on the non-lanthionine bacteriocins of the LAB, especially those containing disulfide (–S–S) bridge(s).

A system used to classify non-lanthionine bacteriocins is presented in Table 4.1. Overall, these peptides can be divided into the class II group, which contains the heat stable, small (<10 kDa) peptides, and class III, which contains the heat labile, relatively large (20 kDa) peptides. Class II peptides have been isolated more frequently and currently are divided into four subgroups. Notably, different researchers have used other terms to name these subgroups. Furthermore, these groupings are tentative. In the future, as more bacteriocins are discovered and characterized, a better classification system can be adopted. The bacteriocins in class IIa, which are also called pediocin-like, cystibiotic, or anti-listerial type peptides, are composed of a single cationic peptide with an N-terminal YGNGV consensus sequence and at least one disulfide bridge formed

Table 4.1 Classification of non-lanthione bacteriocins of the LAB

Class (other names)	Important characteristics	Representative types
Class II	Small heat stable peptides	
IIa (cystibiotics, pediocin-like, antilisterial)	Single peptide, YGNGV sequence, ABC-dependent secretion	(a) Four cysteines with two disulfide bridges: pediocin PA-1/AcH (b) Two cysteines with one disulfide bridge: leucocin A
IIb (two-peptide, or two-component)	Two separate peptides, both needed for optimum activity	Plantaricin S
IIc (*sec*-dependent)	*Sec*-dependent secretion	Divergicin A
IId (miscellaneous)	Differ with types	(a) Thiolbiotics: one SH: lactococcin A; two SH: carnobacteriocin A (b) No thiol group: lactococcin A (c) No leader peptide: enterocin I
Class III	Large heat labile peptides	Pediocin A

Compiled from: Klaenhammer (1993), Jack *et al.* (1995), Moll *et al.* (1999), Ray *et al.* (2001b), Nes and Holo (2002).

Table 4.2 Cystibiotics of the LAB

Initial names	Producing species and strains	Source
Pediocin PA-1+	*Pediococcus acidilactici* Pac 1.0	Culture collection
	Pediococcus parvulus AOT 77	Vegetables
Pediocin AcH	*Pediococcus acidilactici* H	Fermented meat
	Pediococcus pentosaceus S34	Buffalo milk
	Lactobacillus plantarum WHE92	Munster cheese
Enterocin A	*Enterococcus faecium* CTC 492	Fermented sausage
	Enterococcus faecium DPC 1146	Fermented sausage
Divercin 41	*Carnobacterium divergens* V41	Fish viscera
Leucocin A*	*Leuconostoc gelidum* UAL 187	Processed meat
Leucocin B-TA11a*	*Leuconostoc carnosum* TA11a	Processed meat
Leucocin A-TA33a*	*Leuconostoc mesenteroides* TA33a	Processed meat
Leucocin C-TA33a	*Leuconostoc mesenteroides* TA33a	Processed meat
Mesentericin Y105	*Leuconostoc mesenteroides* Y105	Goat milk
Sakacin A*	*Lactobacillus sake* Lb 706	Raw meat
Curvacin A*	*Lactobacillus curvatus* LHT 1174	Fermented sausage
Sakacin P*	*Lactobacillus sake* LTH 673	Meat
Sakacin 674*	*Lactobacillus sake* Lb 674	Meat
Bavaricin A*	*Lactobacillus bavaricus* MI 401	Sourdough
Bavaricin MN	*Lactobacillus sake* MN	Meat
Acidocin A	*Lactobacillus acidophilus* TK 9201	Culture collection
Bifidocin B	*Bifidobacterium bifidus* NCFB1454	Culture collection
Carnobacteriocin B2	*Carnobacterium piscicola* LV17B	Fresh pork
Carnobacterium BM1*	*Carnobacterium piscicola* LV17B	Fresh pork
Piscicocin VI6*	*Carnobacterium piscicola* VI	Fish
Piscicolin 126*	*Carnobacterium piscicola* JG 126	Ham
Piscicocin VIa*	*Carnobacterium piscicola* VI	Fish
Mundticin ATO 6	*Enterococcus mundtii* ATO 6	Vegetables
Enterocin P13	*Enterococcus faecium* P13	Fermented sausage
Bacteriocin 31	*Enterococcus faecalis* 31	Not known

*Bacteriocins that were originally named differently, but have the same amino acids sequence.
+Pediocin PA-1 and Pediocin AcH are called pediocin PA-1/AcH in this chapter. Pediocins designated by other names, but having the same amino acid sequence as pediocin PA-1/AcH have been reported in the literature.

by two cysteines (Jack *et al.*, 1995). These cystibiotics are the focus for the remainder of this chapter.

A list of currently known cystibiotics is presented in Table 4.2. Among the class II bacteriocins, the pediocin-like cystibiotics generally have a relatively wide antibacterial spectrum, and therefore have generated greater interest as potential food biopreservatives and topical therapeutic agents (Ray *et al.*, 2001b). In many instances, researchers have isolated producer strains from different sources and named bacteriocins prior to determining amino acid sequences. This has resulted in the assignment of different names to the same bacteriocin. For example, two independent research groups working with different strains of *Pediococcus acidilactici* discovered what they originally thought were two different peptides and named them pediocin PA-1 and pediocin AcH. After the amino acid sequence of each peptide was determined, it was recognized that the two bacterial strains produced the same peptide. In this review, we call this peptide pediocin PA-1/AcH. Currently, 17 cystibiotics are known (Nes and Holo, 2002).

Several general comments can be made about the production of cystibiotics (as well as other bacteriocins):

- A given strain can produce more than one bacteriocin, or even class of bacteriocin (e.g. leucocins produced by the same *Leuconostoc* strain).
- Different strains of the same species can produce the same bacteriocin (e.g. pediocin PA-1/AcH production by different *Ped. acidilactici* strains).
- Strains of different species and genera can produce the same bacteriocin (e.g. pediocin PA-1/AcH produced by *Ped. acidilactici, Ped. pentosaceus, Ped. parvulus, Lab. plantarum,* and *B. coagulans*).
- Strains of the same species can produce different bacteriocins (e.g. sakacin A and sakacin P produced by *Lab. sake* strains).

4.4 Mode of bactericidal action of cystibiotics

Cystibiotics kill Gram-positive bacteria by forming ion conductance pores in the cytoplasmic membrane. After prolonged exposure to at least some cystibiotics, cell lysis occurs, possibly because of the induction of autolytic enzymes within the cell wall. Cystibiotics initially bind electrostatically to anionic molecules in the cell wall (teichoic and teichouronic acids and perhaps lipid II) and accumulate there prior to partitioning onto the cell membrane. In addition, some molecules may bind directly to anionic phospholipids and negatively charged proteins in the cell membrane. The amphiphatic properties of these peptides enable them to remain soluble in the polar cell wall environment yet bind to and attack the non-polar membrane. It has been proposed that specific receptors (possibly protein in nature) interact with cystibiotics such as pediocin PA-1/AcH (Chikindas *et al.*, 1993). However, pediocin PA-1/AcH also forms ion conductance pores in artificial membrane vesicles prepared from anionic phospholipids, and this suggests that at least for this cystibiotic, a protein receptor is not necessary for activity (Chen *et al.*, 1997a, 1998; Montville and Chen, 1998).

Cystibiotics dissipate the membrane potential and pH gradient across the membrane, and cells die as a result of inhibition of ATP synthesis and solute transport (Jack *et al.*, 1995; Montville and Chen, 1998; Moll *et al.*, 1999; Ennaher *et al.*, 2000; Ray and Miller, 2000; Nes and Holo, 2002). Pores are not highly selective and many types of small molecules and ions pass through the membrane. The damage created is not extensive, as large molecules such as β-galactosidase do not leak out of dead cells (Bhunia *et al.*, 1991; Ray and Daeschel, 1992; Jack *et al.*, 1995).

Following exposure of sensitive Gram-positive bacteria to a cystibiotic, cells lose their viability almost immediately. However, the addition of bacteriocins such as pediocin PA-1/AcH to *Leu. mesenteroides* Ly cells not only causes cell death but ultimately also leads to lysis of cells (Bhunia *et al.*, 1991). This shows that pediocin-induced viability loss and cell lysis are two separate events. Death is caused first because of membrane permeabilization and depletion of cellular energy. Subsequently the imbalance in control mechanisms that regulate an autolytic system results in uncontrolled degradation of the cell wall and cell lysis (Fig. 4.1, Kalchayanand *et al.*, 2002a,b).

Although Gram-negative bacteria are normally resistant to the cystibiotics, cells sub-lethally injured by different types of treatments become sensitive. It is postulated that lipopolysaccharide (LPS) molecules in the outer membrane normally act as a barrier that prevents bacteriocin molecules from contacting anionic phospholipids in the inner membrane. In cells damaged by freeze–thaw

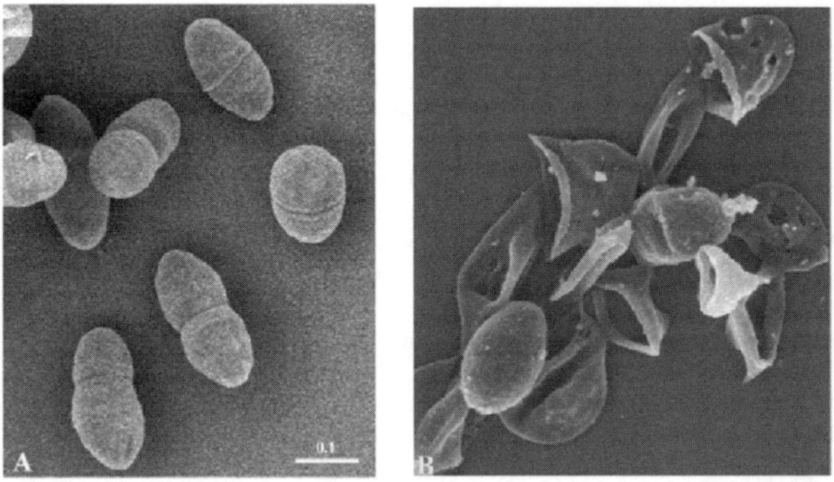

Fig. 4.1 Scanning electron photomicrographs of *Leuconostoc mesenteroides* Ly cells: (A) untreated control; (B) following incubation with pediocin PA-1/AcH (5000 AU/ml) for 30 min at 25 °C. Pedocin treatment reduced viable counts by >2 log CFU/ml in 30 min. Bar in μm.

or high-pressure treatment, for example, the outer membrane barrier is breached, and bacteriocin molecules pass through. Once cystibiotics are in contact with the cytoplasmic membrane, pore formation and cell death are thought to occur via the same mechanisms as discussed above for Gram-positive cells (Ray, 1993; Kalchayanand *et al.*, 1992; Ray and Daeschel, 1992).

4.5 Antibacterial potency and spectrum of activity

The antibacterial potency and spectrum of activity of bacteriocins against Gram-positive bacteria vary considerably. The lantibiotic nisin has a wider spectrum of activity than any non-lanthionine bacteriocin. Within the non-lanthionine groups of bacteriocins, the cystibiotics generally have the broadest antibacterial spectrum. All cystibiotics are highly bactericidal against *L. monocytogenes* and other *Listeria* species.

Purified pediocin PA-1/AcH is active against *L. monocytogenes*, *B. cereus*, *Clostridium perfringens*, *Cl. botulinum* B, *Brochothrix thermosphacta,* and many species and strains of *Leuconostoc, Lactobacillus, Pediococcus, Enterococcus* and others (Bhunia *et al.*, 1988; Ray and Daeschel, 1992; Ray and Miller, 2000). For example, nine strains of *L. monocytogenes* were reduced by 1.5 to 4.0 log CFU/ml after exposure to 1500 activity units (AU)/ml of pediocin PA-1/AcH (Motlagh *et al.*, 1991). In separate studies, six strains from six genera were sensitive to the same concentration of purified pediocin PA-1/AcH in the order: *L. innocua* > *Leu. mesenteroides* > *Ent. faecalis* > *Lab. plantarum* > *Ped. acidilactici* > *Lac. lactis* (Bhunia *et al.*, 1988, 1991; Yang *et al.*, 1992; Yang and Ray, 1994a; Ray and Miller, 2000).

The minimum inhibitory concentrations (MICs) of four purified cystibiotics, pediocin PA-1/AcH, sakacin A, sakacin P, and enterocin A were compared against 61 strains of 21 bacterial species (one to seven strains in each species) by Eijsink *et al.* (1998). Among the 61 strains, the MICs varied from >2 μg/ml to 4×10^{-5} μg/ml for the four bacteriocins. The results demonstrated a very wide range of specificity and potency of cystibiotics against different target species. These differences can be explained by variations in the structure of cystibiotics, the cell wall and membrane structure of target bacterial cells, the age of the cells, the growth conditions and the inherent ability of some strains to produce variants or mutants (Jack *et al.*, 1995; Ennahar *et al.*, 2000).

Pediocin PA-1/AcH is also effective against stressed cells of Gram-negative species such as *Salmonella*, *Escherichia coli* O157:H7, *Pseudomonas fluorescens*, and *Serratia liquefaciens* (Kalchayanand *et al.*, 1994, 1998a,b). In addition, pediocin PA-1/AcH shows even greater bactericidal potency and activity spectrum against both Gram-positive and Gram-negative bacteria when used in combination with nisin A and/or sakacin A, lysozyme, organic acids, sodium dodecyl sulfate (SDS) and ethylenediaminetetraacetic acid (EDTA) (Hanlin *et al.*, 1993; Kalchayanand *et al.*, 1998a,b; Ray and Miller, 2000).

4.6 Immunity and resistance to cystibiotics

The ability of bacterial cells to survive in the presence of a cystibiotic (or other bacteriocin) is attributable to a number of mechanisms, including the production of specific immunity proteins by producer strains, the prevention of bacteriocins from reaching or binding to the cell wall or membrane, and the hydrolysis of bacteriocins by cellular proteases. The most important contributor to resistance in producer cells is the specific immunity protein whose synthesis is linked to production of the bacteriocin. The immunity provided by an immunity protein is strongest against its cognate bacteriocin, and producers typically remain sensitive to other bacteriocins. However, cross-protection has been observed in some cases. For example, the pediocin PA-1/AcH immunity protein provides limited protection against sakacin A, sakacin P, and enterocin A, as well as against pediocin PA-1/AcH itself (Moll *et al.*, 1999; Nes and Holo, 2002).

Within a population of cells of a sensitive bacterial strain, resistant cells are commonly observed. In this case, resistance may be related to physiological alteration of cell structures involved in binding or sensing of bacteriocins in the environment (Motlagh *et al.*, 1992). Physiological changes in cell composition may be triggered in old populations of cells and in cells maintained for prolonged periods at low pH. Variations in the levels of anionic molecules in the cell wall can influence sensitivity to bacteriocins. It has been reported that a pediocin-resistant strain of *L. monocytogenes* had a lower proportion of anionic phospholipids in the membrane than sensitive strains (Chikindas *et al.*, 1993; Chen *et al.*, 1997a,b). Recent studies also have shown that *Listeria* can become resistant to pediocin PA-1/AcH by insertion of transposon Tn*917* into genes encoding proteins anchored to the cell wall (Miller *et al.*, 2002). Resistance can also result from the production of exoproteinases and exopeptidases by bacteria such as *Bacillus* spp. and *Staphylococcus* spp. Interestingly, when the immunity gene is lost from a producer strain, cells become sensitive to the bacteriocin, although sensitivity does not appear immediately (Noerlis and Ray, 1994). For example, producer cells remain resistant to pediocin PA-1/AcH for several generations after losing the production/immunity plasmid. These results suggest that immunity protein may trigger physiological changes in the cell that lead to relatively long-term alterations in the levels of anionic components in the cell wall and membrane.

The resistance of Gram-negative bacteria to cystibiotics and other bacteriocins of the LAB appears to be mainly due to the presence of LPS molecules in the outer membrane. LPS molecules are known to make Gram-negative bacteria resistant to compounds such as bile salts and SDS, and to charged molecules such as dinitrophenol. Bacteriocins of LAB, being hydrophilic and cationic, could be prevented from reaching the inner membrane of the cells by the barrier mechanism of LPS molecules. It has been shown that sublethally stressed Gram-negative bacterial cells are sensitive to bile salts due to disruption of the outer membrane barrier (Ray, 1993). They may become sensitive to pediocin PA-1/AcH for the same reason (Kalchayanand *et al.*, 1994, 1998a,b; Ray, 1993).

4.7 Production and purification of cystibiotics

The influences of different growth parameters, such as media composition, time and temperature of incubation, and initial and terminal pH of the growth media on production, have been studied for only a few cystibiotics. These include pediocin PA-1/AcH (produced by *Ped. acidilactici*), sakacin A (produced by *Lab. sake*), and the leuconocins (produced by several *Leuconostoc* spp. and strains) (Biswas *et al.*, 1991; Yang *et al.*, 1992; Yang and Ray, 1994a,b). In all of these studies, production levels were directly proportional to cell mass at the end of growth.

Conditions that facilitate rapid and abundant growth of a producer strain result in proportionately high levels of bacteriocins. The primary nutrients needed by LAB for production are amino acids and peptides (contained in tryptone, trypticase, and beef extract), readily metabolizable mono- or disaccharides (glucose being the most effective), B-vitamins, and growth factors (contained in yeast extract). Depending upon the producer species and strain, small amounts of other supplements can increase production greatly. These include Tween-80, Mg and Mn salts, acetate, phosphate, and citrate (Yang *et al.*, 1992; Yang and Ray, 1994b; Yang *et al.*, 1999). Other conditions that enhance production of bacteriocins include pH control and incubation at temperatures that are close to the optimum for rapid growth.

Following fermentation, cultures are generally heat-treated to kill the producer cells and inactivate proteolytic enzymes. The culture can also be centrifuged and the supernatant liquid used to measure production. However, depending upon the pH, variable amounts of bacteriocin may remain attached to the surface of the cells and be lost during centrifugation.

A large number of methods have been developed for purification of cystibiotics. For analytical work such as structure–function analysis, high-purity preparations can be obtained by precipitation with ammonium sulphate followed by cation exchange chromatography (Bhunia *et al.*, 1988; Jack *et al.*, 1995; Carolissen-Mackey *et al.*, 1997; Guyonnet *et al.*, 2000). In contrast, only partially purified samples are needed for measurement of potency and for applications work. The latter preparations can be obtained economically in large amounts (Jack *et al.*, 1995).

4.8 Applications

The potential applications for cystibiotics are summarized in Table 4.3. Specific examples of applications in foods are shown in Table 4.4.

Three approaches have been used to study cystibiotics in food preservation. In the first, the producing strain is used as a starter culture for fermentation of the product. For example, pediocin-producing strains have been used to control *L. monocytogenes* during fermentation of sausages and Cheddar cheese (Table 4.4). Foods made using the pediocin PA-1/AcH-producing

Table 4.3 Potential applications of cystibiotics in foods and food processing operations

(a) Meats: low-heat-processed refrigerated products, specialty products, ground meat chubs
(b) Fish: low-heat-processed refrigerated products, raw specialty products
(c) Dairy:
  (i) Soft cheeses, cheese-spread, yogurt
  (ii) Control of *Listeria* spp. during cheese fermentation
  (iii) Accelerate cheese ripening by lysing starter culture cells
(d) Miscellaneous: non-thermally processed foods, carcass wash, beverages, condiments, probiotics, packaging materials, equipment sanitation, sugar processing

Adapted from Ray *et al.* (2001b).

starter culture contained 2 to 3 log CFU/g fewer *L. monocytogenes* than those made with non-producing starters. In the second approach, high numbers of the producer strain are added to food products prior to refrigerated storage. For example, Sakacin A-producing strains have been added to minced meat, bavaricin-producing strains to meat cubes, leucocin A-producing strains to vacuum packaged meat, and pediocin PA-1/AcH-producing strains to frankfurters, fresh chicken and Munster cheese (Table 4.4). Depending on the cystibiotic and food product, *Listeria* populations have been reduced by 1 to 4 log CFU/g during storage using this approach. Furthermore, spoilage was delayed by 5 days compared with control foods containing strains that did not produce cystibiotics. In the third approach, cystibiotics are added to foods inoculated with spoilage and pathogenic bacteria. For example, *L. monocytogenes* was effectively controlled by piscicolin 126 in devilled ham paste, by enterocin A in ham, pork, chicken breast, paté, and sausage, and by pediocin PA-1/AcH in cottage cheese, half-and-half, cheese sauce, fresh meat, ground beef, sausage mix, milk, and liquid egg (Table 4.4). Pediocin PA-1/AcH also limited the growth of several Gram-positive and Gram-negative spoilage bacteria during refrigerated storage of vacuum packaged beef by-products and several types of low heat-processed meat products. An even greater reduction in counts was observed when nisin, lactic, acetic, or propionic acids or lysozyme were added with pediocin PA-1/AcH.

In separate studies, processed meat products with and without pediocin PA-1/AcH were inoculated with spoilage and pathogenic bacteria, vacuum packaged and subjected to hydrostatic pressure at 345 MPa for 5 min at 50 °C. A >7 log CFU/g reduction in several pathogenic and spoilage bacteria was observed when the foods were pressurized in the presence of pediocin. In the absence of the bacteriocin, pressurization produced only a 3 to 4 log CFU/g reduction in numbers. These results suggest that this method should be effective for control of post-heat contaminating bacteria in vacuum packaged refrigerated meat products, and probably in other low-acid food products as well.

The bactericidal efficiency of cystibiotics can be enhanced when they are used in combination with other antimicrobial agents and treatments. For

Table 4.4 Examples of applications of cystibiotics as food preservatives

Cystibiotic	Food type	Target microorganisms	Mode of application	Reference
Bavaricin MN	Meat	*Listeria* and *Lactobacillus*	Producer cells added	Winkowski *et al.* (1993)
Enterocin A	Ham, pork, chicken, paté and sausage	*Listeria*	Bacteriocin added	Aymerich *et al.* (2000a,b)
Leucocin A	Meat	Spoilage organisms	Producer cells added	Leistner *et al.* (1996)
Piscicolin 126	Devilled ham	*Listeria*	Bacteriocin added	Jack *et al.* (1996)
Sakacin A	Minced meat	*Listeria*	Producer cells added	Schillinger *et al.* (1991)
Pediocin PA-1/AcH	Sausage, fermented	*Listeria*	Producer cells as starters	Berry *et al.* (1990); Foegeding *et al.* (1992); Luchansky *et al.* (1992); Baccus-Taylor *et al.* (1993)
	Cheddar cheese	*Listeria*	Producer cells as starters added to milk	Buyong *et al.* (1998)
	Frankfurter package	*Listeria*	Producer cells in package	Berry *et al.* (1991); Degnan *et al.* (1992)
	Chicken, fresh	*Listeria*	Producer cells added	Goff *et al.* (1996)
	Munster cheese	*Listeria*	Producer cells on surface of cheese	Ennahar *et al.* (1998)
	Cottage cheese, half-and-half, cheese sauce	*Listeria*	Bacteriocin added	Pucci *et al.* (1988)
	Meat, fresh	*Listeria*	Bacteriocin added	Nielson *et al.* (1990)

Ground beef, sausage and milk	*Listeria*	Bacteriocin added	Motlagh *et al.* (1992)
Liquid whole egg	*Listeria*	Bacteriocin added	Muriana (1996)
Beef, refrigerated	Spoilage bacteria	Bacteriocin added	Rozbeh *et al.* (1993)
Meat paste, refrigerated	Spoilage bacteria	Bacteriocin added	Coventry *et al.* (1995)
Beef liver, kidney, tongue, hotdogs, roast beef and ham, refrigerated	*Lab. sake, Leuc.* sp., *Pseudomonas* sp., *Serratia* sp., *L. monocytogenes, Staph. aureus, Salmonella, E. coli* O157:H7, *Cl. botulinum*	Pediocin in combination with nisin and organic acids	Ray (1992, 1993, 1994, 1995, 1996); Ray and Hoover (1993); Ray and Miller (2000); Ray *et al.* (2001b)
Roast beef, hotdogs, summer sausage, cotto salami	As above except for *Cl. botulinum*	Pediocin in combination with nisin, lysozyme and organic acids followed by vacuum packing and hydrostatic pressure processing	Kalchayanand *et al.* (1998a,b); Ray *et al.* (2001a)

example, growth of a wide range of bacteria can be inhibited in foods by combining pediocin PA-1/AcH with sakacin, and especially with nisin A (Hanlin *et al.*, 1993; Mulet-Powell *et al.*, 1998; Ray and Miller, 2000; Ray *et al.*, 2001b). In addition, the effectiveness of combinations of pediocin PA-1/AcH and nisin against many bacteria can be enhanced by combining them with one or more of the following compounds: organic acids, such as lactic, acetic or propionic acids, lysozyme, EDTA and SDS, and chitosan (Jack *et al.*, 1995; Ray and Miller, 2000; Ray *et al.*, 2001b; Kalchayanand *et al.*, 1992, 1994, 1998a,b). While some of these compounds (organic acids and lysozyme) can be used in food preservation, others (SDS and EDTA) may not be permitted in foods.

4.9 Safety and legal status

Owing to their natural association with foods, humans have consumed the viable cells and metabolites of most, if not all, cystibiotic-producing LAB for a long time without known adverse effects. Lactic acid bacteria that have been used as starter cultures, namely *Ped. acidilactici*, *Ped. pentosacenes*, *Leu. mesenteroides*, *Lab. plantarum*, *Lab. sake*, and *Lab. curvatus*, can probably be regarded as food grade and safe. *Lab. acidophilus* and *Bif. bifidum* can also be included in this group. Limited toxicity studies conducted with cell cultures and laboratory animals have shown that at least one cystibiotic, pediocin PA-1/AcH, is quite harmless. In addition, fermented sausages made with pediocin-producing strains have been consumed by people without adverse effects (Ray, unpublished results; Ray and Miller, 2000). In contrast, the safety of those LAB currently not used as starter cultures, such as *Carnobacterium* spp. and even *Leu. gelidum, Leu. carnosum* and *Enterococcus* spp. has not been proven. Before such producer strains or their cystibiotics can be used in foods, their food grade and safety status will need to be clarified by appropriate regulatory agencies.

At present, only nisin is approved for use in certain foods in several countries (see Chapters 2, 3, and 15). None of the cystibiotics, including pediocin PA-1/AcH, has been approved for food use, although some commercial preparations containing pediocin PA-1/AcH are available in the USA. In the USA, the use of bacteriocins, even from food grade starter bacteria, is regulated by the Federal Food, Drug and Cosmetic Act and requires integrated reviews by the FDA and Food Safety Inspection Service of the Department of Agriculture, depending on the nature of the food (Field, 1996; Post, 1996). Purified bacteriocins and producer cells are classified as additives and are evaluated accordingly. Under these circumstances, it is rather doubtful that cystibiotics and other bacteriocins, even those produced by food grade starter bacteria, will be used as food preservatives in the near future.

4.10 Conclusions

Although pediocin PA-1/AcH is now commercially produced, it will take some time before it or other cystibiotics are approved for use in food preservation. In the meantime, other uses of these bacteriocins in non-food applications should be investigated. The regulatory agencies also need to be flexible in evaluating possible uses of bacteriocins in foods, especially those cystibiotics that are produced by starter culture bacteria, in order to avoid delay in the resolution of problems that currently face the food industry, such as control of *L. monocytogenes* and other psychrotrophic bacteria in foods.

4.11 References

AXELSON L (1998), 'Lactic acid bacteria: classification and physiology', in Salminen S and von Wright A, *Lactic Acid Bacteria*, 2nd ed., New York, Marcel Dekker, 1–72.

AYMERICH T, ARTIGAS MG, GARRIGA M, MONTFORT JM and HUGAS M (2000a), 'Effect of sausage ingredients and additives on the production of enterocin A and B by *Enterococcus faecium* CTC 492. Optimization of *in vitro* production and antilisterial effect in dry fermented sausages', *J Appl Microbiol*, **88**, 686–694.

AYMERICH T, GARRIGA M, YLLA J, VALLIER J, MONFORT JM and HUGAS M (2000b), 'Application of enterocins as biopreservatives against *Listeria innocua* in meat products', *J Food Prot*, **63**, 721–726.

BACCUS-TAYLOR G, GLASS KA, LUCHANSKY JB and MAURER AJ (1993), 'Fate of *Listeria monocytogenes* and pediococcal starter cultures during the manufacture of chicken summer sausage', *Poultry Sci*, **72**, 1772–1778.

BERRY ED, LIEWEN MB, MANDIOG RW and HUTKINS RW (1990), 'Inhibition of *Listeria monocytogenes* by bacteriocin-producing *Pediococcus* during the manufacture of fermented semidry sausage', *J Food Prot*, **53**, 194–197.

BERRY ED, HUTKINS RW and MANDIGO RW (1991), 'The use of bacteriocin-producing *Pediococcus acidilactici* to control post-processing *Listeria monocytogenes* contamination of frankfurters', *J Food Prot*, **54**, 681–686.

BHUNIA AK, JOHNSON MC and RAY B (1988), 'Purification, characterization and antimicrobial spectrum of a bacteriocin produced by *Pediococcus acidilactici*', *J Appl Bacteriol*, **65**, 261–268.

BHUNIA AK, JOHNSON MC, RAY B and KALCHAYANAND N (1991), 'Mode of action of pediocin AcH from *Pediococcus* H', *J Appl Bacteriol*, **70**, 25–33.

BISWAS SR, RAY P, JOHNSON MC and RAY B (1991), 'Influence of growth conditions on the production of bacteriocin, pediocin AcH, by *Pediococcus acidilactici* H', *Appl. Environ Microbiol*, **52**, 1265–1267.

BUYONG N, KOK J and LUCHANSKY JB (1998), 'Use of genetically enhanced pediocin-producing starter culture, *Lactococcus lactis* subsp *lactis* MM217, to control *Listeria monocytogenes* in cheddar cheese', *Appl Environ Microbiol*, **64**, 4842–4845.

CAROLISSEN-MACKEY V, ARENDSE G and HASTINGS JW (1997), 'Purification of bacteriocins of lactic acid bacteria: problems and pointers', *Int J Food Microbiol*, **34**, 1–16.

CHEN Y, LUDESCHER RD and MONTVILLE TJ (1997a), 'Electrostatic interactions, but not the YGNGV consensus motif, govern the binding of pediocin PA-1 and its fragments to phospholipid vesicles', *Appl Environ Microbiol*, **63**, 4770–4777.

CHEN Y, LUDESCHER RD and MONTVILLE TJ (1998), 'Influence of lipid composition on pediocin PA-1 binding to phospholipid vesicle', *Appl Environ Microbiol*, **64**, 3530–3532.

CHEN Y, SHAPIRA R, EISENSTEIN M and MONTVILLE TJ (1997b), 'Functional characterization of pediocin PA-1 binding to liposomes in the absence of a protein receptor and its relationship to a predicted tertiary structure', *Appl Environ Microbiol*, **63**, 524–531.

CHIKINDAS ML, GARCIA-GARCERA MJ, DRIESSEN AJM, LEDEBOER AM, NISSEN-MEYER J, NES IF, ABEE T, KONINGS WN and VENEMA G (1993), 'Pediocin PA-1, a bacteriocin of *Pediococcus acidilactici* PAC 1.0, forms hydrophilic pores in the cytoplasmic membrane of target cells', *Appl Environ Microbiol*, **59**, 3577–3584.

COVENTRY MJ, MUIRHEAD K and HICKEY MM (1995), 'Partial characteristics of pediocin $P0_2$ and comparison of nisin for biopreservation of meat products', *Int J Food Microbiol*, **26**, 133–145.

DEGNAN AJ, YOUSEF AE and LUCHANSKY JB (1992), 'Use of *Pediococcus acidilactici* to control *Listeria monocytogenes* in temperature-abused vacuum-packaged wieners', *J Food Prot*, **55**, 98–103.

EIJSINK VG, SKEIE M, MIDDELHOVEN PH, BRURBERG MB and NES IF (1998), 'Comparative studies of class IIa bacteriocins of lactic acid bacteria', *J Appl Microbiol*, **64**, 3275–3281.

ENNAHAR S, ASSOBHEL D and HASSELMANN C (1998), 'Inhibition of *Listeria monocytogenes* in a smear-surface soft cheese by *Lactobacillus plantarum* WHE92, a pediocin AcH producer', *J Food Prot*, **61**, 186–191.

ENNAHAR S, SASHIHARA T, SONOMOTO K and ISHIZAKI A (2000), 'Class IIa bacteriocins: biosynthesis, structure and activity', *FEMS Microbiol Rev*, **24**, 85–106.

FIELD FA (1996), 'Use of bacteriocins in food: regulatory considerations', *J Food Prot, Supplement*, 72–77.

FOEGEDING PM, THOMAS AB, PILKINGTON DH and KLAENHAMNER TR (1992), 'Enhanced control of *Listeria monocytogenes* by *in situ*-produced pediocin during dry fermented sausage production', *Appl Environ Microbiol*, **58**, 884–890.

GOFF JH, BHUNIA AK and JOHNSON MG (1996), 'Complete inhibition of low levels of *Listeria monocytogenes* on refrigerated chicken meat with pediocin AcH bound to heat-killed *Pediococcus acidilactic* cells', *J Food Prot*, **59**, 1187–1192.

GUYONNET D, FREMAUX C, CENATIEMPO Y and BERJEAUD JM (2000), 'Method for rapid purification of class IIa bacteriocins and comparison of their activities', *Appl Environ Microbiol*, **66**, 1744–1748.

HANLIN MB, KALCHAYANAND N, RAY P and RAY B (1993), 'Bacteriocins of lactic acid bacteria in combination have greater antibacterial activity', *J Food Prot*, **56**, 252–255.

JACK RW, TAGG JR and RAY B (1995), 'Bacteriocins of Gram-positive bacteria', *Microbiol Rev*, **59**(2), 171–200.

JACK RW, WAN J, GORDON J, HARMARK K, DAVIDSON BE, HILLIER AJ, WETTENHALL REH, HICKEY MW and COVENTRY MJ (1996), 'Characterization of the chemical and antimicrobial properties of piscicolin 126, a bacteriocin produced by *Carnobacterium piscicola* JG 126', *Appl Environ Microbiol*, **62**, 2897–2903.

KALCHAYANAND N, HANLIN MB and RAY B (1992), 'Sublethal injury makes Gram-negative and resistant Gram-positive bacteria sensitive to the bacteriocins, pediocin AcH and nisin', *Lett Appl Microbiol*, **15**, 239–243.

KALCHAYANAND N, SIKES T, DUNNE CP and RAY B (1994), 'Hydrostatic pressure and electroporation have increased bactericidal efficiency in combination with bacteriocins', *Appl Environ Microbiol*, **60**, 4174–4177.

KALCHAYANAND N, SIKES A, DUNNE CP and RAY B (1998a), 'Factors influencing death and injury of foodborne pathogens by hydrostatic pressure pasteurization', *Food Microbiol*, **15**, 207–214.

KALCHAYANAND N, SIKES A, DUNNE CP and RAY B (1998b), 'Interaction of hydrostatic pressure, time and temperature of pressurization and pediocin AcH on inactivation of foodborne bacteria', *J Food Prot*, **61**, 425–431.

KALCHAYANAND N, DUNNE CP, SIKES A and RAY B (2002a), 'Viability loss and morphology changes of *Listeria monocytogenes*, *Salmonella* Typhimurium and *Escherichia coli* O157:H7 cells following exposure to high hydrostatic pressure in the presence and absence of bacteriocins', *Appl Environ Microbiol*, (accepted).

KALCHAYANAND N, FRETHEM C, DUNNE P, SIKES A and RAY B (2002b), 'Hydrostatic pressure and bacteriocin triggered cell wall lysis of *Leuconostoc mesenteroides* Ly', *Innovation Fd Sci & Emerging Technol*, (accepted).

KLAENHAMMER TR (1993), 'Genetics of bacteriocins production by lactic acid bacteria', *FEMS Microbiol Rev*, **12**, 39–86.

LEISTNER JJ, GREER GG and STILES ME (1996), 'Control of beef spoilage by a sulfide-producing *Lactobacillus sake* strain with bacteriocinogenic *Leuconostoc gelidum* UAL187 during anaerobic storage at 2°C', *Appl Environ Microbiol*, **62**, 2610–2614.

LUCHANSKY JB, GLASS KA, HARSONO KD, DEGNAN AJ, FAITH NG, CAUVIN B, BUCCUS-TAYLOR G, ARIHARA K, BATER B, MAURER AJ and CASSENS RG (1992), 'Genomic analysis of *Pediococcus* starter cultures used to control *Listeria monocytogenes* in turkey summer sausage', *Appl Environ Microbiol*, **58**, 3053–3059.

MILLER KW, STEINMETZ T, HUNTER I, PETERS A, AMARENDRAN V and RAY B (2002), 'Genes involved in resistance of *Listeria* to pediocin AcH', Am Soc Microbiol, Annual Meeting, Abstract (accepted).

MOLL GN, KONINGS WN and DRIESSEN AJM (1999), 'Bacteriocins: mechanism of membrane insertion and pore formation', *Antonie van Leeuwenhoek*, **76**, 185–189.

MONTVILLE TJ and CHEN Y (1998), 'Mechanistic action of pediocin and nisin: recent progress and unresolved questions', *Appl Microbiol Biotechnol*, **50**, 511–519.

MOTLAGH AM, HOLLA S, JOHNSON MC, RAY B and FIELD RA (1992), 'Inhibition of *Listeria* spp, in sterile food systems by pediocin AcH, a bacteriocin produced by *Pediococcus acidilactici* H', *J Food Prot*, **55**, 337–343.

MOTLAGH AM, JOHNSON MC and RAY B (1991), 'Viability loss of foodborne pathogens by starter culture metabolites', *J Food Prot*, **54**, 873–878, 884.

MULET-POWELL N, LOCOSTE-ARMYNOT AM, VINAS M and SIMEON DEBUOCHBERG M (1998), 'Interactions between pairs of bacteriocins from lactic acid bacteria', *J Food Prot*, **61**, 1210–1212.

MURIANA PM (1996), 'Bacteriocins for control of *Listeria* spp. in food', *J Food Prot, Supplement*, 54–63.

NES IF and HOLO H (2002), 'Unmodified peptide-bacteriocin (class II) produced by lactic acid bacteria', in Dutton CJ, Haxell MA, McArthur HAI and Wax RG, *Peptide Antibiotics*, New York, Marcel Dekker, 81–115.

NIELSON JW, DICKSON JS and CROUSE JD (1990), 'Use of bacteriocin produced by *Pediococcus acidilactici* to inhibit *Listeria monocytogenes* associated with fresh meat', *Appl Environ Microbiol*, **56**, 2142–2145.

NOERLIS Y and RAY B (1994), 'Factors influencing immunity and resistance of *Pediococcus acidilactici* to the bacteriocin, pediocin AcH', *Lett Appl Microbiol*, **18**, 138–143.

POST RC (1996), 'Regulatory perspective of the USDA on the use of antimicrobials and inhibitors in food', *J Food Prot, Supplement*, 78–81.

PUCCI MJ, VEDAMUTHU ER, KUNKA BS and VANDENBERGH PA (1988), 'Inhibition of *Listeria monocytogenes* by using bacteriocin PA-1 produced by *Pediococcus acidilactici* PAC1.0', *Appl Env Microbiol*, **54**, 2349–2353.

RAY B (1992), 'Pediocin of *Pediococcus acidilactici* as a food biopreservative', in Ray B and Daeschel M, *Food Biopreservatives of Microbial Origin*, Boca Raton, CRC Press, 265–322.

RAY B (1993), 'Sublethal injury, bacteriocins, and food microbiology', *ASM-News*, **59**, 285–291.

RAY B (1994), 'Pediocins of *Pediococcus* species', in DeVuyst L and Vandamme EJ, *Bacteriocins of Lactic Acid Bacteria*, New York, Blackie Academic and Professional, 465–496.

RAY B (1995), '*Pediococcus* in fermented foods', in Hui YH and Khachatourians GG, *Food Biotechnology Microorganisms*, New York, VCH Publishers, 745–796.

RAY B (1996), 'Characteristics and applications of pediocin(s) of *Pediococcus acidilactici*: pediocin PA-1/AcH', in Bozoglu TF and Ray B, *Lactic Acid Bacteria*, New York, Springer-Verlag, 98 NATO-ASI series, 155–204.

RAY B (2000), *Fundamental Food Microbiology*, 2nd edn, Boca Raton, CRC Press, 109, 222, 269.

RAY B and DAESCHEL M, Eds. (1992), *Food Biopreservatives of Microbial Origin*, Boca Raton, CRC Press.

RAY B and HOOVER DG (1993), 'Pediocins', in Hoover DG and Steenson LR, *Bacteriocins of Lactic Acid Bacteria*, San Diego, Academic Press, 108–210.

RAY B and MILLER KW (2000), 'Pediocin', in Naidu AS, *Natural Food Antimicrobial Systems*, Boca Raton, CRC Press, 525–566.

RAY B, KALCHAYANAND N, DUNNE P and SIKES A (2001a), 'Microbial destruction during hydrostatic pressure processing of food', in Bozoglu F, Deak T and Ray B, *Novel Processes and Control Technologies in the Food Industry*, Oxford, IOS Press, 95–122.

RAY B, MILLER KW and JAIN MK (2001b), 'Bacteriocins of lactic acid bacteria', *Indian J Microbiol*, **41**, 1–21.

ROZBEH M, KALCHAYANAND N, FIELD RA, JOHNSON MC and RAY B (1993), 'The influence of biopreservatives on the bacterial level of refrigerated vacuum-packaged beef', *J Food Safety*, **13**, 99–111.

SALMINEN S and VON WRIGHT A (1998), *Lactic Acid Bacteria*, 2nd edn, New York, Marcel Dekker, Inc.

SCHILLINGER U, KAYA M and LÜCKE F-K (1991), 'Behaviour of *Listeria monocytogenes* in meat and its control by a bacteriocin producing *Lactobacillus sake*', *J Appl Bacteriol*, **70**, 473–478.

SNEATH PHA, ED. (1986), *Bergy's Manual of Systemic Bacteriology*, Vol. 2, Baltimore, William and Wilkins. 1065, 1071, 1075, 1209, 1346, 1418.

WINKOWSKI K, CRANDALL AD and MONTVILLE TJ (1993), 'Inhibition of *Listeria monocytogenes* by *Lactobacillus bavaricus* MN in beef systems at refrigeration temperatures', *Appl Environ Microbiol*, **59**, 2552–2557.

YANG R and RAY B (1994a), 'Prevalence and biological control of bacteriocin-producing psychrotphic leuconostocs associated with spoilage of vacuum-packaged processed meats', *J Food Prot*, **57**, 209–217.

YANG R and RAY B (1994b), 'Factors influencing production of bacteriocins by lactic acid bacteria', *Food Microbiol*, **11**, 281–291.

YANG R, CHEN Y and RAY B (1999), 'Enhanced bacteriocin production by lactic acid bacteria in a dairy-based medium, supplemented with β-galactosidase', *Indian J Microbiol*, **39**, 235–240.

YANG R, JOHNSON MC and RAY B (1992), 'Novel method to extract large amounts of bacteriocins from lactic acid bacteria', *Appl Environ Microbiol*, **58**, 3355–3359.

5

Natamycin: an effective fungicide for food and beverages

J. Stark, DSM Food Specialties, The Netherlands

5.1 Introduction

Natamycin was discovered in the 1950s and was described for the first time by Struyck as follows:

> A new crystalline antibiotic, pimaricin, has been isolated from the fermentation broth of a culture of a *Streptomyces* species, isolated from a soil sample obtained near Pietermaritzburg, State of Natal, Union of South Africa. This organism has been named *Streptomyces natalensis* (Struyck *et al.*, 1957–58).

Today, almost half a century later, the antimycotic natamycin (the name used in earlier literature was pimaricin) is still widely used for the prevention of mould and yeast growth on food products.

Natamycin is produced by fermentation using *Strep. natalensis*. It can be recovered using extraction, filtration and drying. The dry powder is sufficiently stable to be stored for years without substantial loss of activity. Besides its broad-spectrum activity against moulds and yeasts, the biopreservative natamycin has some unique characteristics, which make it particularly suitable for preventing fungal growth on the surface of food products and in beverages. It is safe for the consumer, is effective at low concentrations, has no negative effects on the quality of food products, remains on the surface of the cheese or sausage and has a prolonged working time. Although natamycin has been used for almost 40 years in the food industry, the development of resistant strains has not been reported to date. As natamycin has no antibacterial activity, the bacterial ripening processes of food products such as cheese and sausages are not influenced in a negative way.

5.2 Chemical and physical properties

Natamycin belongs to the group of polyene macrolide antimycotics. This group is characterised by its macrocyclic lactone-ring with a number of conjugated carbon–carbon double bonds (Fig. 5.1). The molecular weight of natamycin is 665.75. Its empirical formula is $C_{33}H_{47}NO_{13}$. Its CAS registry number is 7681-93-8. The correct gross structure was determined using proton- and ^{13}C-NMR (nuclear magnetic resonance) spectroscopy (Ceder *et al.*, 1977). The complete three-dimensional structure was determined later (Lancelin and Beau, 1990, 1995; Duplantier and Masamune, 1990). The isoelectric point of natamycin is 6.5 (Raab, 1974).

Natamycin is a white or creamy-white powder with almost no odour or taste. Its most common form, the trihydrate, can be stored for many years at room temperature, provided the powder is protected from moisture and light. Even after several years of storage under these conditions only a few per cent loss of activity is observed. Neutral aqueous suspensions of natamycin are also very stable. A suspension containing 0.5% natamycin at pH 6.5 lost hardly any activity at room temperature after two years of storage in the dark (Clark *et al.*, 1964). Under these conditions, most of the natamycin is in the stable crystalline form. Only a minor fraction (about 40 ppm) is in solution. Although dissolved natamycin is relatively unstable, these results demonstrate that under optimal storage conditions it can be quite stable. The amphoteric character of natamycin is responsible for its low solubility in most solvents. The solubility is increased at high or low pH values (Brik, 1981).

At extreme pH values, natamycin is completely soluble. Since only the dissolved fraction has antifungal activity, extremes of pH can be used to

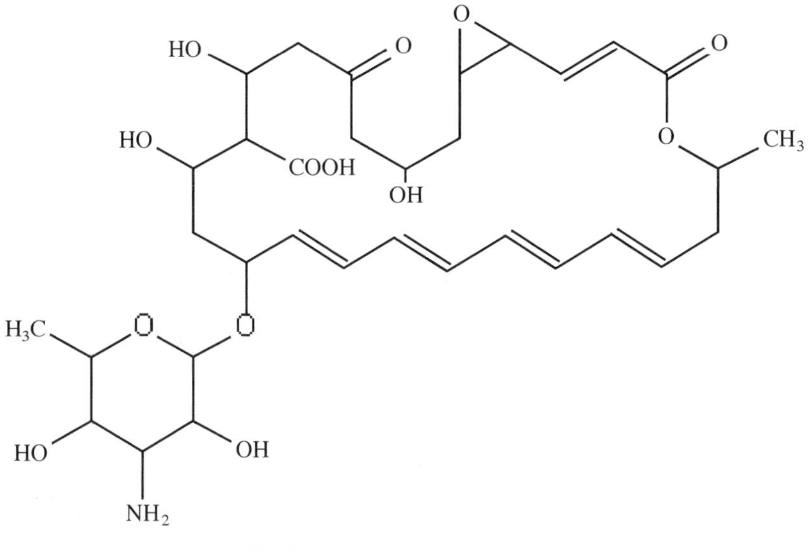

Fig. 5.1 Structure of natamycin.

optimise/control the availability of active natamycin. However, natamycin is rapidly decomposed under these conditions. Therefore such solutions are effective only when used immediately after preparation. In practice, this way of applying natamycin might be beneficial in cases of extreme contamination, but under normal hygienic conditions there is no added value in this approach. Therefore, such treatment is not normally recommended.

Solid natamycin and natamycin suspensions are quite stable to heat and it has been reported that natamycin suspensions could withstand a temperature of 50 °C for several days without losing much activity. Even heating for several hours at 100 °C causes only a slight decrease of its activity.

Natamycin is decomposed by ultraviolet light with loss of the tetraene structure so it is best stored in the dark. Gamma irradiation decomposes natamycin as well. Natamycin is inactivated by chlorine, peroxides, antioxidants, sulphites and sodium formaldehyde sulphoxidate (Brik, 1981). Some of these compounds are applied as cleaning agents in the food industry. Uncontrolled use of chlorine in a factory may indirectly induce mould growth on cheese. In the presence of deactivating compounds or less optimal conditions, solid natamycin is the most stable form. Dissolved natamycin is much more susceptible to inactivation. Under neutral aqueous conditions most of the natamycin on the surface of a food product or in a stock suspension is present as the more stable crystal. Under extremely unfavourable conditions, rapid elimination of the dissolved fraction will ultimately lead to complete decomposition of natamycin through enhanced dissolution.

Natamycin trihydrate, crystallised from aqueous solvents, contains approximately 7.5% water. Methanol extracts can be obtained by crystallisation from a saturated methanolic solution. As soon as the methanol extract is in contact with water, it rapidly converts to the trihydrate form. The anhydrous form of natamycin can be prepared by drying the trihydrate in vacuum at room temperature over phosphorus pentoxide. The anhydrate is unstable; it loses 15% of its activity when stored for 48 hours under optimal conditions (Brik, 1981). However, when stored under nitrogen or at low temperatures it is more stable (van Rijn *et al.*, 1995). The barium and calcium salts of natamycin and a new crystalline form, also a trihydrate, have been described by van Rijn *et al.* (1995). The stability of the new trihydrate is somewhat less than the stability of the regular type. The salts in their solid form are stable but, once in contact with water, they dissolve rapidly and are converted into the more stable regular trihydrate. In principle, this phenomenon could be used to obtain a highly active preparation of dissolved natamycin for a short period of time.

5.3 Mechanism of action

The eukaryotic cell membrane contains lipids, phospholipids, proteins and sterols. The sterols play an important role in the selective action of polyene antimycotics. Ergosterol is the major sterol in the cell membrane of moulds and yeasts, while cholesterol is the major sterol in the mammalian cell membrane. Natamycin has a high affinity for ergosterol and binds irreversibly to the ergosterol in the fungal cell membrane. This perturbs the permeability of the cell membrane, which leads to rapid leakage of essential ions and small peptides and so causes cell lysis.

The detailed mechanism of action of natamycin is thought to resemble that of another polyene antimycotic, amphotericin B. The latter forms chiral aggregates at the bilayer–water interface. In the absence of ergosterol, the aggregates remain at the surface of the bilayer (Milhaud *et al.*, 2002). In the presence of ergosterol, the aggregates are embedded in the phospholipid bilayers. The embedding in the hydrophobic bilayer requires a rearrangement of the aggregates that is, in turn, responsible for the hollowing of the aggregates. These structures are thought to be the precursors of pore formation.

Vesicles of *Acholeplasma laidlawii* have been prepared with either ergosterol or cholesterol as the membrane sterol. When the vesicles contained 20–40 mol% of ergosterol, which is equivalent to the ergosterol content of a fungal cell membrane, natamycin induced leakage of potassium ions. Natamycin caused no damage when only cholesterol was present in the vesicles (De Kruijff *et al.*, 1974; Teerlink *et al.*, 1980). It can be concluded that ergosterol is the molecular target for natamycin on the fungal cell membrane. Since bacteria do not contain sterols in their cell membrane, they are not susceptible to lysis by natamycin.

Studies with amphotericin B suggest that the cell wall forms a natural barrier to the fungicide. This protective effect increases as the culture ages (Malewicz and Borowski, 1979). It is possible that a similar protective effect occurs in the case of natamycin. This is supported by observations made in practice, which indicate that natamycin is very effective in preventing fungal growth, but ineffective in killing mature fungal tissue. Therefore natamycin cannot be used as a disinfectant. Most moulds grow by extension of the hyphal tip and the cell wall is formed just behind the new tissue of the hyphal tip. If the cell wall indeed prevents interaction of natamycin with the cell membrane, this should mean that natamycin stops the growth of the fungal mycelium only at the tip of the growing hypha by interacting with the fungal cell membrane. Thus the cell wall prevents direct interaction of natamycin with the older tissue in the lower part of the hyphae. The older cells are not killed directly but are also unlikely to survive. Natamycin also inhibits the germination of fungal spores by an unknown mechanism.

Mycotoxins produced by fungi can cause health problems and some are carcinogenic. Removal or elimination of moulds after growth on a food does not affect mycotoxins, which can migrate deep into the foodstuff. Therefore, it is extremely important to prevent growth of toxigenic fungi on food products. The growth-inhibiting mode of action of natamycin lends itself well to this application.

5.4 Sensitivity of moulds and yeasts to natamycin

Natamycin is active against most fungi at low concentrations. The minimum inhibitory concentration (MIC) for natamycin against almost all foodborne fungi is less than 20 ppm, while the solubility of natamycin in aqueous food systems is around 40 ppm. It has been demonstrated in practice that under acceptable hygienic conditions, this concentration of dissolved natamycin is sufficient to prevent fungal growth.

Struyk *et al.* (1957–58) determined the MICs of natamycin against 66 moulds and yeasts, including food spoilage strains and those that were pathogenic to people and plants. The majority of yeasts and moulds were inhibited at concentrations of 1–15 ppm of natamycin. Some species of *Trichophyton* and *Phythium* were less sensitive but these were irrelevant for the food industry. Struyk concluded that natamycin has a broad spectrum of activity against fungi, is fungicidal and is not effective against bacteria. Over the past 50 years, the original conclusions of Struyk have been confirmed in many studies using a wide range of species of relevance to the food industry (Tables 5.1–5.3).

Several well-known methods for determining the MICs of antimicrobials against bacteria and yeasts are recognised and used widely. By contrast, there are no standardised methods for the determination of MICs for filamentous fungi. The quantities of mould spores used by different investigators can vary or are not

Table 5.1 Sensitivity to natamycin of fungi isolated in Dutch cheese factories and warehouses

Microorganism	MIC* (ppm)
Aspergillus flavus & parasiticus	10–20
Aspergillus fumigatus, penicillioides & versicolor	< 10
Cladosporium candidum	
Debaryomyces hansenii	
Eurotium herbariorum	
Geotrrichum candidum	
Mucor racemosus	
Penicillium brevicompactum, camembertii, commune, corylophilum, glabrum, nalgiovense, roquefortii & solitum	
Scopulariopsis brevicaulis & fusca	
Syncephalastrum racemosum	
Wallemia sebi	
Penicillium discolor+	20–30

* 5μl of a suspension of 10^5 spores/ml was spotted on agar plates containing 0, 10, 20, 30 or 40 ppm of natamycin and incubated at 24 °C for 6 days. The minimum inhibitory concentration (MIC) was defined as the lowest concentration of natamycin at which no growth was observed. The MIC of natamycin for most moulds and yeasts is most likely < 5 ppm (see Tables 5.2 and 5.3).
+ At a concentration of 10^4 spores/ml the MIC for *Pen. discolor* is 10–20 ppm.
Source: from DSM Food Specialties Research Laboratorium Delft, The Netherlands; Dr R.A. Samson, Centraal Bureau voor Schimmelcultures, Utrecht, The Netherlands.

Table 5.2 Sensitivity to natamycin of fungi occurring on sausages*

Microorganism	MIC (ppm)⁺	Source of microorganism
Moulds		
Alternaria alternata	< 5	Blood
Aspergillus flavus	10–20	Air
Aspergillus niger	< 5	Fruit
Aspergillus versicolor	< 5	Soil
Cladosporium cladosporioides	< 5	Meat stamp
Eurotium appendiculatum	< 5	Smoked sausage
Eurotium herbariorum	< 5	Board
Geotrichum candidum	5–10	Soil
Mucor racemosus	< 5	Sausage
Penicillium chrysogenum	< 5	Meat
Penicillium glabrum	5–10	Soil
Penicillium nalgiovense	< 5	Sausage
Penicillium verrucosum	< 5	Meat
Rhizopus stolonifer	5–10	Bread
Wallemia sebi	< 5	Sea salt
Yeasts		
Candida zeylandoides	< 5	Sausage
Cryptococcus laurentii	< 5	Air
Debaryomyces hansenii	< 5	Sausage
Rhodutorula mucilaginosa	< 5	Air
Trichosporon pullulans	< 5	Frozen beef

* The selection of strains was based on their occurrence on sausages and meat products. When not available from meat sources, representative isolates were selected.
⁺ 0.1 ml of a suspension of 10^4 spores/ml (moulds) or CFU/ml (yeasts) was spread on agar plates containing natamycin and incubated at 24 °C for 7 days. The minimum inhibitory concentration (MIC) was defined as the lowest concentration of natamycin at which no growth was observed.
Source: from DSM Food Specialties Research Laboratorium Delft, The Netherlands; Dr R.A. Samson, Centraal Bureau voor Schimmelcultures, Utrecht, The Netherlands.

reported at all, although this is just as important as the concentration of fungicide added to the medium. Nevertheless, in most cases, the MIC of natamycin is independent of spore concentration. The manufacturer DSM Food Specialties recommends spreading 0.1 ml of a suspension of at least 10^4 spores/ml on an agar plate. This concentration of spores is high but reflects the levels of contamination that can be expected in practice. The use of fungal mycelia for determining MICs is not advised, as it is impossible to obtain reliable cell counts.

It is recommended that freshly prepared spores are used in MIC determinations. Moulds are grown on appropriate media such as malt extract agar (MEA) and incubated in darkness at 24 °C for 7–10 days. The spores are gently scraped from the agar surface and suspended in sterile water with 0.05% Tween 80. Stock suspensions are diluted using sterile water. A spore suspension of 0.1 ml is spread on suitable agar plates containing different concentrations of natamycin. Alternatively, a droplet of 5 μl of the concentrated spore suspension may be spotted onto the agar. After incubation for 4–6 days in darkness at 24 °C, the

Table 5.3 Sensitivity to natamycin of fungi isolated from beverages and fruit products

Microorganism	MIC (ppm)[+]	Source of microorganism
Yeasts		
Saccharomyces cerevisiae	1.5	Grape juice
Saccharomyces cerevisiae	1.5	Apple juice
Saccharomyces exiguus	< 1.0	Soft drink
Saccharomyces carlsbergensis	< 1.0	Brewers yeast
Zygosaccharomyces microellipsoides	1.0	Apple juice
Zygosaccharomyces microellipsoides	2.5	Orange juice
Zygosaccharomyces bailii	1.0	Apple juice
Zygosaccharomyces bailii	1.0	Pear juice
Zygosaccharomyces rouxii	1.0	Strawberry juice
Schizosaccharomyces pombe	< 1.0	Grape juice
Brettanomyces bruxellensii	2.5	Wine
Brettanomyces intermedius	< 1.0	Wine
Dekkara anomala	2.5	Stout
Dekkara anomala	5.0	Soft drink
Candida etchellsii	< 1.0	Concentrated lemon juice
Candida glabrata	4.0	Concentrated orange juice
Debaromyces hansenii	1.0	White wine
Pichia membranaefaciens	1.0	Lemonade
Pichia kluyverii	1.0	Fruit
Pichia angusta	< 1.0	Orange juice
Lodderomyces elongisporus	2.0	Concentrated orange juice
Torulaspora dellbreuckii	< 1.0	Lemonade
Hanseniaspora osmophila	< 1.0	Cider
Williopsis saturnus	1.0	Fruit juice
Rhodutorula rubra	< 1.0	Yoghurt
Moulds		
Byssochlamys nivea	2.0	Fruit juice
Trichosporonoides nigrescens	5.0	Melon jam
Mucor circinelloides	2.0	Strawberries
Stemphiliomma valparadisiacum	4.0	Apple juice
Talaromyces macrosporus	2.0	Pineapple juice
Cladosporium tenuissimum	2.0	Fruit
Zygosporium mycophilum	2.0	Apple pulp
Gilbertella persiacaria	10.0	Fruit

* MICs were determined as described in Table 5.2.
Source: from DSM Food Specialties Research Laboratorium Delft, The Netherlands.

MIC is determined as the lowest concentration of natamycin at which no growth is observed.

During storage of the agar plates, the fully dissolved natamycin is partly inactivated. After 20 days at 24 °C, the concentration of natamycin in MEA plates can be reduced by up to 50% (unpublished results, DSM Food Specialties). Therefore, it is recommended that only freshly prepared agar plates are used and that the concentration of natamycin in control agar plates is checked analytically at the end of prolonged incubation times.

Natamycin is poorly soluble in water and its crystals have no antifungal activity. Therefore, experiments to determine the MIC using agar plates containing natamycin suspensions higher than its maximum solubility in neutral aqueous environments are scientifically meaningless. However, such experiments might be useful to examine the prolonged working time of natamycin, e.g. in a cheese model.

Several surveys on the mycoflora in the cheese industry have been reported (De Boer and Stolk-Horsthuis, 1977; Fente-Sampayo, 1995; Hoekstra *et al.*, 1998). Natamycin sensitivity of fungi from Dutch cheese warehouses was determined by De Boer and Stolk-Horsthuis (1977). The main fungi detected were *Penicillium* and *Aspergillus* species although *Cladosporium*, *Scopulariopsis* and *Acremomium* species were also isolated. MICs of less than 10 ppm were reported for 26 species. More recently, Hoekstra *et al.* (1998) reported that a stable 'home flora' was present in all cheese factories. Environmental conditions, especially humidity, determine which species becomes predominant. In the more humid areas of a cheese production site, the composition of the mycoflora differed from that in the dryer areas of the factory. The predominant fungi detected in cheese factories were *Penicillium brevicompactum*, *Pen. corylophilum*, *Aspergillus penicillioides*, *Cladosporium* species and yeast species. The yeast *Debaryomyces hansenii* was predominant in the brines of several factories. The predominant fungi in cheese warehouses were *A. penicillioides* and *A. versicolor*. The sensitivity to natamycin of the most important fungi isolated in this study is shown in Table 5.1. Only three organisms, *A. flavus*, *A. parasiticus* and *Pen. discolor*, had a MIC greater than 10 ppm. In practice, the amount of active natamycin of about 40 ppm present on cheese surfaces is sufficient to inhibit all these species.

Fente-Sampayo *et al.* (1995) determined the composition of the fungal flora on soft cheeses and the production environment in 10 farm-level cheese production plants in Spain. In total, 35 species were isolated. Species of the genera *Penicillium*, *Aspergillus*, *Cladosporium*, *Rhizopus* and *Geotrichum* were most frequently isolated. The MIC for natamycin of all isolates was 10 ppm or less, except for a *Geotrichum* species that had a MIC of 12.5 ppm. A survey of 16 Dutch factories producing dry sausages was carried out to isolate less sensitive moulds and yeast using agar plates containing 0, 1, 2, 3 or 4 ppm of natamycin (De Boer *et al.*, 1979). Airborne yeasts and moulds were isolated using uncovered Petri dishes. For each natamycin concentration, 20 plates were used per factory. Yeasts and moulds from various objects in manufacturing and storage areas were isolated using Rodac plates (30 per factory) or swabs (25 per factory). Growth was observed on very few plates containing 2 ppm natamycin. More recently, the sensitivity to natamycin of fungi isolated from sausages was determined, as shown in Table 5.2. Again, a concentration of 10 ppm of natamycin was sufficient to inhibit most species.

Yeasts are predominant in the spoilage of acidic beverages such as fruit juices, lemonades, wine and beer. Some moulds, especially species producing heat-resistant spores or with thicker heat-resistant hyphae, are also associated with the spoilage of beverages and pasteurised fruit products. Important heat-resistant spoilage moulds are species belonging to the genera *Byssochlamys* and *Talaromyces*. Some strains are also a health risk, since the mycotoxin patulin is produced. Ascospores of *Byssochlamys* can survive several minutes at 90 °C. This mould is also able to grow at low O_2 or elevated CO_2 concentrations. Its unique characteristics give it a selective advantage in products such as fruit juices and canned fruits.

In a German study the sensitivity of 83 yeasts to natamycin was examined (Henninger, 1977). None of the yeasts isolated from food products were able to grow at a concentration of 3 ppm of natamycin. More recently, the MICs for natamycin of yeasts associated with spoilage of beverages and fruit products were determined, as shown in Table 5.3. The MICs of most species was less than 2.5 ppm. Notably, several sorbate-resistant yeast species, which can cause severe spoilage problems in the food industry, were also inhibited at these low concentrations of natamycin. *Zygosaccharomyces bailii*, a well-known sorbate-resistant yeast was inhibited by 1 ppm of natamycin. In the same study, many heat-resistant moulds of relevance to the food industry were also inhibited by as little as 2.5 ppm of natamycin.

5.5 Resistance

After decades of continuous use of natamycin, no resistant moulds and yeasts have been reported. This is quite remarkable, since most preservatives induce resistance. For example, several moulds and yeasts resistant to sorbate have been isolated. However, it is unlikely that fungi will develop resistance to natamycin because it interacts with ergosterol, a major constituent of the fungal cell membrane. Although low-ergosterol or ergosterol-free mutants that are also resistant to natamycin have been produced in the laboratory, they have slower growth rates and are unlikely to survive in nature (Ziogas *et al.*, 1983; Hamilton-Miller, 1974).

Furthermore, there is no separation between fungistatic and fungicidal concentrations of natamycin. Natamycin appears to have an all-or-none effect, which destroys the cell membrane without noticeable prior damage (Kotler-Brajtburg *et al.*, 1979). An interesting additional explanation is the single-hit theory. It is suggested that in aqueous solutions natamycin always occurs as micelles. If such a micelle comes into contact with a fungal cell, the local concentration of natamycin is always high enough to kill the cell immediately. Therefore selection of resistant mutants is unlikely.

Attempts have been made to induce natamycin tolerance in moulds from cheese warehouses by serial transfer onto media with increasing concentrations of natamycin (De Boer and Stolk-Horsthuis, 1977). After 25–30 transfers, none of the 26 strains studied became obviously less sensitive to natamycin. A study with *Penicillium discolor* (Frisvad *et al.*, 1997), a relatively natamycin-tolerant mould (see Table 5.1), showed that on agar plates inoculated with 10^2 spores/ml, the MIC was 10–20 ppm. However, when the agar plate was inoculated with 10^3–10^4 spores/ml, the MIC was 20–30 ppm. These results illustrated the importance of control of inoculum size in MIC determinations. Moreover, attempts to induce resistance in *Pen. discolor* have not been successful (unpublished results, DSM Food Specialties). It is concluded that *Pen. discolor* is a more tolerant mould species due to natural variation.

5.6 Applications

5.6.1 Surface treatments

One of the advantages of natamycin over other preservatives such as sorbate is that natamycin crystals remain on the surface of the product. Also, the dissolved fraction of natamycin hardly penetrates the product. In a study with Gouda cheese, no migration of natamycin was observed further than 3–4 mm into the rind (Daamen and van den Berg, 1985). In another study, a maximal penetration depth of 2.6 mm in Tilsiter cheese was reported (Kiermeier and Zierer, 1975). As another example, natamycin can be used to ensure a clean surface on blue cheese without affecting the internal mould growth in the punch holes of the cheese (Morris and Castberg, 1980). Growth of spoilage moulds on the surface of blue cheese must be avoided as it closes the punch holes and inhibits the development of *Penicillium roquefortii*, which give the cheese its characteristic appearance and flavour. Dipping the blue cheese in a suspension of 1 g/l of natamycin was sufficient to prevent fungal growth on the surface for three weeks, while the development of the blue mould in the punch holes was not affected.

Another advantage of natamycin is its prolonged working time. In practice, cheeses or sausages are treated with coatings, casings or suspensions containing 100–2000 ppm of natamycin. The initial concentration of natamycin on the surface of the product depends on the treatment. Usually this concentration is 1–2 mg/dm^2. At these concentrations, only a fraction of 30–40 ppm is in solution. Most of the natamycin is present in the form of stable crystals. Disappearance of dissolved natamycin, e.g. due to interaction with fungal cell membranes, diffusion or chemical decomposition, is compensated for by dissolution from the crystals and by diffusion over the surface of the product. This protects the product for a longer period of time against fungal growth.

Many scientific studies have been published about the successful application of natamycin on almost every type of cheese: Gouda (Lück and Cheesman, 1978; Engel *et al.*, 1983; De Ruig and van den Berg, 1985; Daamen and van den Berg, 1985), Edam (Engel *et al.*, 1983), Cheddar (Lück and Cheesman, 1978; Sachdeva *et al.*, 1994); Tilsiter (Engel *et al.*, 1983), Italian cheeses such as Caciotta (Neviani *et al.*, 1981) and Fontina, Tallegio, Montasio, Asiago, Provolone, Pecorino and Romano (Lodi *et al.*, 1989); Swedish hard cheeses (Mattson, 1977); blue cheese (Morris and Castberg, 1980); Indian cheeses (Verma *et al.*, 1988; Pugazhenthe *et al.*, 1999). For the surface treatment of cheese, natamycin is usually added to the aqueous polymer dispersion, the so-called plastic emulsion that is applied to the cheese rind as a coating. On cheese types that are not treated with a coating, natamycin can be applied by dipping or spraying. Natamycin can also be added to the brine as an alternative treatment for round cheeses such as Edam (Zuthof and Isidorus, 1981).

The antifungal effect of natamycin on sausages has also been extensively studied. Natamycin prevents fungal growth during normal ageing and storage without affecting the quality of Dutch raw sausages (Moerman, 1972), German raw sausages (Hechelman and Leistner, 1969; Stiebing *et al.*, 2001) and Italian sausages such as dry salami (Cattaneo *et al.*, 1978; Baldini *et al.*, 1979; Holley, 1981, 1986). Sausages are treated with natamycin by dipping (e.g. at 500–1000 ppm) or spraying (e.g. at 2000 ppm). Natamycin can also be applied by treatment of sausage casings before stuffing by soaking for 1–2 h in a suspension containing 500–1000 ppm of natamycin.

The efficacy of natamycin and sorbate in the prevention of mould growth on raw German sausages has been compared (Stiebing *et al.*, 2001). Sausages were dipped in suspensions containing 0.2% natamycin or a solution containing 10% sorbate. It was demonstrated that natamycin treatment was superior to that afforded by sorbate. Sorbate gave less protection against mould growth, inhibited the bacterial starter cultures, causing a delay of the ripening process, which led to defects in the colour development of the product and penetrated into the sausages. In the case of natamycin no negative effects on the quality of the product were observed.

Natamycin can also be used to prevent fungal growth on fruit. Strawberries, cranberries and raspberries can be sprayed in the field. A more effective protection is obtained when the berries are dipped in a suspension of 10–100 ppm natamycin after harvest. The shelf-life of the berries was prolonged by several days (Ayres and Denisen, 1958). Natamycin was shown to be effective in preventing fungal growth on the surface of apples and pears. Treatment with a lecithin emulsion containing 200 ppm of natamycin was the most effective way to prevent fungal growth (Staden and Witmondt, 1967).

5.6.2 Beverages

Owing to its optimal availability, low concentrations of natamycin (1–5 ppm) are sufficient to prevent fungal spoilage of juices, lemonades, beer, wine, iced tea and fruit yoghurts. Natamycin reduces the initial mould and yeast population rapidly. This treatment is sufficient since after production the packaging of most beverages is usually well sealed and contamination before the consumer opens the packaging is unlikely. It has also been demonstrated that low concentrations of natamycin (1–20 ppm) are quite stable in orange juice. After 12 weeks of refrigerated storage, 70% of the initial natamycin remained active (Shirk and Clark, 1963). This means that natamycin is also suitable to protect 'fresh' beverages with a short shelf-life, such as chilled orange juice, against fungal spoilage. Natamycin is also effective in protecting iced tea, a beverage with higher pH values. Sorbic acid is currently used in iced tea but the pH of the product has to be reduced for the sorbic acid to be effective. Both the low pH and the taste of the sorbic acid itself can have adverse effects on the organoleptic properties of iced tea.

5.7 Toxicology

The mutagenicity, absorption, distribution, excretion and metabolism of natamycin have been investigated in several studies. Mutagenicity has been examined with the Ames test using two *Salmonella* Typhimurium strains and two *Escherichia coli* strains. Another mutagenicity test was executed using a wild type *Bacillus subtilis* and a mutant deficient in its DNA repair system. If the mutant was more sensitive to natamycin than its wild-type parent, it could be concluded that natamycin acted on DNA. The degradation products aponatamycin, dinatamycinolidediol and mycosamine and the commercial product Delvocid® were included in these studies. All the mutagenicity tests were negative.

In several studies it was demonstrated that natamycin and its degradation products are hardly absorbed by the body. After oral administration of ^{14}C-labelled natamycin to rats, none of the label was detected outside the gastrointestinal tract. After a very long exposure time of 150 days, small amounts were detected in the liver, kidney and fatty tissue. Similar results were found with dogs. Acute intraperitoneal toxicity of natamycin and its degradation products was determined in mice. Neither natamycin nor its degradation products presented a toxic risk. Finally, the very long history of safe use of natamycin as food additive confirms that natamycin is a safe fungicide for preventing growth of moulds and yeasts in foods and beverages.

5.8 Regulatory status for use in foods

Natamycin is allowed for use in many countries as a food additive in a variety of foods to control the growth of yeasts and moulds. The primary food uses are in the surface treatment of cheese and sausages. These uses are covered in the Food and Agriculture Organization/World Health Organization (FAO/WHO) Codex Alimentarius (see Chapter 15 for details of international food regulations). Several Codex standards for cheese permit the presence of natamycin on the cheese surface/rind at a level of 2 mg/dm^2 to a maximum depth of 5 mm. Recently, the use of natamycin was incorporated in a new Codex group standard for unripened cheese and fresh cheese. The Joint FAO/WHO Expert Committee on Food Additives (JECFA) reviewed the safety of natamycin in 1968, 1976 and 2001. JECFA assigned an Acceptable Daily Intake value (ADI) of 0.3 mg/kg body weight per day. This ADI was reaffirmed by JECFA during the latest review of natamycin in 2001.

In the European Union, natamycin is permitted as a food additive for the surface treatment of hard, semi-hard and semi-soft cheese and dried, cured sausages. The maximum permitted level is 1 mg/dm^2 surface and it must not be present at a depth greater than 5 mm (EU Directive 95/2/EC). In the USA, natamycin is an approved food additive. It may be applied on cheese in amounts not to exceed 20 mg/kg in the finished product as determined by International Dairy Federation Standard 140A: 1992 (21 CFR 172.155). In Canada, natamycin is permitted on 47 named cheeses up to 20 ppm and in shredded/grated cheese up to 10 ppm residual level (*Canada Gazette* Part II, Vol. 116, no. 7, 14 April 1982).

In Australia, natamycin is approved in cheese and cheese products at a level of 15 mg/kg on cheese surfaces based on individual cheese weight. It is also allowed on fermented, uncooked processed comminuted meat products when determined in a surface sample taken to a depth of not less than 3 mm and not more than 5 mm including the casing, applied to the surface of food (ANZA Food Standard 1.3.1). In South Africa, natamycin is allowed in a broad range of food products and beverages (*Government Gazette* no. 8436, 1982, and no. 5729, 1977). Elsewhere, the permitted uses differ from country to country. In most non-EU European countries, natamycin is allowed for the surface treatment of cheeses and/or sausages or other meat products. In Brazil, Argentina and Venezuela, it is allowed for the application on cheese and sausages. In Colombia it is only allowed for the treatment of meat products. In countries such as Mexico, Chile, Costa Rica and some Arab countries, it is permitted as a general food additive. Since regulations may change, one always has to check the regulation in a country before applying natamycin in food products.

5.9 Future prospects

Natamycin has a long history of safe use in the prevention of fungal growth on the surface of cheese and sausages. In the future, economic losses due to spoilage and health risks associated with the growth of pathogenic or mycotoxin-producing moulds will remain important issues. Therefore, it is expected that natamycin will remain the most important antifungal agent for the surface treatment of cheese and sausages. Although natamycin has also been shown to inhibit moulds and yeasts in other foods and beverages at very low concentrations (1–5 ppm), its use in other food products is still not permitted in most countries. The recent inclusion of natamycin in a new Codex group standard for unripened cheese and fresh cheese can be considered as a first step towards a more general approval in the future.

The increasing occurrence of sorbate-resistant yeasts and heat-resistant moulds will require improved preservation systems. Also, the trend towards more fresh products such as fruit juices represents a severe risk of increasing mould problems. Sterilization processes will be replaced by milder heat treatments or by new mild preservation techniques such as ultra high pressure or pulsed electric field processing. These techniques will improve the organoleptic and nutritional properties of the products; however, they will also introduce new spoilage problems by selecting resistant strains. Therefore, it is to be expected that in the future natamycin will be accepted in more countries for application in products containing fresh fruit, beverages and other products sensitive to fungal spoilage.

5.10 References

AYRES, J.C. and DENISEN, E.L. (1958), 'Maintaining freshness of berries using selected packaging materials and antifungal agents', *Food Technology*, **12** (10), 562–567.

BALDINI, P., PALMIA, F., RACZYNSKI, G. *et al.* (1979), 'Impiego della pimaricina nella prevenzione della crescita della muffe sui prodotti di salumeria italiani', *Industria Conserve*, **54** (4), 305–307.

BRIK, H. (1981), 'Natamycin', in K. Florey, *Analytical Profiles of Drug Substances* Vol. 10, Academic Press, New York, 513–561.

CATTANEO, P., D'AUBERT, S. and RIGHETTI, A. (1978), 'Attivita antifungina della pimaricina in salami crudi stagionati. *Industrie Alimentari*', **17** (9), 658–664.

CEDER, O., HANSSON, B. and RAPP, U. (1977), 'Pimaricin. VIII. Structural and configurational studies by electron impact and field desorption mass spectrometry, carbon-13 (25.2 Mhz) and proton (270 Mhz)-NMR spectroscopy', *Tetrahedron,* **33** (20), 2703–2714.

CLARK, W.L., SHIRK, R.J. and KLINE, E.F. (1964), 'Pimaricin, a new food fungistat', in Molin, N., *Microbial Inhibitors in Food*, Almquist & Wiksell, Gotenborg, 167–184.

DAAMEN, C.B.G. and VAN DEN BERG, G. (1985), 'Prevention of mould growth on cheese by means of natamycin', *Voedingsmiddelentechnologie*, **18** (2), 26–29.

DE BOER, E. and STOLK-HORSTHUIS, M. (1977), 'Sensitivity to natamycin (pimaricin) of fungi isolated in cheese warehouses', *Journal of Food Protection*, **40** (8), 533–536.

DE BOER, E., LABOTS, M., STOLK-HORSTHUIS, M. *et al.* (1979), 'Sensitivity to natamycin of fungi in factories producing dry sausage', *Fleischwirtschaft*, **59** (12), 1868–1869.

DE KRUIJFF, B., GERRITSEN, W.J., OERLEMANS, F. *et al.* (1974), 'Polyene antibiotic-sterol interactions in membranes of *Acholeplasma laidlawii* cells and lecithin liposomes. II. Temperature dependence of the polyene antibiotic-sterol complex formation', *Biochimica et Biophysica Acta*, **339** (1), 44–56.

DE RUIG, W.G. and VAN DEN BERG, G. (1985), 'Influence of the fungicides sorbate and natamycin in cheese coatings on the quality of the cheese', *Netherlands Milk and Dairy Journal*, **39** (3), 165–172.

DUPLANTIER, A.J. and MASAMUNE, S. (1990), 'Pimaricin. Stereochemistry and synthesis of its aglycon (pimarolide) methyl ester', *Journal of the American Chemical Society* **112** (19), 7079–7081.

ENGEL, G., HERTEL, K. and TEUBER M. (1983), 'Nachweis und Abbau von Natamycin (Pimaricin) auf der Käseoberfläche', *Milchwissenschaft*, **38** (3), 145–147.

FENTE-SAMPAYO, C.A., VAZQUEZ-BELDA, B., FRANCO-ABUIN, C. *et al.* (1995), 'Distribution of fungal genera in cheese and dairies. Sensitivity to potassium sorbate and natamycin', *Archiv für Lebensmittelhygiene*, **46** (3), 62–65.

FRISVAD, J.C., SAMSON, R.A., RASSING, B.R. *et al.* (1997), '*Penicillium discolor*, a new species from cheese, nuts and vegetables', *Antonie van Leeuwenhoek*, **72**, 119–126.

HAMILTON-MILLER, J.M.T. (1974), 'Fungal sterols and the mode of action of the polyene antibiotics', *Advances in Applied Microbiology*, **17**, 109–134.

HECHELMANN, H. and LEISTNER, L. (1969), 'Hemmung von unerwünschtem Schimmelpilzwachstum auf Rohwürsten durch Delvocid (Pimaricin)', *Fleischwirtschaft*, **49** (12), 1639–1641.

HENNINGER, W. (1977), 'Die Abtötung von getränkeverderbenden Hefen durch Pimaricin', *Das Erfrischungsgetränk Mineralwasser-Zeitung*, **30** (13), 1–6.

HOEKSTRA, E.S., VAN DER HORST, M.I., SAMSON, R.A. *et al.* (1998), 'Survey of the fungal flora in Dutch cheese factories and warehouses', *Journal of Food Mycology*, **1** (1), 13–22

HOLLEY, R.A. (1981), 'Prevention of surface mould growth on Italian dry sausage by natamycin and potassium sorbate', *Applied and Environmental Microbiology*, **41** (2), 422–429.

HOLLEY, R.A. (1986), 'Effect of sorbate and pimaricin on surface mould and ripening of Italian dry salami', *Lebensmittel Wissenschaft und Technologie*, **19** (1), 59–65.

KIERMEIER, F. and ZIERER, E. (1975), 'Zur Wirkung von Pimaricin auf Schimmelpilze und deren Aflatoxinbildung bei Käsen', *Zeitschrift für Lebensmitteluntersuchung und Forschung*, **157** (5), 253–262.

KOTLER-BRAJTBURG, J., MEDOFF, G., KOBAYASHI, G.S. *et al.* (1979), 'Classification of polyene antibiotics according to chemical structure and biological effects', *Antimicrobial Agents and Chemotherapy*, **15** (5), 716–722.

LANCELIN, J.M. and BEAU, J.M. (1990), 'Stereostructure of pimaricin', *Journal of the American Chemical Society* **112** (10), 4060–4061.

LANCELIN, J.M. and BEAU, J.M. (1995), 'Stereostructure of glycosylated polyene macrolides: the example natamycin', *Bulletin Societé Chimie de France* **132**, 215–223.

LODI, R., TODESCO, R. and BOZZETTI, V. (1989), 'Nouvelles applications de la natamycine sur des fromages typiques Italiens', *Microbiologie-Aliments-Nutrition*, **7** (1), 81–84.

LÜCK, H. and CHEESMAN, C.E. (1978) 'Mould growth on cheese as influenced by pimaricin or sorbate treatments', *South African Journal of Dairy Technol.*, **10** (3), 143–146

MALEWICZ, B. and BOROWSKI, E. (1979), 'Energy dependence and reversibility of membrane alterations induced by polyene macrolide antibiotics in *Chlorella vulgaris*', *Nature* **281**, 80–82.

MATTSSON, N. (1977), 'Mould control on cheese. Experience with potassium sorbate and pimaricin for surface treatment of hard cheese', *Svenska Mejeritidningen*, **69** (3), 14–15.

MILHAUD, J., PONSINET, V., TAKASASHI, M. *et al.* (2002) 'Interactions of the drug amphotericin B with phospholipid membranes containing or not ergosterol: new insight into the role of ergosterol', *Biochimica et Biophysica Acta*, **1558**, 95–108.

MOERMAN, P.C. (1972), 'Schimmelwering op vleeswaren door Pimaricine', *Voedingsmiddelentechnologie*, **3** (51/52), 261–264.

MORRIS, H.A. and CASTBERG, H.B. (1980), 'Control of surface growth on blue cheese using pimaricin', *Cultures Dairy Products Journal*, **15** (2), 21–23.

NEVIANI, E., EMALDI, G.C. and CARINI, S. (1981), 'L'impiego di pimaricina come antifungino sulle croste dei formaggi: technologia e microflora di superficie', *Latte*, **6** (5), 335–343.

PUGAZHENTHI, T.R., DHANALAKSHMI, B., NARASIMHAN, R. *et al.* (1999), 'Effect of antimycotic agents on *Penicillium citrinum* in cheese', *Indian Veterinary Journal*, **76**,

537–539.

RAAB, W.P. (1974), '*Natamycin (Pimaricin). Properties and Medical Applications*', Georg Thieme, Stuttgart.

SACHDEVA, S., SING, S., TIWARI, B.D. *et al.* (1994), 'Effect of processing variables on the quality and shelf-life of processed cheese from buffalo milk Cheddar cheese', *The Australian Journal of Dairy Technology*, **49** (2), 75–78.

SHIRK, R.J. and CLARK, W.L. (1963), 'The effect of pimaricin in retarding the spoilage of fresh orange juice', *Food Technology*, **17** (8), 1062–1066.

STADEN, O.L. and WITMONDT, M. (1967), 'Lecithine-pimaricine-perspectieven op hard fruit', *De Fruitteelt*, **57** (38), 1180–1182.

STIEBING, A., OBERHAUS, T. and BAUMGART J. (2001), 'Natamycin-Verhinderung des Schimmelpilzwachtums bei Rohwurst', *Fleischwirtschaft*, **81** (8), 97–100.

STRUYK, A.P., HOETTE, I., DROST, G. *et al.* (1957–58), 'Pimaricin, a new antifungal antibiotic', *Antibiotics Annual*, 878–885.

TEERLINK, T., DE KRUYFF, B. and DEMEL, R.A. (1980), 'The action of pimaricin, etruscomycin and amphotericin B on liposomes with varying sterol content', *Biochimica et Biophysica Acta*, **599** (2), 484–492.

VAN RIJN, F.T.J., STARK, J., TAN, H.S. *et al.* (1995), 'A novel antifungal composition', US Patent Application No. 08/446,782.

VERMA, H.S., YADIV, J.S. and NEELAKANTAN S. (1988), 'Preservative effect of selected antifungal agents on butter and cheese', *Asian Journal of Dairy Research*, **7** (1), 34–38.

ZIOGAS, B.N., SISLER, H.D. and LUSBY, W.R. (1983), 'Sterol content and other characteristics of pimaricin-resistant mutants of *Aspergillus nidulans*, *Pesticide Biochemistry and Physiology*, **20** (3), 320–329.

ZUTHOF, J.B.I. and ISIDORUS, J.B. (1981), 'Cheese manufacturing – Flowing brine over mould pressed cheese in racks', GB Patent 2072481.

6

Organic acids

J. Samelis, National Agricultural Research Foundation, Greece and J. N. Sofos, Colorado State University, USA

6.1 Introduction

Organic acids (acetic, lactic, malic, citric, etc.) are natural constituents of many foods. They are also used widely as additives in food preservation. The antimicrobial action of organic acids is primarily based on their ability to reduce the pH in the water phase of foods. When pH values are lower than 4.0, the acids restrict bacterial growth, while yeasts and molds become competitive against bacteria at pH values below 5.0. Weak lipophilic organic acids (acetic, propionic, sorbic, benzoic) penetrate the cell membrane in the undissociated form to inhibit growth or cause death by dissociating and acidifying the cytoplasm. Mechanisms of antimicrobial activity are based on inhibition of enzymes, membrane function, nutrient transport and overall metabolic activity. Organic acids in foods can act as acidulants or preservatives, while their salts or esters can be effective antimicrobials at pH values near neutrality.

Foods preserved by means of acid/low pH can be differentiated into acidic (most fruits, some vegetables, juices, beverages), fermented (yogurt, cheese, pickles and sausages) or acidified (sauces, mayonnaise, salad dressings) products. The reduction in pH by fermentation or addition of organic acids is among the most widely used of food preservation methods, often combined with salting, curing or drying. Lactic acid is the main product of food fermentations, followed by acetic, propionic, malic and other acids at various concentrations, depending on the product and the microorganisms involved. Glucose, sucrose, lactose and other sugars are converted into organic acids by microorganisms in fermented foods of dairy, meat or plant origin. Vinegar (acetic acid) and lemon juice (citric acid) have been used as flavoring and preservative agents in traditional foods since antiquity. Citric, malic, acetic and tartaric acids are present in lemons, oranges, apples, grapes, other fruits and juices or beverages produced from them. Commercially available acetic, adipic, citric, fumaric, gluconic, lactic, malic, succinic and tartaric acids are permitted for use as acidulants in foods such as pickles, salad dressings, desserts, soft drinks, jams, soups and margarine. Propionic, sorbic and benzoic acids and their salts are permitted and used extensively in foods including fruit-based drinks, bakery, cheese, meat, fish and egg products. Although present naturally in certain plant tissues, the lipophilic organic acids are manufactured for commercial use by chemical synthesis.

The antimicrobial effectiveness of organic acids in foods depends on the type of acid used, its concentration and method of application. Efficacy is also affected by temperature, pH, water activity, oxygen, salt and other antimicrobials. Implicit factors, such as the numbers and types of microorganisms present, their metabolic activity and microbial interactions may increase or decrease the antimicrobial action of acids. In general, the activity of organic acids is enhanced at low pH and/or when they are combined with additional antimicrobial factors.

In this chapter, a brief overview of the antimicrobial action of organic acids in laboratory media is followed by a more detailed review of their application in foods. Special emphasis is given to the use of organic acids for decontamination of meat surfaces, and to the potential for the development of acid resistance in pathogenic bacteria. The chemically manufactured lipophilic acids and their salts (sorbates, benzoates) are not covered as extensively as those organic acids considered as acidulants (acetic, lactic, malic, citric, etc.). For more information on lipophilic acids, the reader is referred to reviews by Chipley (1993), Sofos (1989) and Sofos and Busta (1993, 1999). The important chemical and physical properties of organic acids are summarized in Table 6.1.

6.1.1 Organic acids in laboratory media

Numerous studies have been published on the effects of pH and organic acids on survival and growth of foodborne microorganisms in laboratory media. When HCl is used to adjust the pH of culture media, most foodborne pathogens have a minimum pH limit for growth in the range of 4.0 to 5.0 with *Escherichia coli* O157:H7, *Yersinia enterocolitica*, *Staphylococcus aureus*, *Aeromonas hydrophila* and *Salmonella* showing the lowest growth limit at 4.0 (Table 6.2). Low pH by itself extends the lag phase and reduces the growth rate and cell biomass, reflecting the expenditure of energy by the cells for intracellular pH maintenance (Lund and Eklund, 2000). When an organic acid is involved, however, the pH-dependent effect of the undissociated acid needs to be considered. The inhibitory effects of organic acids at various molar concentrations at constant pH values adjusted with a strong inorganic acid have been compared (Sorrells *et al.*, 1989; Buchanan *et al.*, 1993; Buchanan and Edelson, 1999a; Young and Foegeding, 1993; Ouattara *et al.*, 1997). In general, the pH where inhibition of a microorganism by an organic acid occurs has been

Table 6.1 Main physical and chemical properties of organic acids used as food acidulants or preservatives

Common acid name	Molecular formula	Molecular weight	pK_a	Specific gravity at 20/4 °C	Melting point (°C)	Boiling point (°C)	Solubility
Acetic	$C_2H_4O_2$	60.06	4.75	1.049	16.7	118.1	Miscible in water, alcohol, glycerol, ether
Adipic	$C_6H_{10}O_4$	146.14	4.43	1.360 at 25/4 °C	152	337.5	g/100 ml in water: 1.4 at 20 °C, 160 at 100 °C, very soluble in ethanol
Ascorbic	$C_6H_8O_6$	176.12	4.17	1.650	190–192		g/100 ml in water: 33 at 20 °C, 80 at 100 °C
Benzoic	$C_7H_6O_2$	122.12	4.19	1.266–1.321	122.4	249.2	g/l in water: 2.9 at 20 °C, 12.0 at 60 °C, 68.0 at 95 °C
Citric	$C_6H_8O_7$	192.14	3.14	1.665 (anhydrous)	153	Decomposes	g/100 ml in water: 59.2 at 20 °C, 73.5 at 60 °C, 84.0 at 100 °C
Fumaric	$C_4H_4O_4$	116.07	3.03	1.635	287	290, sublimes	g/100 ml in water: 0.63 at 25 °C, 2.4 at 60 °C, 9.4 at 100 °C
Lactic	$C_3H_6O_3$	90.08	3.86	1.249 at 15 °C	16.8	122 at 15 mm	Very soluble in water and alcohol, less soluble in ether, practically insoluble in chloroform
Malic	$C_4H_6O_5$	134.09	3.40	1.595 (D or L) 1.601 (DL)	100 (D or L) 128 (DL)	140 (D or L, decomposes) 150 (DL)	g/100 ml at 20 °C: 55.8 (DL), 36.3 (L) in water, 45.5 (DL), 86.6 (L) in alcohol
Propionic	$C_3H_6O_2$	74.09	4.87	0.998 at 15/4 °C	−21.5	141.1	Miscible in water, alcohol, ether, chloroform
Sorbic	$C_6H_8O_2$	112.14	4.76		134.5	228 (decomposes)	0.16 g/100 ml water at 20 °C, 14.8 g/100 ml ethanol at 25 °C
Succinic	$C_4H_6O_4$	118.09	4.16	1.564	185–188	235	1 g/13 ml cold water or 1 ml boiling water
Tartaric	$C_4H_6O_6$	150.09	2.98	1.759	168–170		g/100 ml in water: 139 at 20 °C, 217 at 60 °C, 343 at 100 °C

Modified from Deshpande *et al.* (1995).

Table 6.2 Survival and growth pH limits of some foodborne pathogens in laboratory media

Organism	Minimum pH for growth*	Inhibitory pH of acetic acid⁺	Lethal pH of acetic acid
Aeromonas hydrophila	4.0	NR	NR
Campylobacter jejuni	4.9	NR	NR
Bacillus cereus	4.4–5.0	4.9	4.9
Clostridium botulinum	4.6	NR	NR
Escherichia coli O157:H7	4.0-4.4	4.5–5.0	<3.0
Listeria monocytogenes	4.3–4.5	4.59	4.37
Salmonella spp.	4.0–4.5	4.9–5.4	4.5
Staphylococcus aureus	4.0	5.0	4.9
Yersinia enterocolitica	4.0–4.2	5.0	4.75

* The lower pH allowing the initiation of growth in media adjusted with strong acid (i.e. HCl).
⁺ The pH that inhibits growth when acetic acid is the acidulant.
NR, not reported.
Modified from Lund and Eklund (2000) and Smittle (2000).

higher than the minimum pH allowing initiation of growth (Table 6.2). The minimum inhibitory concentration (MIC) of an organic acid against microorganisms has to be calculated under constant conditions because interacting factors (salt, sugar, antimicrobials other than acids, water activity) and the environment (temperature, oxygen, etc.) enhance the acid/pH effects (Houtsma *et al.*, 1993; Lund and Eklund, 2000).

In a pioneering study, Nunheimer and Fabian (1940) reported that organic acids inhibited *Staph. aureus* in the order acetic > lactic > citric > malic > tartaric > HCl. Another early study showed that volatile acids were effective against *Salmonella* Typhimurium in the order formic > acetic > propionic > butyric > HCl (Geopfert and Hicks, 1969). Based on equal pH, *L. monocytogenes* was affected by organic acids at 25 or 35 °C in the order acetic > lactic > citric > malic > HCl; however, based on an equal molar concentration the order changed to citric > malic > lactic > acetic > HCl (Sorrells *et al.*, 1989). Similar results were reported by Young and Foegeding (1993), who noted that pathogen inhibition based on initial undissociated acid concentrations was citric > lactic > acetic. Acetic, lactic and citric acids at 0.1% in tryptose broth inhibited growth of *L. monocytogenes* in that order; inhibition increased as the temperature increased in the range 7 to 35 °C (Ahamad and Marth, 1989). Conner and Kotrola (1995) reported that the order of effectiveness of organic acids against *E. coli* O157:H7 in tryptic soy broth (TSB) supplemented with yeast extract and acidified to pH 4.0–5.5 was mandelic = acetic > citric > lactic > malic > tartaric. When organic acids were added (0.5%) to brain heart infusion (BHI) broth adjusted to pH 3.0 with HCl, the relative inactivation rates of previously acid-adapted or non-adapted strains of *E. coli* O157:H7 were as shown in Table 6.3. Similar results have also been reported for meat spoilage bacteria in laboratory media; on a weight basis, the order of inhibition was acetic > propionic > lactic > citric, whereas on a molar basis, the above order was reversed (Ouattara *et al.*, 1997).

Many predictive models of the combined effects of temperature, pH and organic acids (mainly lactic acid) on the survival and growth of *Salmonella* spp., *Listeria* spp., *E. coli* and other bacteria have been developed on the basis of data obtained in laboratory media (Gibson *et al.*, 1988; McClure *et al.*, 1994; George *et al.*, 1996; Houtsma *et al.*, 1996; Buchanan *et al.*, 1997; Razavilar and Genigeorgis, 1998; Presser *et al.*, 1997, 1998).

6.2 Organic acids in complex food systems

Under constant environmental conditions, microbial growth and death rates are slower in complex food systems than in culture media of similar pH due to compositional factors that frequently protect the cells. Temperature has the most important influence on the antimicrobial efficacy of organic acids: storage at refrigeration temperatures delays microbial inactivation in low-pH foods.

Table 6.3 Effect of acid type on the relative rate of inactivation at pH 3.0 when enterohemorrhagic *Escherichia coli* were cultured in tryptic soy broth with 1% glucose (acid-adapted) or without glucose (non-adapted) at 37 °C for 18 h

E. coli strain	Strain serotype and source	Relative rate of inactivation by acid of cells when being:	
		Acid-adapted	Non-adapted
S1, B1409	O157:H7, clinical isolate, CDC	HCl < malic < citric < acetic <lactic	HCl < citric < malic < acetic = lactic
S3, 30-2C4	O157:H7, clinical isolate from dry salami outbreak, CDC via FSIS	HCl = malic = citric < acetic <lactic	HCl < malic < citric < acetic < lactic
S5, Ent-C9490	O157:H7, clinical isolate from Jack-in-the-Box outbreak, CDC	HCl = malic = citric = acetic <lactic	HCl < citric = malic < acetic = lactic
S7, 95JB1	O111:H-, clinical isolate from mettwurst outbreak, Australia	HCl = malic < citric < acetic <lactic	HCl < malic = citric < acetic < lactic
S8, ATCC 25922	Not available, reference strain	HCl < citric < acetic < malic <lactic	HCl < citric = malic = acetic = lactic
S9, SEA13B88	O157:H7, apple juice isolate from Odwalla outbreak, FDA	HCl = malic = citric = acetic <lactic	HCl = citric < malic < acetic < lactic

Modified from Buchanan and Edelson (1999a).

6.2.1 Organic acids as antimicrobials in naturally acidic foods

Fruits and fruit juices naturally contain organic acids such as citric (citrus fruits), malic (apples) and tartaric (grapes) acids. These are characterized by a lower pK_a than those of acetic and lactic acids, the main organic acids in acidified and fermented foods (Table 6.1). The lower antimicrobial activity of citric and malic acids compared with lactic and acetic acids at the same pH mean that fruits and their juices may allow survival and growth of microorganisms despite their low pH. The pH of fruit juices ranges from 2.9 to 4.3, depending on fruit variety (Lund and Snowdon, 2000). At this pH range, citric, malic and other organic acids present in fruits inhibit many contaminating microorganisms but not lactic acid bacteria (LAB) and yeasts. Spoilage of fresh and processed fruits and fruit juices by LAB and yeasts is well documented (Deak and Beuchat, 1993; Lund and Snowdon, 2000; Alwazeer *et al.*, 2002). Pasteurization remains the most effective method of inactivating spoilage bacteria in fruit juices but high pressure and irradiation have also been applied (Parish, 1998; Foley *et al.*, 2002; Alwazeer *et al.*, 2002). The lower the pH of a fruit juice, the greater the killing effect of heat, high pressure and irradiation on microorganisms. Silva *et al.* (1999) observed a linear decrease in thermal *D*-values of *Alicyclobacillus acidoterrestris*, a common spoilage agent of heat-treated fruit juices, with decreasing pH and soluble solids in fruit juices. Thus, while thermal inactivation is enhanced in more acidic juices, increased concentrations of solids protect spoilage bacteria from heat.

An increasing number of hemorrhagic colitis outbreaks linked with the consumption of unpasteurized apple juice and cider have prompted regulatory action by the US Food and Drug Administration (Besser *et al.*, 1993; Centers for Disease Control and Prevention, 1996). A warning statement has been mandated on packaged fruit juices not treated to reduce target pathogen populations by 5 log CFU/ml (Food and Drug Administration, 1998). In response, research has been intensified to find new intervention treatments resulting in the required reduction in numbers (Fisher and Golden, 1998; Ingham and Uljas, 1998; Uljas and Ingham, 1999). The survival of *E. coli* O157:H7 in apple cider with or without preservatives is summarized in Table 6.4. It is evident that lactic, propionic and sorbic acids, potassium sorbate and sodium benzoate significantly contribute to population reductions. Reductions were always faster and greater at abuse temperatures. The antimicrobial effect was also greater in fruit juices when organic acids or salts were combined with each other (Zhao *et al.*, 1993) or with other preservatives, such as dimethyl-carbonate or sodium bisulfite (Fisher and Golden, 1998). Also, freezing/thawing and/or short-term storage of apple juice or cider at fluctuating temperatures increased the effectiveness of organic acids or that of subsequent heat treatments (Ingham and Uljas, 1998).

Treatment of sliced fruits such as apples with organic acids (e.g. dipping in a 3.4% solution of ascorbic acid or 1.7% citric acid) enhanced destruction of *E. coli* O157:H7 during domestic drying (Burnham *et al.*, 2001; Lakkakula *et al.*, 2001). Citric acid at 0.5–1% applied on the surface of peeled oranges (pH $\geq$ 6.0)

Table 6.4 Survival of *Escherichia coli* O157:H7 in apple cider with or without organic acid preservatives

pH of apple cider	Inoculation level (log CFU/ml)	Experimental conditions	Survival/Reduction in counts	Reference
3.6–4.0	5.0	Fresh-pressed; preservative-free; stored at 25 °C	Survived for 2–3 days; undetectable at 7 days	Besser *et al.* (1993) and Zhao *et al.* (1993)
		0.1% sodium benzoate added; stored at 25 °C	Survived for <1–2 days; undetectable at 7 days	
		0.1% potassium sorbate added; stored at 25 °C	Survived for 1–3 days; undetectable at 7 days	
		Fresh pressed; preservative-free; stored at 8 °C	Slight growth (1 log CFU/ml); survived for 10 to 31 days depending on batch	
		0.1% sodium benzoate added; stored at 8 °C	No growth; undetectable within 7 days	
		0.1% potassium sorbate added; stored at 8 °C	Little effect compared to preservative-free cider; survived for 15–20 days	
3.7–4.1	4.5	Preservative-free	91–98% survival after 21 days at 4 °C	Miller and Kaspar (1994)
		Commercially added 0.05% potassium sorbate	90–93% survival after 21 days at 4 °C	
		Commercially added 0.1% potassium sorbate	84–91% survival after 21 days at 4 °C	
		Commercially added 0.1% sodium benzoate	43–98% survival after 21 days at 4 °C	
3.3–3.5	7.0	Preservative-free; unpasteurized; stored at 4 °C	3.2 and 6.0 log reduction at 7 and 21 days	Roering *et al.* (1999)
		Preservative-free; pasteurized; stored at 4 °C	2.8 and 5.2 log reduction at 7 and 21 days	
		Preservative-free; unpasteurized; stored at 10 °C	3.4 and 6.0 log reduction at 7 and 21 days	
		Preservative-free; pasteurized; stored at 10 °C	2.5 and 5.1 log reduction at 7 and 21 days	
3.3, 3.7 or 4.1; adjusted	7.0	Preservative-free; unpasteurized; treated with short-term storage at 4, 25 or 35 °C and/or freeze-thawing (48 h at −20 °C; 4 h at 4 °C)	≥5.0 log reduction at pH 3.3 in freeze-thawed cider (no acids required), but ≤4.6 log reduction at pH 3.7 or 4.1 without acids	Uljas and Ingham (1999)
		As above, plus 0.1% lactic acid	≥5.0 log reduction at pH 3.7 when combined with 4 h at 35 °C; however, in 4 °C-treated cider lactic acid slightly enhanced survival	
		As above, plus 0.1% sorbic acid	≥5.0 log at pH 4.1 when combined with 6 h at 35 °C, or 4 h at 35 °C plus freeze-thawing, or 12 h at 25 °C plus freeze-thawing.	
		As above, plus 0.1% propionic acid	Depending on cider temperature at inoculation, 4.0 to >5.0 log reduction at pH 3.7 when combined with freeze-thawing.	

extended product shelf-life at 4 to 21 °C (Pao and Petracek, 1997). Acids combined with low storage temperatures (4–8 °C) may inhibit growth of pathogens on peeled oranges, where *Salmonella, E. coli* O157:H7, *L. monocytogenes* and *Staph. aureus* can grow at abuse temperatures (Pao *et al.*, 1998).

6.2.2 Organic acids as antimicrobials in acidified foods

Acetic acid is traditionally used in the form of vinegar as a natural acidulant and flavoring agent in acidified foods. Entani *et al.* (1998) reported that growth of *E. coli* O157:H7 and other foodborne pathogens was inhibited by vinegar containing 0.1% acetic acid, and that the bactericidal effect of acetic acid was synergistically enhanced by sodium chloride but attenuated by glucose. Mayonnaise, salad dressings and sauces owe their stability and safety to the antimicrobial action of acetic acid and to a lesser extent of citric acid (reviewed comprehensively by Smittle, 2000). In general, pathogens die off quickly when inoculated in mayonnaise or other acidified products such as ketchup, mustard or sweet pickle relish, with *E. coli* O157:H7 being the most resistant (Table 6.5). However, refrigerated storage of such foods significantly delays pathogen inactivation, especially when the initial inoculation is >3 log CFU/g and the product (e.g. mayonnaise) is full fat. The product pH is also critical, as death of *E. coli* O157:H7 in ketchup (pH 3.6) stored at 5 or 23 °C was slower than that in mustard or sweet pickle relish (Table 6.5).

Mixing of mayonnaise with vegetables or meat in salads increases the pH above 4.0 and enhances the potential for pathogen survival or growth. For example, *E. coli* O157:H7 can survive in coleslaw (pH 4.3–4.5) and in salads containing beef and up to 40% mayonnaise (pH $\geq$ 5.4) stored at 5 °C for 72 h (Wu *et al.*, 2002). Furthermore, the pathogen grew within 24 h in beef-containing mayonnaise salads at pH > 5.5 stored at 21 to 30 °C (Abdul-Raouf *et al.*, 1993a).

The pH of vegetable salads is generally reduced during storage, especially under modified atmospheres, owing to formation of lactic and other organic acids by spoilage bacteria (Abdul-Rauf *et al.*, 1993a). Although raw ripe tomatoes are naturally acidic (pH 3.4–4.9) (Lund, 1992), *Salmonella* Montevideo has been shown to grow in chopped tomatoes (pH 4.1) stored at 20 to 30 °C, while numbers remained unchanged at 5 °C for 9 days (Zhuang *et al.*, 1995). The authors recommended that tomatoes be chilled to 10 °C as rapidly as possible after harvesting, and that treatments with chlorine (200 ppm) should be applied to reduce populations of *Salmonella* potentially present on tomato surfaces (Zhuang *et al.*, 1995).

Following several foodborne disease outbreaks associated with fresh produce, interest in the application of various disinfectants on the surface of raw fruit or vegetables has also increased (Beuchat and Ryu, 1997). Acetic, lactic and citric acids have been shown to be of variable effectiveness for decontamination of fresh produce (Nguyen-the and Carlin, 1994; Francis and O'Beirne, 1997; Beuchat and Ryu, 1997). For example, 2–3 log CFU/g reductions of *E. coli*

Table 6.5 Survival of *Escherichia coli* O157:H7 in acidified foods

Type of food	Product pH	Inoculation level (log CFU/ml or g)	Storage temperature (°C)	Survival/reduction in counts	Reference
Mayonnaise and mayonnaise-based dressings and sauces	3.65–4.44	7	7 and 25	Rapid death at 25 °C, but survival up to 35 days at 7 °C	Wcagent *et al.* (1994)
Mayonnaise	3.6–3.9	3.8	5 and 20	Survival for 8 and 21 days at 20 °C and 34 to 55 days at 5 °C, depending on batch	Zhao and Doyle (1994)
Real mayonnaise	3.86–3.97	0.2–0.3	5, 20 and 30	Undetectable within 1–2 days at 5, 20 or 30 °C	Hathcox *et al.* (1995)
		2.2		Undetectable after 4, 7 and 58 days at 30, 20 and 5 °C, respectively	
		6.2		Undetectable after 7 days at 30 °C and 21 days at 20 °C. Approaching undetectable levels after 93 days at 5 °C	
Reduced-calorie mayonnaise dressing	4.08	0.2–0.3	5, 20 and 30	Undetectable within 1–2 days at 5, 20 or 30 °C	Hathcox *et al.* (1995)
		2.2		Undetectable after 4, 7 and 28 days at 30, 20 and 5 °C, respectively	
		6.2		Undetectable after 4, 11 and 58 days at 30, 20 and 5 °C, respectively	
Ketchup	3.6	5	5 and 23	Survival for 1 day at 23 °C and for up to 7 days at 5 °C depending on the strain	Tsai and Ingham (1997)
Mustard, sweet pickle relish	2.8–3.1	5	5 and 23	Death within 1 h, irrespective of storage temperature or acid adaptation of the inoculum	Tsai and Ingham (1997)
Coleslaw salad	4.3–4.5	5.3	4, 11 and 21	0.4–0.5 log reduction within 3 days at 21 °C, but only 0.1–0.2 log reduction at 4 or 11 °C	Wu *et al.* (2002)

O157:H7 on apples were achieved with 5% acetic acid alone or in combination with 3% hydrogen peroxide (Wright *et al.*, 2000). However, antimicrobial treatments of alfalfa seeds with acetic, lactic or citric acids were not effective in reducing numbers of inoculated *Salmonella* (Weissinger and Beuchat, 2000).

6.2.3 Organic acids in fermented foods

Organic acids produced during food fermentations inhibit many microorganisms by lowering the pH and acting as direct antimicrobials in their undissociated form. Additional factors, which may include sodium chloride, curing salts, spices and herbs, bacteriocins, other natural antimicrobials, low oxygen and water activity, may enhance the antimicrobial action of organic acids, and vice versa. Production of fermented foods is dependent on LAB, mainly species of *Lactococcus, Streptococcus, Pediococcus, Lactobacillus* and *Leuconostoc* (Caplice and Fitzgerald, 1999). For more information on fermented foods, the reader is referred to comprehensive reviews by Teuber (2000), Lucke (2000) and Nout and Rombouts (2000).

The ability of pathogens to survive in fermented foods has been investigated extensively. Following the *E. coli* O157:H7 outbreaks associated with fermented sausages in the mid-1990s (Centers for Disease Control and Prevention, 1995; Tilden *et al.*, 1996), the United States Department of Agriculture Food Safety and Inspection Service (UDSA-FSIS) announced a mandate to require sausage manufacturers to assure a 5-log CFU/g reduction in the final product (Reed, 1995). The traditional fermentation processes for producing dry or semi-dry sausages are sufficient to obtain a 1–3 log CFU/g reduction in counts of inoculated pathogens with *E. coli* O157:H7 being the most resistant (Glass *et al.*, 1992; Hinkens *et al.*, 1996; Calicioglu *et al.*, 1997; Faith *et al.*, 1997; Ihnot *et al.*, 1998; Ellajosyula *et al.*, 1998; Riordan *et al.*, 1998; Getty *et al.*, 2000). Reductions were greater in the presence of starter cultures and lower fermentation temperatures, which require a concurrent increase in fermentation time (Tomicka *et al.*, 1997). Furthermore, inactivation of inoculated pathogens was enhanced when the ripened product was stored at abuse temperatures in air rather than refrigerated in vacuum packages (Nissen and Holck, 1998; Getty *et al.*, 2000). However, to achieve the recommended 5 log CFU/g reduction, a post-fermentation heat treatment is required. Temperatures from 45 to 63 °C for a few seconds and up to 3 h have been tested (see review by Getty *et al.*, 2000). Nevertheless, it is unlikely that *E. coli* O157:H7 would survive in salami stored at 5 °C for 32 days if present at $\leq$100 CFU/g post-processing; however, cross-contamination of salami with the pathogen at 10^4–10^5 CFU/g after processing would pose a health risk (Clavero and Beuchat, 1996). *Listeria monocytogenes* and other *Listeria* spp. have been shown to die off in naturally fermented, unheated Greek salami by the end of fermentation (7th day) under industrial conditions (Samelis *et al.*, 1998; Samelis and Metaxopoulos, 1999).

6.2.4 Organic acids as food preservatives in high-pH foods

The addition of organic acids or their salts as preservatives in high-pH foods has been investigated and numerous commercial applications are in place. For example, propionic acid (0.1–0.4%) is added to bread and puddings, flour and bakery products in some countries (Lund and Eklund, 2000). Benzoic acid at 0.2 to 0.5% is used in semi-preserved fish products or liquid egg. Sorbic acid (0.1 to 0.6% depending on the food) is another important food preservative (Sofos, 1989; Sofos and Busta, 1993, 1999).

There has been much recent interest in the use of sodium lactate, acetate and diacetate as additives in processed meat formulations to inhibit *L. monocytogenes*. Post-processing contamination of vacuum packaged ready-to-eat meat products with *L. monocytogenes* represents a serious health risk (Glass and Doyle, 1989; Centers for Disease Control and Prevention, 1999; Tompkin *et al.*, 1999). Sodium lactate (2–3%), sodium acetate (0.25–0.5%), sodium diacetate (0.25–0.5%), potassium sorbate (0.26%) and glucono-D-lactone (GDL) (0.25–0.5%) have been shown to control *L. monocytogenes* (Weaver and Shelef, 1993; Qvist *et al.*, 1994; Wederquist *et al.*, 1994; Blom *et al.*, 1997; Bedie *et al.*, 2001). The antimicrobial efficacy of these additives depends on pH, water activity, moisture, fat, nitrite and salt content of the product, as well as storage conditions (temperature, packaging atmosphere) (Chen and Shelef, 1992; Shelef and Addala, 1994; Buncic *et al.*, 1995; Houtsma *et al.*, 1996; Nebrink *et al.*, 1999). The antimicrobial action increases when the additives were combined, even at reduced concentrations (Qvist *et al.*, 1994; Blom *et al.*, 1997; Stekelenburg and Kant-Muermans, 2001; Mbandi and Shelef, 2001).

We have recently shown complete growth inhibition (120 days) of *L. monocytogenes* inoculated on frankfurters formulated with 1.8% sodium lactate and 0.25% sodium acetate, sodium diacetate or GDL and stored at 4 °C under vacuum (Samelis *et al.*, 2002c). Lactic acid, acetic acid, sodium diacetate, potassium benzoate and potassium sorbate have also shown promise for use as dipping or spraying solutions (at 2.5–5%) for sliced cooked meats to control *L. monocytogenes* during refrigerated storage (Palumbo and Williams, 1994; Samelis *et al.*, 2001c). Another effective method of application is the use of packaging films containing organic acids; a recent study has shown >3 log CFU/g reductions of inoculated *L. monocytogenes, E. coli* O157:H7 and *Salmonella* Typhimurium DT104 on sliced Bologna and summer sausage packaged in films containing 0.5 to 1% *p*-aminobenzoic acid and sorbic acid (Cagri *et al.*, 2002).

Organic acids or salts have also been tested in fresh muscle foods. For example, combinations of sodium lactate (2–4%) with sodium acetate (0.1–0.3%), sodium propionate (0.1–0.2%), or sodium citrate (0.1 to 0.3%) have been effectively applied in ground beef (Shelef and Yang, 1991; Maca *et al.*, 1997). Growth of *Clostridium sporogenes* and *L. monocytogenes* was inhibited on microwave-ready beef roasts containing sodium lactate (Unda *et al.*, 1991). *Escherichia coli* O157:H7 populations declined in ground roasted beef slurries acidified to pH 4.7 with lactic or citric acid or to pH 4.7–5.4 with acetic acid at 21 °C for 24 h; at pH 5 and 5.4, lactic and citric acids failed to inhibit the pathogen. At 30 °C, *E. coli* O157:H7 declined only in beef slurries acidified to pH 4.7 and 5.0 with acetic acid (Abdul-Raouf *et al.*, 1993b). Lactic acid has also been evaluated at pH 3.8–5.4 in combination with common salt and diacetyl against *E. coli* O157:H7 in skim milk at 4 and 12 °C. The pathogen did not survive at pH 3.8, declined 3 log CFU/ml at pH 4.1, survived for up to 35 days at pH 4.4, and grew by 2.2 log CFU/ml at pH 5.0 after 35 days at 12 °C. Inactivation rates of *E. coli* O157:H7 in sour milk, yogurt and buttermilk were similar or greater (Guraya *et al.*, 1998).

6.3 Organic acids in meat decontamination

The need for meat decontamination and the effectiveness of current practices in the meat industry are discussed in Chapter 3. In this chapter, we focus on the efficacy of decontamination treatments based on organic acids. Although organic acid decontamination applied commercially has been shown to reduce the prevalence of *E. coli* O157:H7 on beef carcasses, the inherent ability of the pathogen to resist acidity remains an important safety concern (Conner and Kotrola, 1995; Lin *et al.*, 1996; Diez-Gonzalez and Russell, 1997; Elder *et al.*, 2000; Huffman, 2002). Furthermore, low population reductions (<1 log CFU/cm^2) of this pathogen following decontamination with organic acid sprays have been reported as shown in Table 6.6. Although the efficacy of spraying treatments is increased by increasing acid concentrations from 1.5 to 5% and temperature from 20 to 55 °C, the pathogen may not be fully inactivated (Cutter and Siragusa, 1994; Hardin *et al.*, 1995).

Binding of *E. coli* O157:H7 to collagen has been suggested as a mechanism of survival for the organism (Fratamico *et al.*, 1996). According to Calicioglu *et al.* (2002a), this problem may be reduced if carcasses are first sprayed with a non-ionic surfactant such as Tween 20 (5% vol/vol) prior to spraying with 2% lactic acid alone or in combination with 0.5% sodium benzoate (Table 6.6). Reductions of *E. coli* O157:H7 by lactic or acetic acid (2–3%) treatments have been greater (>2.5 log CFU/cm^2) when beef tissues were inoculated with bovine feces spiked with the pathogenic strains rather than aqueous cell suspensions (Table 6.6). Apparently, reductions in the former case were enhanced by the 'washing' effect (removal of inoculated feces) of acid spray treatments. Nevertheless, at equal concentrations (e.g. 2%), lactic acid was more effective than acetic acid in inactivating *E. coli* O157:H7 attached onto fresh meat or suspended in the decontamination waste fluids due to its lower pH in solution (lower pK_a of lactic acid) (Hardin *et al.*, 1995; Samelis *et al.*, 2001a, 2002a; Alakomi *et al.*, 2000).

Salmonella spp. are generally more sensitive to organic acid decontamination than *E. coli* O157:H7 (Dickson, 1991; Hardin *et al.*, 1995). However, *Salmonella* reductions may also be low under certain conditions. For example, treatment with 2% acetic acid at ambient temperature reduced counts of attached *Salmonella* Typhimurium on beef tissue previously sterilized with irradiation by

Table 6.6 Organic acid decontamination of meat inoculated with *Escherichia coli* O157:H7*

Tissue	Level and method of inoculation	Acid decontamination method	Pathogen reductions (log CFU/cm^2) after treatment	Reference
Beef, lean and adipose		1, 3, 5% lactic, acetic, citric; pilot scale washer	1.0–2.0	Cutter and Siragusa (1994)
Beef, raw sirloin tips	3 or 6 log CFU/g; surface inoculated with cells in phosphate buffer	1.5% lactic or acetic, sprayed at 20 or 55 °C	<0.30	Brackett *et al.* (1994)
Beef, lean	Dipped into a 9 log CFU/ml inoculum in broth	1% lactic, acetic or fumaric, 55 °C dip	0.62 (lactic), 0.40 (acetic), 1.76 (fumaric)	Podolak *et al.* (1995)
Beef tenderloin and adipose	Dipped into 2 log to 9 log CFU/ml inocula in 0.1% peptone water	2% acetic acid rinse	No reduction	Fratamico *et al.* (1996)
Beef trim with *ca.* 20% fat	3 log CFU/g; spraying of 1 ml broth culture	2 or 4% lactic or acetic, sprayed at 55 °C	0.10–0.64	Conner *et al.* (1997)
Beef, lean and fat	Surface inoculated with 9 log CFU/ml in 0.85% saline	5% acetic acid, spraying at 20 °C	<1.0	Delazari *et al.* (1998)
Beef brisket	5 log CFU/cm^2; faeces spiked with the pathogen	2% acetic or lactic, spraying at 55 °C	3.0 (acetic) 4.2 (lactic)	Hardin *et al.* (1995)
Beef carcass tissue, lean	5 or 7 CFU/cm^2; sterile faeces spiked with the pathogen	2% acetic at 56 °C; model spray washer	4.49–4.70	Cutter *et al.* (1997)
Beef carcass tissue, lean	4 log CFU/cm^2; faeces spiked with the pathogen	3% acetic or lactic, spraying at 32±2 °C	>2.7	Dorsa *et al.* (1997)
Beef carcass quarters and fat-covered sub-primal cuts	4 to 5 log CFU/cm^2; bovine manure slurry	2% lactic acid alone or combined with 0.5% sodium benzoate	2.3–2.7, but increased to 3.1–3.4 when beef was pre-sprayed with 5% Tween 20	Calicioglu *et al.* (2002a)

* For more information on decontamination of meat, see Chapter 3.

only 0.5 to 0.8 log (Dickson, 1992). Reductions of *Salmonella* Typhimurium on beef treated with 2% acetic acid at 37 °C (Tinney *et al.*, 1997) or with 1% lactic or acetic acid at 55 °C (Podolak *et al.*, 1995) were 1 and 0.2–0.8 log, respectively. However, *Salmonella* Typhimurium DT104 and non-O157:H7 enterohemorrhagic *E. coli* on beef surfaces were reduced by 2–3 log CFU/cm^2 after spraying with lactic or acetic acid (Cutter and Rivera-Betancourt, 2000). Also, treatment of beef trimmings with hot water (95 °C for 3 s) followed by 2% lactic acid at 55 °C for 11 sec reduced *Salmonella* Typhimurium by 1.8 log CFU/g (Ellebracht *et al.*, 1999). Using an *in vitro* pork skin model, van Netten *et al.* (1994a) reported immediate death of 2 log CFU/cm^2 of *Salmonella* Typhimurium exposed to 2% lactic acid at 21 °C for 90 s. *Salmonella* Wentworth was reduced by 3.4 log CFU/cm^2 on lean and 3.7 log CFU/cm^2 on beef adipose tissue following treatment with 1% acetic acid combined with 3% hydrogen peroxide (Bell *et al.*, 1997). Several authors have also evaluated the efficacy of spraying of poultry skin with lactic, acetic and other organic acids (Tamblyn and Conner, 1997; Bautista *et al.*, 1998; Yang *et al.*, 1998).

Beef inoculated with *L. monocytogenes* and then treated with lactic acid resulted in reductions and inhibition of growth of the organism for 21–42 days at 4 °C under vacuum (Ariyapitipun *et al.*, 2000; Dickson and Siragusa, 1994). Likewise, beef treated with lactic acid first and then inoculated with *L. monocytogenes* inhibited growth of the pathogen for 2 days (El-Khateib *et al.*, 1993). However, high survival of *L. monocytogenes* on fresh beef trimmings following spraying with 2 or 4% acetic and lactic acid has also been reported (Conner *et al.*, 1997). Also, naturally occurring *Listeria* survived on beef strip loins sprayed with 1.5% lactic acid and stored under vacuum at −1.1 to 2 °C (Prasai *et al.*, 1997). Notably, *L. monocytogenes* inoculated onto lactic acid-treated pork bellies, grew after two days of aerobic storage at 4 °C (van Netten *et al.*, 1997).

Despite variable results, it has been suggested by many authors that *L. monocytogenes* and other pathogens on acid-decontaminated meat did not represent an increased health hazard due to the residual effect of the acids remaining on the meat after treatment (van Netten *et al.*, 1997, 1998; Dorsa *et al.*, 1997, 1998a,b; Dickson, 1991; Dickson and Siragusa, 1994; Podolak *et al.*, 1995). We evaluated survival and growth of *L. monocytogenes* inoculated onto whole beef slices treated with warm water (55 °C) or hot water (75 °C), and 2% acetic or lactic acid at 55 °C, and stored at 4 or 10 °C in vacuum packages (Ikeda *et al.*, 2003). Dipping of inoculated meat for 30 s in hot water and acid resulted in immediate reductions of 1.4–2.6 log CFU/cm^2 of the pathogen compared with <1 log CFU/cm^2 reduction with warm water alone. During storage at 10 °C for 28 days, *L. monocytogenes* increased by 1.6–1.8 and 3.6–4.6 log CFU/cm^2 following treatment with warm and hot water, respectively. Growth occurred as early as Day 14 in meat decontaminated with hot water and stored at 10 °C, while growth occurred after 21 days in water-treated samples stored at 4 °C. In contrast, *L. monocytogenes* survived (1.6–2.8 log CFU/cm^2) but did not grow in acid-treated meat for 28 days at 4 or 10 °C (Ikeda *et al.*, 2003). These results demonstrate a residual activity of acids on fresh lean meat during vacuum-packed storage.

Little is known about the long-term effects of acid decontamination on the microbial ecology of meats. It has yet to be ascertained whether acid interventions cause changes in the original flora that may compromise safety of the meat (Jay, 1996). For example, lactic acid decontamination of a meat model system has been shown to reduce mesophilic enterobacteriaceae by 1–3 log CFU/cm^2 resulting in a shift in the predominant microflora to Gram-positive bacteria and yeasts (van Netten *et al.*, 1994b). A possible risk associated with a change in microflora is that pathogens may gain a foothold and grow to large numbers given that they are generally more resistant to organic acids than meat spoilage species (Greer and Dilts, 1995). It has been demonstrated that growth of *E. coli* O157:H7 on beef decontaminated by steam vacuuming and spraying with 0.2 M lactic acid was 2 to 3 log CFU/cm^2 greater than on untreated beef stored at 10 °C in air and vacuum (Nissen *et al.*, 2001). The background flora in untreated meat grew to high numbers and suppressed the growth of *E. coli* O157:H7 (Vold *et al.*, 2000; Nissen *et al.*, 2001). Other pathogens more cold-tolerant than *E. coli* O157:H7, such as *Salmonella* Enteritidis and *Y. enterocolitica*, increased to high levels on both decontaminated and untreated chicken or pork skin, indicating increased competitiveness with the natural flora at 10 °C (Nissen *et al.*, 2001).

Possible long-term effects of acid decontamination on the microbial ecology of the meat plant environment have been largely overlooked. The runoff and aerosol dispersion from acid spray washings may settle and collect on meat contact surfaces of equipment. Such wet surfaces may provide an environment favorable for colonization and proliferation leading to attachment and biofilm formation. In our laboratory, we have monitored survival, growth and biofilm formation of pathogens in waste fluids from acidic and plain water meat washings at refrigeration and abuse temperatures (Samelis *et al.*, 2001a,b, 2002a,b; Stopforth *et al.*, 2002). It was shown that *E. coli* O157:H7 had greater potential than *L. monocytogenes* or *Salmonella* Typhimurium DT104 for survival (up to seven days) in 2% organic acid meat washings (mainly acetic rather than lactic acid) when stored at 4 °C compared with 10 °C (Samelis *et al.*, 2001a). In water washings at 4 and 10 °C, *E. coli* O157:H7 survived but the low storage temperatures and the predominance of a *Pseudomonas*-like natural meat flora (>10^8 CFU/ml at two to four days) synergistically inhibited its growth (Samelis *et al.*, 2001a, 2002a,b). This growth inhibition was not exerted against *L. monocytogenes* at 4 or 10 °C, *Salmonella* Typhimurium DT104 at 10 °C (Samelis *et al.*, 2001a), or *E. coli* O157:H7 at 15 °C (unpublished data), reflecting the differences in psychrotrophy among the three pathogens. Mixing of acidic and water washings, as may happen in meat plants, enhanced the survival of pathogens, particularly *E. coli* O157:H7, because of sub-lethal pH (>3.5 to about 4.5, Table 6.7). Acid washings diluted with water at pH 4.0 suppressed growth of the *Pseudomonas*-like natural flora while selecting for lactic acid bacteria and yeasts. This selection was not observed in acid-containing washings at pH 4.5 or higher, in which the Gram-negative flora grew

Table 6.7 Fate of *Escerichia coli* O157:H7* (inoculated at 10^5 CFU/ml) and natural flora in meat washings at different pH values and with organic acid contents stored at 10 °C for 2 weeks

Acid content in meat washings	pH of washings			Reduction in *E. coli* O157:H7 (log CFU/ml)		Counts (log CFU/ml) and predominant organisms in natural flora	
	Inoculation	1 week	2 weeks	1 week	2 weeks	1 week	2 weeks
No acid (water)	6.98	7.33	7.43	0.7	1.1	8.8 (*Pseudomonas*-like bacteria	9.1 (*Pseudomonas*-like bacteria
1% lactic acid	2.63	2.62	2.71	>3.9	>3.9	1.8 (*Rhodotorula*-like yeasts)	4.5 (*Rhodotorula*-like yeasts)
0.2% lactic acid	3.20	3.09	3.28	1.9	>3.9	5.9 (mixed yeast flora)	6.9 (mixed yeast flora)
0.02% lactic acid	4.56	6.58	6.98	0.7	1.0	8.8 (*Pseudomonas*-like bacteria	9.0 (*Pseudomonas*-like bacteria
1% acetic acid	3.20	3.21	3.27	0.8	3.2	4.1 (*E. coli* survivors only)	1.9 (*E. coli* survivors only)
0.2% acetic acid	3.76	3.57	3.65	0.1	0.3	4.8 (*E. coli* survivors only	5.1 (LAB mixed with *E. coli* survivors
0.02% acetic acid	4.66	4.62	6.68	0.1	0.3	5.9 (LAB mixed with yeasts)	7.3 (LAB and Gram-negative bacteria mixed with yeasts

*Strain ATCC 43895, which is of permanent acid resistance, after culturing (30 °C, 24 h) in broth with 1% glucose.
Source: modified from Samelis *et al.* (2002a).

to levels similar to those in water washings (8 log CFU/ml). These findings clearly demonstrated that acid decontamination has the potential to alter the microbial ecology of meat plant environments. Most importantly, organic acid decontamination may lead to development of acid-resistant bacteria in meat plants giving rise to a potential food safety risk, which is discussed in more detail in the next section.

6.4 Development of acid resistance in microorganisms

Acid resistance (AR) and acid tolerance (AT) are important physiological characteristics that contribute to the survival and growth of microorganisms in acidic, acidified or fermented foods. They may be also important virulence determinants by assisting foodborne pathogens to survive the acid barriers in the stomach and small intestine and cause disease at very low (<10 cells) infectious doses (Bearson *et al.*, 1997; Gahan and Hill, 1999; Park *et al.*, 1999; Foster, 2000). Thus, the potential for microorganisms to develop AR or AT is of primary importance in food safety (Sheridan and McDowell, 1998; Bower and Daeschel, 1999).

It is necessary to distinguish between AR and AT, two terms that are closely related to each other and are sometimes used interchangeably. AR reflects the inherent ability of a microorganism for survival at lethal pH as a result of hardening to acid by frequent and long-term exposures to adverse pH conditions. For example, microorganisms that prevail naturally in low-pH food environments or in the animal or human gastrointestinal tract, such as LAB or enteric bacteria, can survive sudden shifts in pH (high AR) even if their previous natural environment has been non-acid. Conversely, AT reflects enhanced survival at lethal pH of any microorganism, including species of typically low AR, after brief exposure and induced adaptation to moderately acidic conditions.

The acid tolerance response (ATR) is one of the most important and complex food-related stress resistances of bacterial pathogens (Kroll and Patchett, 1992; Foster, 1995; Rowbury, 1997, 2001). Pathogenic bacteria increase their ATR upon exposure to sub-lethal pH while they are growing exponentially (Foster, 1995; Davis *et al.*, 1996; O'Driscoll *et al.*, 1996; Jordan *et al.*, 1999a; Plan-Thanh *et al.*, 2000). Stationary phase cells may also exhibit an inducible pH-dependent AT, which further increases their pH-independent AR (Lee *et al.*, 1994; Davis *et al.*, 1996; Buchanan and Edelson, 1996, 1999a). In turn, the pH-independent AR is part of the generalized stress resistance (GSR), which is always expressed upon entry of bacterial cells into stationary phase, and is regulated by RpoS (Hengge-Aronis, 1993; Arnold and Kaspar, 1995; Lin *et al.*, 1995; Cheville *et al.*, 1996; Davis *et al.*, 1996). Consequently, bacteria are more resistant to acid in the stationary than in the exponential phase of growth (Davis *et al.*, 1996; Jordan *et al.*, 1999a). Both the stationary-phase ATR and the pH-independent AR vary and depend on species and strain, type of acidulant, growth substrate and presence of oxygen (Buchanan and Edelson, 1996, 1999a;

Lin *et al.*, 1995, 1996; Ryu *et al.*, 1999). A detailed description of the mechanisms and the genes involved in induction, activation or expression of the ATRs and ARs is outside the scope of this chapter but can be found in comprehensive reviews by Foster (2000) and Rowbury (1997, 2001).

Most bacteria require an adaptation period in an acidic environment to exhibit an increased ATR and to enhance their AR. A common technique used in the laboratory to prepare acid-adapted cells is acid shock, i.e. the sudden shift of an exponentially growing culture to pH 5.0–5.8 for a few growth cycles (Davis *et al.*, 1996; Leyer and Johnson, 1992, 1993; Gahan *et al.*, 1996; Lou and Yousef, 1997). Other techniques include culturing of bacteria to stationary phase in media pre-adjusted to pH 5.0 (Dickson and Kunduru, 1995) or in media containing up to 1% glucose, which is converted to acid by the growing organism (Buchanan *et al.*, 1994; Buchanan and Edelson, 1996; Wilde *et al.*, 2000). Regardless of the preparation method, acid-adapted bacteria may exhibit cross protection to other common stresses such as heat (Farber and Pagotto, 1992; Leyer and Johnson, 1993; Lou and Yousef, 1997; Buchanan and Edelson, 1999b; Ryu and Beuchat, 1999). Most importantly, significant increases in virulence have been shown in acid-adapted or acid-stressed laboratory cultures of *L. monocytogenes*, *Salmonella* and *E. coli* O157:H7 (Humphrey *et al.*, 1996; O'Driscoll *et al.*, 1996; Buncic and Avery, 1998; Gahan and Hill, 1999).

Although acid adaptation of laboratory cultures is normally transient, there is concern that long-term survival of acid-adapted pathogens may trigger mutations of permanent stress resistance and increased virulence (Archer, 1996). Recent epidemiological data indicate continuous adaptation and development of resistance by microorganisms to antibiotics and food-related stresses including acid (Glynn *et al.*, 1998; Bower and Daeschel, 1999; Jung and Beuchat, 2000; Threlfall *et al.*, 2000; Bacon *et al.*, 2002). Therefore, intensified research is needed to understand and monitor bacterial stress responses in foods further in order to better assess and manage risks (Lammerding and Paoli, 1997; Doores, 1999; Sofos, 2002).

Acid adaptation has been shown to enhance survival of *E. coli* O157:H7, *Salmonella* and *L. monocytogenes* in acidic foods (Gahan *et al.*, 1996; Leyer *et al.*, 1995; Leyer and Johnson, 1992). Acid adaptation also enhanced survival of *E. coli* O157:H7 on meat decontaminated with acetic acid and in organic acid-containing washings from meat decontamination (Berry and Cutter, 2000; Samelis *et al.*, 2002a). However, others have found acid adaptation to have no effect, or a negative influence, on survival of *Salmonella* in beef treated with organic acids, or of *E. coli* O157:H7 during post-process heating of pepperoni, or under host-simulated conditions following exposure to apple juice (Dickson and Kunduru, 1995; Riordan *et al.*, 2000; Uljas and Ingham, 1998). Likewise, acid-adapted *E. coli* O157:H7 were more sensitive than non-adapted cells to drying (60 °C, 10 h) of beef jerky previously treated with marinades containing 5% acetic acid (Calicioglu *et al.*, 2002b). When jerky was untreated or treated in a traditional marinade, acid-adapted and non-adapted *E. coli* O157:H7 were of similar sensitivity to drying, indicating that acid adaptation did not offer any cross-protection (Calicioglu *et al.*, 2002b). Also, acid adaptation did not promote survival or growth of *L. monocytogenes* on beef jerky marinated as above (Calicioglu *et al.*, 2002c) or on fresh beef stored at 4 or 10 °C following decontamination with water (55 and 75 °C) or 2% lactic or acetic acid at 55 °C (Ikeda *et al.*, 2003).

The above findings suggest that acid stressing may enhance tolerance or resistance of pathogens to acid or other stresses, most probably on a temporary basis. However, the extent to which recovery and repair of acid injury occurs in bacteria during food processing or in the host is rather unpredictable at present. Studies with substrate-limited *E. coli* in laboratory media have shown that acid stress imposes larger energetic burdens on cells compared to other food-related stresses such as heat or osmolarity (Krist *et al.*, 1998; Shadbolt *et al.*, 2001). When an acid stress (pH 3.5/HCl) was applied before an osmotic stress (water activity, a_w 0.90), acid survivors were inactivated rapidly and decreased below the detection limit within a few hours of subsequent exposure to low a_w. However, rapid inactivation was not observed when the order of the stresses was reversed, suggesting that acid stress may be more high-energy-demanding than osmotic stress (Shadbolt *et al.*, 2001). Increasing the level of acid stress against *E. coli* O157:H7 in meat decontamination washing mixtures by using lactic instead of acetic acid at equal concentrations, or by increasing the concentration of either acid resulted in low to minimal survival upon subsequent exposure of cells to broth acidified at pH 3.5 with lactic acid (Samelis *et al.*, 2002a). However, when the acid stress in meat washing mixtures was reduced at pH > 4.0 or in the presence of acetic acid, subsequent survival of *E. coli* O157:H7 exposed to pH 3.5 with lactic acid for 120 min was high (Samelis and Sofos, 2003). Thus, whether acid-stressed pathogens in food environments are sensitized to subsequent stresses may depend on the magnitude (type of acid and pH) and duration of the initial acid stress.

Environmental or processing factors such as temperature, oxygen, salt and nitrite can also influence bacterial responses to acid. For example, low incubation temperatures ($\leq$15 °C) have been shown to reduce the acid tolerance of *E. coli* O157:H7 in addition to delaying or preventing growth (Cheng and Kaspar, 1998; Semanchek and Golden, 1998). By contrast, others have shown that the lower the temperature of stored foods or of the plant environment, the greater and longer the survival of *E. coli* O157:H7, particularly when the food or its processing environment are acidic (Tsai and Ingham, 1997; Roering *et al.*, 1999; Getty *et al.*, 2000; Samelis *et al.*, 2002a). In broth cultures, lactate, ethanol or a combination of both reduced the AT of *E. coli* O157:H7 (Jordan *et al.*, 1999b). Exposure to ethanol also sensitized *L. monocytogenes* to low pH, organic acids or osmotic stress (Barker and Park, 2001). By supplementing a traditional marinade with lactate and ethanol, we confirmed improved inactivation of both pathogens during drying of jerky (Calicioglu *et al.*, 2002b,c). Casey and Condon (2000, 2002) showed that nitrite acted synergistically with acidic pH to increase inactivation of *E. coli* O157:H7 in fermented sausage while sodium chloride had the opposite effect (Casey and

Condon, 2002). Emerging food preservation technologies such as high pressure appear to enhance acid sensitivity of *E. coli* O157:H7 (Pagan *et al.*, 2001).

Recent research in our laboratory has revealed dramatic sensitization of stationary phase *E. coli* O157:H7 to lactic or acetic acid (pH 3.5–3.7) following exposure to non-acid (water) washings for 2–7 days at 10 °C. A similar trend was noted for *L. monocytogenes* and *Salmonella* Typhimurium DT104 but was not as prominent as for *E. coli* O157:H7. In another study, we examined the interaction of the natural meat flora, mainly consisting of *Pseudomonas*-like bacteria, with *L. monocytogenes* in non-acid (water) meat washings at 35 °C. The pathogen increased in number (1–2 log CFU/ml) irrespective of the natural flora, which, when present, predominated (>10^8 CFU/ml) by Day 1. After 8 days, *L. monocytogenes* declined (1 log CFU/ml) or remained constant in the absence or presence of the natural flora, respectively. When the natural flora was removed by filtering the washings prior to inoculation, the ATR of *L. monocytogenes* increased upon storage. In contrast, when the natural flora was present, the pathogen was acid-sensitized after 8 days at 35 °C (Samelis *et al.*, 2001b,d, 2002b).

Acid sensitization of bacteria in non-acid meat washings may result from intracellular changes triggered by several external factors. A relatively high pH in the washings and accumulation of alkaline metabolites by the pseudomonads in the background flora may have led the pathogens to adapt to conditions of neutral pH (Rowbury, 2001; Samelis *et al.*, 2001a,b). Thus, the genes required for expression of high ATR may have been switched off. According to Rowbury (2001), *E. coli* possess extracellular sensing components, which are converted by stress, particularly pH stress, to extracellular induction components. These compounds act as alarmones and pre-warn unstressed cells to be prepared to cope with upcoming acid stress. If the environmental conditions suppress the secretion of such sensors by *E. coli* O157:H7 or those that are secreted are inactivated by competitive flora, *E. coli* O157:H7 would be eventually acid-sensitized.

A classical strategy to control AR and AT in bacteria involves the continuing application of high, lethal levels of organic acids as preservatives in foods. The risk in such a strategy is that inadvertent sub-lethal doses of acids may, in the long term, enhance generation or evolution of pathogenic strains with increased resistance and virulence. Resistance may be avoided if acids are combined with other stresses and applied in the right sequence during food processing (Samelis and Sofos, 2003). Many synthetic and natural antimicrobials (e.g. nisin, herbs and spices, essential oils, lactoferrin, chitosan) and new technologies (e.g. high pressure) could be combined with organic acids to control microorganisms in foods (see related chapters in this book). Another approach may be to make pathogens misinterpret the environmental signals by avoiding, reducing or rotating the use of acid-based technologies, such as acid meat decontamination. Bacteria may become physiologically unprepared to cope with a sudden acid stress during food processing, in the kitchen or in the host (Sheridan and McDowell, 1998).

In conclusion, factors that induce acid sensitization must be suitably monitored, while simultaneously taking into account any potential effects on survival or growth. Conversely, factors that have the potential to enhance ATR must be avoided, if possible. Directed subsequent exposure of sensitized pathogens to the target stress may result in rapid cell inactivation.

6.5 Legislation, labeling and consumer acceptance

According to US regulations, organic (including acetic, lactic, citric, malic, propionic and tartaric) acids and their salts are Generally Recognized As Safe (GRAS) for miscellaneous and general purpose usage (CFR, Title 21, Doores, 1993). The maximum permitted levels depend on the type of acid (salt) relative to the type of food product. For several applications of organic acids in foods, the *quantum satis* rather than a maximum permissible level is accepted. Unless an organic acid is naturally present or microbially produced in a food, its usage as an acidulant or preservative must be labeled on the product. Further details on regulations are available in Chapter 15 of this book.

6.5.1 Consumer acceptance

Unless the sensory quality of a food is negatively influenced by the amount of acid formed or added, consumers generally accept the usage of organic acids and their salts in foods. Most consumers regard organic acids as food-grade compounds, and recognize their use in households as flavorings or natural food acidulants from ancient times. Thus, compared with emerging technologies (e.g. irradiation) that may be faced with skepticism and probably fear by consumers, food treatments with organic acids are 'consumer-friendly.' Recent developments in acid-based decontamination for fresh meat and produce have not been opposed by consumers in the US. Meat decontamination has not yet been approved for commercial use in Europe because of regulatory constraints rather than consumer objections. Acid foods are generally liked for their taste and flavor and some are considered minimally processed.

Processors must guarantee a high sensory quality of acid-preserved foods and this has not always been achievable. For example, Uljas and Ingham (1999) reported that apple ciders with 0.1% lactic and sorbic acids held for 6 h at 35 °C and freeze-thawed were less liked by consumers than pasteurized cider, but were better liked when the acids were omitted from the combination treatments.

6.6 Future trends

Organic acids will continue to be important as food acidulants, flavorings and preservatives since they are of natural origin and occur in foods as natural components or products of fermentation. Any efforts to increase the amounts of organic acids added to foods are limited by their sensory properties. Future research efforts will be focused on extending the uses of organic acids in additional or novel foods, as well as on exploiting less well-known or novel organic acid derivatives in foods. Monitoring the efficacy of acid decontamination treatments in different products and elucidating the effects of these interventions on the microbial ecology of foods are future research challenges. Another interesting field is the determination of novel combinations of organic acids with other natural antimicrobials, at defined concentrations, in order to obtain highly synergistic antimicrobial effects and increase pathogen control and product shelf-life. However, since organic acid treatments have the potential to create acid-adapted pathogens and may lead to mutants with permanent acid (stress) resistance and increased virulence, research to evaluate the acid tolerance response of bacteria in foods must be intensified (Sofos, 2002; Samelis and Sofos, 2003).

6.7 References

ABDUL-RAOUF U M, BEUCHAT L R and AMMAR M S (1993a) 'Survival and growth of *Escherichia coli* O157:H7 on salad vegetables', *Appl Environ Microbiol*, **59**, 1999–2006.

ABDUL-RAOUF U M, BEUCHAT L R and AMMAR M S (1993b) 'Survival and growth of *Escherichia coli* O157:H7 in ground, roasted beef as affected by pH, acidulants, and temperature', *Appl Environ Microbiol*, **59**, 2364–2368.

AHAMAD N and MARTH E H (1989) 'Behavior of *Listeria monocytogenes* at 7, 13, 21 and 35 °C in tryptose broth acidified with acetic, citric, and lactic acids', *J Food Prot*, **52**, 688–695.

ALAKOMI H-L, SKYTTA E, SAARELA M, MATTILA-SANHOLM T, LATVA-KALA K and HELANDER I M (2000) 'Lactic acid permeabilizes Gram-negative bacteria by disrupting the outer membrane', *Appl Environ Microbiol*, **66**, 2001–2005.

ALWAZEER D, CACHON R and DIVIES C (2002) 'Behavior of *Lactobacillus plantarum* and *Saccharomyces cerevisiae* in fresh and thermally processed orange juice', *J Food Prot*, **65**, 1586–1589.

ARCHER D L (1996) 'Preservation microbiology and safety: evidence that stress enhances virulence and triggers adaptive mutations', *Trends Food Sci Technol*, **7**, 91–95.

ARIYAPITIPUN T, MUSTAPHA A and CLARKE A D (2000) 'Survival of *Listeria monocytogenes* Scott A on vacuum-packaged raw beef treated with polylactic acid, lactic acid, and nisin', *J Food Prot*, **63**, 131–136.

ARNOLD K W and KASPAR C W (1995) 'Starvation and stationary-phase-induced acid tolerance in *Escherichia coli* O157:H7', *Appl Environ Microbiol*, **61**, 2037–2039.

BACON R T, SOFOS J N, BELK K E, HYATT D R and SMITH G C (2002) 'Prevalence and antibiotic susceptibility of *Salmonella* isolated from beef animal hides and carcasses', *J Food Prot*, **65**, 284–290.

BARKER C and PARK S F (2001) 'Sensitization of *Listeria monocytogenes* to low pH, organic acids and osmotic stress by ethanol', *Appl Environ Microbiol*, **67**, 1594–1600.

BAUTISTA D A, CHEN J, BARBUT S and GRIFFITHS M W (1998), 'Use of an autobioluminescent *Salmonella* Hadar to monitor the effects of acid and temperature treatments on cell survival and viability on lactic acid-treated poultry carcasses', *J Food Prot*, **61**, 1439–1445.

BEARSON S, BEARSON B and FOSTER J W (1997) 'Acid stress responses in enterobacteria', *FEMS Microbiol Lett*, **147**, 173–180.

BEDIE G K, SAMELIS J, SOFOS J N, BELK K E, SCANGA J A and SMITH G C (2001) 'Antimicrobials in the formulation to control *Listeria monocytogenes* postprocessing contamination on frankfurters stored at 4 °C in vacuum packages', *J Food Prot*, **64**, 1949–1955.

BELL K Y, CUTTER C N and SUMNER S S (1997) 'Reduction of foodborne microorganisms on beef carcass tissue using acetic acid, sodium bicarbonate, and hydrogen peroxide spray washes', *Food Microbiol*, **14**, 439–448.

BERRY E D and CUTTER C N (2000) 'Effects of acid adaptation of *Escherichia coli* O157:H7 on efficacy of acetic acid spray washes to decontaminate beef carcass tissue', *Appl Environ Microbiol*, **66**, 1493–1498.

BESSER R E, LETT S M, WEBER J T, DOYLE M P, BARRET T J WELLS J G and GRIFFIN P M (1993) 'An outbreak of diarrhea and hemolytic uremic syndrome from *Escherichia coli* O157:H7 in fresh-pressed apple cider', *JAMA*, **269**, 2217–2220.

BEUCHAT L R and RYU J H (1997) 'Produce handling and processing practices', *Emerg Inf Dis*, **3**, 459–465.

BLOM H, NERBRINK E, DAINTY R, HAGTVEDT T, BORCH E, NISSEN H and NESBAKKEN T (1997) 'Addition of 2.5% lactate and 0.25% acetate controls growth of *Listeria monocytogenes* in vacuum-packed, sensory-acceptable servelat sausage and cooked ham stored at 4 °C', *Int J Food Microbiol*, **38**, 71–76.

BOWER C K and DAESCHEL M A (1999) 'Resistance responses of microorganisms in food environments', *Int J Food Microbiol*, **50**, 33–44.

BRACKETT R E, HAO Y-Y and DOYLE M P (1994) 'Ineffectiveness of hot acid sprays to decontaminate *Escherichia coli* O157:H7 on beef', *J Food Prot* **57**, 198–203.

BUCHANAN R L and EDELSON S G (1996) 'Culturing enterohemorrhagic *Escherichia coli* in the presence and absence of glucose as a simple means of evaluating the acid tolerance of stationary-phase cells' *Appl Environ Microbiol*, **62**, 4009–4013.

BUCHANAN R L and EDELSON S G (1999a) 'pH-dependent stationary-phase acid resistance response of enterohemorrhagic *Escherichia coli* in the presence of various acidulants', *J Food Prot*, **62**, 211–218.

BUCHANAN R L and EDELSON S G (1999b) 'Effect of pH-dependent, stationary phase acid resistance on the thermal tolerance of *Escherichia coli* O157:H7', *Food Microbiol*, **16**, 447–458.

BUCHANAN R L, GOLDEN M H and WHITING R C (1993) 'Differentiation of the effects of pH and lactic and acetic acid concentration on the kinetics of *Listeria monocytogenes* inactivation', *J Food Prot*, **56**, 474–478.

BUCHANAN R L, GOLDEN M H, WHITING R C, PHILLIPS J G and SMITH J L (1994) 'Non thermal inactivation models for *Listeria monocytogenes'*, *J Food Sci*, **59**, 179–188.

BUCHANAN R L, GOLDEN M H and PHILLIPS J G (1997) 'Expanded models for the non-thermal inactivation of *Listeria monocytogenes'*, *J Appl Microbiol*, **82**, 567–577.

BUNCIC S and AVERY S M (1998) 'Effects of cold storage and heat-acid shocks on growth and verotoxin 2 production of *Escherichia coli* O157:H7', *Food Microbiol*, **15**, 319–328.

BUNCIC S, FITZGERALD C M, BELL R G and HUDSON J A (1995) 'Individual and combined listericidal effects of sodium lactate, potassium sorbate, nisin and curing salts at refrigeration temperatures', *J Food Safety*, **15**, 247–264.

BURNHAM J A, KENDALL P A and SOFOS J N (2001) 'Ascorbic acid enhances destruction of *Escherichia coli* O157:H7 during home-type drying of apple slices', *J Food Prot*, **64**, 1244–1248.

CAGRI A, USTUNOL Z and RYSER E T (2002) 'Inhibition of three pathogens on bologna and summer sausage using antimicrobial edible films' *J Food Sci*, **67**, 2317–2324.

CALICIOGLU M, FAITH N G, BUEGE D R and LUCHANSKY J B (1997) 'Viability of *Escherichia coli* O157:H7 in fermented semidry low-temperature cooked beef summer sausage', *J Food Prot*, **60**, 1158–1162.

CALICIOGLU M, KASPAR C W, BUEGE D R and LUCHANSKY J B (2002a), 'Effectiveness of spraying with Tween 20 and lactic acid in decontaminating inoculated *Escherichia coli* O157:H7 and indigenous *Escherichia coli* biotype I on beef', *J Food Prot*, **65**, 26–32.

CALICIOGLU M, SOFOS J N, SAMELIS J, KENDALL P A and SMITH G C (2002b) 'Inactivation of acid-adapted and nonadapted *Escherichia coli* O157:H7 during drying and storage of beef jerky treated with different marinades', *J Food Prot*, **65**, 1394–1405.

CALICIOGLU M, SOFOS J N, SAMELIS J, KENDALL P A and SMITH G C (2002c) 'Destruction of acid-adapted and nonadapted *Listeria monocytogenes* during drying and storage of beef jerky', *Food Microbiol*, **19**, 545–559.

CAPLICE E and FITZGERALD G F (1999) 'Food fermentations: role of microorganisms in food production and preservation', *Int J Food Microbiol*, **50**, 131–149.

CASEY P G and CONDON S (2000) 'Synergistic lethal combination of nitrite and acid pH on a verotoxin-negative strain of *Escherichia coli* O157', *Int J Food Microbiol*, **55**, 255–258.

CASEY P G and CONDON S (2002) 'Sodium chloride decreases the bacteriocidal effect of acid pH on *Escherichia coli* O157:H45', *Int J Food Microbiol*, **76**, 199–206.

CENTERS FOR DISEASE CONTROL AND PREVENTION (1995) '*Escherichia coli* O157:H7 outbreak linked to commercially distributed dry-cured salami – Washington and California, 1994', *Morbid Mortal Weekly Rep*, **44**, 157–160.

CENTERS FOR DISEASE CONTROL AND PREVENTION (1996) 'Outbreak of *Escherichia coli* O157:H7 infections associated with drinking unpasteurized commercial apple juice – British columbia, California, Colorado and Washington, October 1996', *Morbid Mortal Weekly Rep*, **45**, 975–982.

CENTERS FOR DISEASE CONTROL AND PREVENTION (1999) 'Update: multistate outbreak of listeriosis – United States, 1998–1999', *Morbid Mortal Weekly Rep*, **47**, 1117–1118.

CHEN N and SHELEF L A (1992) 'Relationship between water activity, salts of lactic acid and growth of *Listeria monocytogenes* in a meat model system', *J Food Prot*, **55**, 574–578.

CHENG C.-M and KASPAR C W (1998) 'Growth and processing conditions affecting acid tolerance in *Escherichia coli* O157:H7', *Food Microbiol*, **15**, 157–166.

CHEVILLE A M, ARNOLD K W, BUCHRIESER C, CHENG C.-M and KASPAR C W (1996) '*rpoS* regulation of acid, heat and salt tolerance in *Escherichia coli* O157:H7', *Appl Environ Microbiol*, **62**, 1822–1824.

CHIPLEY J R (1993) 'Sodium benzoate and benzoic acid', in Davidson P M and Branen A L, *Antimicrobials in Foods*, New York, Marcel Dekker, 11–48.

CLAVERO M R S and BEUCHAT L R (1996) 'Survival of *Escherichia coli* O157:H7 in broth and processed salami as influenced by pH, water activity, and temperature and suitability of media for its recovery', *Appl Environ Microbiol*, **62**, 2735–2740.

CONNER D E and KOTROLA J S (1995) 'Growth and survival of *Escherichia coli* O157:H7 under acidic conditions', *Appl Environ Microbiol*, **61**, 382–385.

CONNER D E, KOTROLA J S, MIKEL W B and TAMPLYN K C (1997) 'Effect of acetic-lactic acid treatments applied to beef trim on populations of *Escherichia coli* O157:H7 and *Listeria monocytogenes* in ground beef', *J Food Prot*, **60**, 1560–1563.

CUTTER C N and RIVERA-BETANCOURT M (2000) 'Interventions for the reduction of *Salmonella* Typhimurium DT 104 and non-O157:H7 enterohemorrhagic *Escherichia coli* on beef surfaces', *J Food Prot*, **63**, 1326–1332.

CUTTER C N and SIRAGUSA G R (1994) 'Efficacy of organic-acids against *Escherichia coli* O157:H7 attached to beef carcass tissue using a pilot-scale model carcass washer', *J Food Prot*, **57**, 97–103.

CUTTER C N, DORSA W J and SIRGAUSA G R (1997) 'Parameters affecting the efficacy of spray washes against *Escherichia coli* O157:H7 and fecal contamination on beef', *J Food Prot*, **60**, 614–618.

DAVIS M J, COOTE P J and O'BYRNE C P (1996) 'Acid tolerance in *Listeria monocytogenes*: the adaptive acid tolerance response (ATR) and growth-phase-dependent acid resistance', *Microbiology*, **142**, 2975–2982.

DEAK T and BEUCHAT L R (1993) 'Yeasts associated with fruit juice concentrates', *J Food Prot*, **56**, 777–782.

DELAZARI I, IARIA S T, RIEMANN H P, CLIVER D O and MORI T (1998), 'Decontaminating beef for *Escherichia coli* O157:H7', *J Food Prot*, **61**, 547–550.

DESHPANDE S S, SALUNKHE D K and DESHPANDE U S (1995) 'Food acidulants', in Maga J A and Tu A T, *Food Additive Toxicology*, New York, Marcel Dekker, 11–87.

DICKSON J S (1991) 'Control of *Salmonella* Typhimurium, *Listeria monocytogenes*, and *Escherichia coli* O157:H7 on beef in a model spray chilling system', *J Food Sci*, **56**, 191–193.

DICKSON J S (1992) 'Acetic acid action on beef tissue surfaces contaminated with *Salmonella* typhimurium' *J Food Sci*, **57**, 297–301.

DICKSON J S and KUNDURU M R (1995) 'Resistance of acid-adapted salmonellae to organic-acid rinses on beef', *J Food Prot*, **58**, 973–976.

DICKSON J S and SIRAGUSA G R (1994) 'Survival of *Salmonella* Typhimurium, *Escherichia coli* O157:H7 and *Listeria* monocytogenes during storage on beef sanitized with organic acids', *J Food Saf*ety, **14**, 313–327.

DIEZ-GONZALEZ F and RUSSELL J B (1997) 'The ability of *Escherichia coli* O157:H7 to decrease its intracellular pH and resist the toxicity of acetic acid', *Microbiology*, **143**, 1175–1180.

DOORES S (1993) 'Organic acids', in Davidson P M and Branen A L, *Antimicrobials in Foods*, New York, Marcel Dekker, 95–136.

DOORES S (1999) *Food Safety: Current Status and Future Needs*, Washington, DC, American Academy of Microbiology.

DORSA W J, CUTTER C N and SIRAGUSA G R (1997) 'Effects of acetic acid, lactic acid and trisodium phosphate on the microflora of refrigerated beef carcass surface tissue inoculated with *Escherichia coli* O157:H7, *Listeria innocua*, and *Clostridium sporogenes*', *J Food Prot*, **60**, 619–624.

DORSA W J, CUTTER C N and SIRAGUSA G R (1998a) 'Long-term effect of alkaline, organic acid, or hot water washes on the microbial profile of refrigerated beef contaminated with bacterial pathogens after washing' *J Food Prot*, **61**, 300–306.

DORSA W J, CUTTER C N and SIRAGUSA G R (1998b) 'Bacterial profile of ground beef made from carcass tissue experimentally contaminated with pathogenic and spoilage bacteria before being washed with hot water, alkaline solution, or organic acid and then stored at 4 or 12 °C', *J Food Prot*, **61**, 1109–1118.

ELDER R O, KEEN J E, SIRAGUSA G R, BARKOCY-GALLAGHER G A, KOOHMARAIE M and LAEGREID W W (2000) 'Correlation of enterohemorrhagic *Escherichia coli* O157 prevalence in feces, hides, and carcasses of beef cattle during processing' *US Proc*

Natl Acad Sci, **97**, 2999–3003.

EL-KHATEIB T, YOUSEF A E and OCKERMAN H W (1993) 'Inactivation and attachment of *Listeria monocytogenes* on beef muscle treated with lactic acid and selected bacteriocins', *J Food Prot*, **56**, 29–33.

ELLAJOSYULA K R, DOORES S, MILLS E D, WILSON R A, ANANTHESWARAN R C and KNABEL S J (1998) 'Destruction of *Escherichia coli* O157:H7 and *Salmonella* Typhimurium in Lebanon bologna by interaction of fermentation pH, heating temperature, and time', *J Food Prot*, **61**, 152–157.

ELLEBRACHT E A, CASTILLO A, LUCIA L M, MILLER R K and ACUFF G R (1999) 'Reduction of pathogens using hot water and lactic acid on beef trimmings', *J Food Sci*, **64**, 1094–1099.

ENTANI E, ASAI M, TSUJIHATA S, TSUKAMOTO Y and OHTA M (1998) 'Antibacterial action of vinegar against food-borne pathogenic bacteria including *Escherichia coli* O157:H7', *J Food Prot*, **61**, 953–959.

FAITH N G, PARNIERE N, LARSON T, LORANG T D and LUCHANSKY J B (1997) 'Viability of *Escherichia coli* O157:H7 in pepperoni during the manufacture of sticks and the subsequent storage of slices at 21, 4, and −20 °C under air, vacuum and CO_2', *Int J Food Microbiol*, **37**, 47–54.

FARBER J M and PAGOTTO F (1992) 'The effect of acid shock on the heat resistance of *Listeria monocytogenes*', *Lett Appl Microbiol*, **15**, 197–201.

FISHER T L and GOLDEN D A (1998) 'Survival of *Escherichia coli* O157:H7 in apple cider as affected by dimethyl dicarbonate, sodium bisulfite and sodium benzoate', *J Food Sci*, **63**, 904–906.

FOLEY D M, PICKETT K, VARON J, LEE J, MIN D B, CAPORASO F and PRAKASH A (2002) 'Pasteurization of fresh orange juice using gamma irradiation: microbiological, flavor and sensory analyses', *J Food Sci*, **67**, 1495–1501.

FOOD AND DRUG ADMINISTRATION (1998) 'Food labeling: warning and notice statements; labeling of juice products', *Fed Regist*, **63**, 20486–20493.

FOSTER J W (1995) 'Low pH adaptation and the acid tolerance response of *Salmonella* Tyrhimurium', *Crit Rev Microbiol*, **21**, 215–237.

FOSTER J W (2000) 'Microbial responses to acid stress', in Storz G and Hengge-Aronis R, *Bacterial Stress Responses*, Washington D C, ASM Press, American Society for Microbiology, Chapter 7, 99–115.

FRANCIS G A and O' BEIRNE D (1997) 'Effect of gas atmosphere, antimicrobial dip and temperature on the fate of *Listeria innocua* and *Listeria monocytogenes* on minimally processed lettuce', *Int J Food Sci Technol*, **32**, 141–151.

FRATAMICO P M, SCHULTZ F J, BENEDICT R C, BUCHANAN R L and COOKE P H (1996) 'Factors influencing attachment of *Escherichia coli* O157:H7 to beef tissues and removal using selected sanitizing rinses', *J Food Prot*, **59**, 453–459.

GAHAN C G M and HILL C (1999) 'The relationship between acid stress responses and virulence in *Salmonella* Typhimurium and *Listeria monocytogenes'*, *Int J Food Microbiol*, **50**, 93–100.

GAHAN C G M, O'DRISCOLL B and HILL C (1996) 'Acid adaptation of *Listeria monocytogenes* can enhance survival in acidic foods and during milk fermentation', *Appl Environ Microbiol*, **62**, 3128–3132.

GEOPFERT J M and HICKS R (1969) 'Effect of volatile fatty acids on *Salmonella* Typhimurium', *J Bacteriol*, **97**, 956–958.

GEORGE S M, RICHARDSON L C C and PECK M W (1996) 'Predictive models of the effect of temperature, pH and acetic and lactic acids on the growth of *Listeria monocytogenes', Int J Food Microbiol*, **32**, 73–90.

GETTY K J K, PHEBUS R K, MARSDEN J L, FUNG D Y C and KASTNER C L (2000) '*Escherichia coli* O157:H7 and fermented sausages: a review', *J Rapid Meth Autom Microbiol*, **8**, 141–170.

GIBSON A M, BRATCHELL N and ROBERTS T A (1988) 'Predicting microbial growth: growth responses of salmonellae in a laboratory medium as affected by pH, sodium chloride and storage temperature', *Int J Food Microbiol*, **6**, 155–178.

GLASS K A and DOYLE M P (1989) 'Fate of *Listeria monocytogenes* in processed meat products during refrigerated storage', *Appl Environ Microbiol*, **55**, 1565–1569.

GLASS K A, LOEFFELHOLZ J M FORD J P and DOYLE M P (1992) 'Fate of *Escherichia coli* O157:H7 as affected by pH or sodium chloride and in fermented, dry sausage', *Appl Environ Microbiol*, **58**, 2513–2516.

GLYNN M K, BOPP C, DEWITT W, DABNEY P, MOKHTAR M and ANGULO F J (1998) 'Emergence of multidrug-resistant *Salmonella enterica* serotype Typhimuiurm DT104 infections in the United States', *New Engl J Med*, **338**, 1333–1338.

GREER G G and DILTS B D (1995) 'Lactic acid inhibition of the growth of spoilage bacteria and cold tolerant pathogens on pork', *Int J Food Microbiol*, **25**, 141–151.

GURAYA R, FRANK J F and HASSAN A N (1998) 'Effectiveness of salt, pH, and diacetyl as inhibitors of *Escherichia coli* O157:H7 in dairy foods stored at refrigeration temperatures', *J Food Prot*, **61**, 1098–1102.

HARDIN M D, ACUFF G R, LUCIA L M, OMAN J S and SAVELL J W (1995) 'Comparison of methods for decontamination from beef carcass surfaces', *J Food Prot*, **58**, 368–374.

HATHCOX A K, BEUCHAT L R and DOYLE M P (1995) 'Death of enterohemorrhagic *Escherichia coli* O157:H7 in real mayonnaise and reduced-calorie mayonnaise dressing as influenced by initial population and storage temperature', *Appl Environ Microbiol*, **61**, 4172–4177.

HENGGE-ARONIS R (1993) 'Survival of hunger and stress: The role of *rpoS* in early stationary phase gene regulation in *E. coli*', *Cell*, **72**, 165–168.

HINKENS J C, FAITH N G, LORANG T D, BAILEY P, BUEGE D, KASPAR C W and LUCHANSKY J B (1996) 'Validation of pepperoni processes for control of *Escherichia coli* O157:H7', *J Food Prot*, **59**, 1260–1266.

HOUTSMA P C, DEWIT J C and ROMBOUTS F M (1993) 'Minimum inhibitory concentration (MIC) of sodium lactate for pathogens and spoilage organisms occurring in meat products', *Int J Food Microbiol*, **20**, 247–257.

HOUTSMA P C, KANT-MUERMANS M L, ROMBOUTS F M and ZWIETERING M H (1996) 'Model for the combined effects of temperature, pH and sodium lactate on growth rates of *Listeria innocua* in broth and Bologna-type sausages', *Appl Environ Microbiol*, **62**, 1616–1622.

HUFFMAN RD (2002), 'Current and future technologies for the decontamination of carcasses and fresh meat', *Meat Sci*, **62**, 285–294.

HUMPHREY T J, WILLIAMS A, MCALPINE K, LEVER M S, GUARD-PETTER J and COX J M (1996) 'Isolates of *Salmonella enterica* Enteritidis PT4 with enhanced heat and acid tolerance are more virulent in mice and more invasive in chicken', *Epidemiol Infect*, **117**, 79–88.

IHNOT A M, ROERING A M, WIERZBA R K, FAITH N G and LUCHANSKY J B (1998) 'Behavior of *Salmonella* Typhimurium DT104 during the manufacture and storage of pepperoni', *Int J Food Microbiol*, **40**, 117–121.

IKEDA J S, SAMELIS J, KENDALL P A, SMITH G C and SOFOS J N (2003) 'Acid adaptation does not promote survival or growth of *Listeria monocytogenes* on fresh beef following acid and nonacid decontamination treatments' *J Food Prot*, in press.

INGHAM S C and ULJAS H E (1998) 'Prior storage conditions influence the destruction of *Escherichia coli* O157:H7 during heating of apple cider and juice', *J Food Prot*, **61**, 390–394.

JAY J M (1996) 'Microorganisms in fresh ground meats: the relative safety of products with low versus high numbers', *Meat Sci*, **43**, S59–S66.

JORDAN K N, OXFORD L and O'BYRNE C P (1999a) 'Survival of low-pH stress by *Escherichia coli* O157:H7: correlation between alterations in the cell envelope and increased acid tolerance', *Appl Environ Microbiol*, **65**, 3048–3055.

JORDAN S L, GLOVER J, MALCOLM L, THOMSON-CARTER F M, BOOTH I R and PARK S F (1999b) 'Augmentation of killing of *Escherichia coli* O157 by combinations of lactate, ethanol, and low-pH conditions', *Appl Environ Microbiol*, **65**, 1308–1311.

JUNG Y S and BEUCHAT L R (2000) 'Sensitivity of multidrug-resistant *Salmonella* Typhimurium DT104 to organic acids and thermal inactivation in liquid egg products', *Food Microbiol*, **17**, 63–71.

KRIST K A, ROSS T and MCMEEKIN T A (1998) 'Final optical density and growth rate; effects of temperature and NaCl differ from acidity', *Int J Food Microbiol*, **43**, 195–203.

KROLL R G and PATCHETT R A (1992) 'Induced acid tolerance in *Listeria monocytogenes*', *Lett Appl Microbiol*, **14**, 224–227.

LAKKAKULA S, KENDALL P A, SAMELIS J. and SOFOS J N (2001) 'Destruction of *Escherichia coli* O157:H7 on apples of different varieties treated with citric acid before drying' in *88th Annual Meeting of International Association for Food Protection (IAFP)*, 5–8 August, 2001, Minneapolis, Minnesota, USA. Abstract P032, 54.

LAMMERDING A M and PAOLI G M (1997) 'Quantitative risk assessment: an emerging tool for emerging foodborne pathogens', *Emerg Infect Dis*, **3**, 483–487.

LEE I S, SLONCZEWSKI J L and FOSTER J W (1994) 'A low-pH-inducible, stationary-phase acid tolerance response in *Salmonella* Typhimurium', *J Bacteriol*, **176**, 1422–1426.

LEYER G J and JOHNSON E A (1992) 'Acid adaptation promotes survival of *Salmonella* spp. in cheese', *Appl Environ Microbiol*, **58**, 2075–2080.

LEYER G J and JOHNSON E A (1993) 'Acid adaptation induces cross-protection against environmental stresses in *Salmonella* Typhimurium', *Appl Environ Microbiol*, **59**, 1842–1847.

LEYER G J, WANG L L and JOHNSON E A (1995) 'Acid adaptation of *Escherichia coli* O157:H7 increases survival in acidic foods', *Appl Environ Microbiol*, **61**, 3752–3755.

LIN J, LEE I S, FREY J, SLONCZEWSKI J L and FOSTER J W (1995) 'Comparative analysis of extreme acid survival of *Salmonella* Typhimurium, *Shigella flexneri* and *Escherichia coli*', *J Bacteriol*, **177**, 4097–4104.

LIN J, SMITH M P, CHAPIN K C, BAIK H S, BENNETT G N and FOSTER J W (1996) 'Mechanisms of acid resistance in enterohemorrhagic *Escherichia coli*', *Appl Environ Microbiol*, **62**, 3094–3100.

LOU Y and YOUSEF A E (1997) 'Adaptation to sublethal environmental stresses protects *Listeria monocytogenes* against lethal preservation factors', *Appl Environ Microbiol*, **63**, 1252–1255.

LUCKE F-K (2000) 'Fermented meats', in Lund B M, Baird-Parker T C and Gould G W, *The Microbiological Safety and Quality of Food*, Maryland, Aspen Publishers, Vol. 1, Chapter 19, 420–444.

LUND B M (1992) 'Ecosystems in vegetable foods', *J Appl Bacteriol Symp Suppl*, **73**,

115S–126S.

LUND B M and EKLUND T (2000) 'Control of pH and use of organic acids', in Lund B M, Baird-Parker T C and Gould G W, *The Microbiological Safety and Quality of Food*, Maryland, Aspen Publishers, Vol. 1, Chapter 8, 175–199.

LUND B M and SNOWDON A L (2000) 'Fresh and processed fruits', in Lund B M, Baird-Parker T C and Gould G W, *The Microbiological Safety and Quality of Food*, Maryland, Aspen Publishers, Vol. 1, Chapter 27, 738–758.

MACA J V, MILLER R K and ACUFF G R (1997) 'Microbiological, sensory and chemical characteristics of vacuum-packaged ground beef patties treated with salts of organic acids', *J Food Sci*, **62**, 591–596.

MBANDI E and SHELEF L A (2001) 'Enhanced inhibition of *Listeria monocytogenes* and *Salmonella* Enteritidis in meat by combinations of sodium lactate and diacetate', *J Food Prot*, **64**, 640–644.

MCCLURE P J, BLACKBURN C W, COLE M B, CURTIS P S, JONES J E, LEGAN J D, OGDEN I D, PECK M W, ROBERTS T A, SUTHERLAND J P and WALKER S J (1994) 'Modelling the growth, survival and death of microorganisms in foods: the UK Food Micromodel approach', *Int J Food Microbiol*, **23**, 265–275.

MILLER L G and KASPAR C W (1994) '*Escherichia coli* O157:H7 acid tolerance and survival in apple cider', *J Food Prot*, **57**, 460–464.

NEBRINK E, BORCH E, BLOM H and NESBAKKEN T (1999) 'A model based on absorbance data on the growth rate of *Listeria monocytogenes* and including the effects of pH, NaCl, Na-lactate and Na-acetate', *Int J Food Microbiol*, **47**, 99–109.

NGUYEN-THE C and CARLIN F (1994) 'The microbiology of minimally processed fresh fruits and vegetables', *Crit Rev Food Sci Nutr*, **34**, 371–401.

NISSEN H and HOLCK A (1998) 'Survival of *Escherichia coli* O157:H7, *Listeria monocytogenes* and *Salmonella kentucky* in Norwegian fermented, dry sausage', *Food Microbiol*, **15**, 273–279.

NISSEN H, MAUGESTEN T and LEA P (2001) 'Survival and growth of *Escherichia coli* O157:H7, *Yersinia enterocolitica* and *Salmonella* enteritidis on decontaminated and untreated meat', *Meat Sci*, **57**, 291–298.

NOUT M J R and ROMBOUTS F M (2000) 'Fermented and acidified plant foods', in Lund B M, Baird-Parker T C and Gould G W, *The Microbiological Safety and Quality of Food*, Maryland, Aspen Publishers, Vol. 1, Chapter 26, 685–737.

NUNHEIMER T D and FABIAN F W (1940) 'Influence of organic acids, sugars, and sodium chloride upon strains of food poisoning staphylococci', *Am J Public Health*, **30**, 1040–1048.

O'DRISCOLL B, GAHAN C G M and HILL C (1996) 'Adaptive acid tolerance response in *Listeria monocytogenes*: isolation of an acid-tolerant mutant, which demonstrates increased virulence', *Appl Environ Microbiol*, **62**, 1693–1698.

OUATTARA B, SIMARD R E, HOLLEY R A, PIETTE G J-P and BEGIN A (1997) 'Inhibitory effect of organic acids upon meat spoilage bacteria', *J Food Prot*, **60**, 246–253.

PAGAN R, JORDAN S, BENITO A and MACKEY B (2001) 'Enhanced acid sensitivity of pressure-damaged *Escherichia coli* O157 cells', *Appl Environ Microbiol*, **67**, 1983–1985.

PALUMBO S A and WILLIAMS A C (1994) 'Control of *Listeria monocytogenes* on the surface of frankfurters by acid treatments', *Food Microbiol*, **11**, 293–300.

PAO S and PETRACEK P D (1997) 'Shelf life extension of peeled oranges by citric acid treatment', *Food Microbiol*, **14**, 485–491.

PAO S, BROWN E and SCHNEIDER K R (1998) 'Challenge studies with selected pathogenic bacteria on freshly peeled Hamlin orange', *J Food Sci*, **63**, 359–362.

PARISH M E (1998) 'Orange juice quality after treatment by thermal pasteurization or isostatic high pressure', *Lebensm-wiss u-Technol*, **31**, 439–442.

PARK S, WOROBO R W and DURST R A (1999) '*Escherichia coli* O157:H7 as an emerging foodborne pathogen: a literature review', *Crit Rev Food Sci Nutr*, **39**, 481–502.

PLAN-THANH L, MAHOUIN F and ALIGE S (2000) 'Acid responses of *Listeria monocytogenes*', *Int J Food Microbiol*, **55**, 121–126.

PODOLAK R K, ZAYAS J F, KASTNER C L and FUNG D Y C (1995) 'Reduction of *Listeria monocytogenes, Escherichia coli* O157:H7 and *Salmonella* Typhimurium during storage on beef sanitized with fumaric, acetic and lactic acids', *J Food Safety*, **15**, 283–290.

PRASAI R K, KASTNER C L, KENNEY P B, KROPF D H, FUNG D Y C, MEASE L E, VOGT L R and JOHNSON D E (1997) 'Microbiological quality of beef subprimals as affected by lactic acid sprays applied at various points during vacuum storage', *J Food Prot*, **60**, 795–798.

PRESSER K A, RATKOWSKY D A and ROSS T (1997) 'Modelling the growth rate of *Escherichia coli* as a function of pH and lactic acid concentration', *Appl Environ Microbiol*, **63**, 2355–2360.

PRESSER K A, ROSS T and RATKOWSKY D A (1998) 'Modelling the growth limits (growth/no growth interface) of *Escherichia coli* as a function of temperature, pH, lactic acid concentration, and water activity', *Appl Environ Microbiol*, **64**, 1773–1779.

QVIST S, SEHESTED K and ZEUTHEN P (1994) 'Growth suppression of *Listeria monocytogenes* in a meat product', *Int J Food Microbiol*, **24**, 283–293.

RAZAVILAR V and GENIGEORGIS C (1998) 'Prediction of *Listeria* spp. growth as affected by various levels of chemicals, pH, temperature and storage time in a model broth', *Int J Food Microbiol*, **40**, 149–157.

REED C A (1995) 'Challenge study – *Escherichia coli* O157:H7 in fermented sausage', US Dept of Agriculture, Food Safety and Inspection Service, Washington, DC, April 28 1995 (letter to plant managers).

RIORDAN D C R, DUFFY G, SHERIDAN J J, EBLEN B S, WHITING R C, BLAIR I S and MCDOWELL D A (1998) 'Survival of *Escherichia coli* O157:H7 during the manufacture of pepperoni', *J Food Prot*, **61**, 146–151.

RIORDAN D C R, DUFFY G, SHERIDAN J J, WHITING R C, BLAIR I S and MCDOWELL D A (2000) 'Effect of acid adaptation, product pH, and heating on survival of *Escherichia coli* O157:H7 in pepperoni', *Appl Environ Microbiol*, **66**, 1726–1729.

ROERING A M, LUCHANSKY J B, IHNOT A M, ANSAY S E, KASPAR C W and INGHAM S C (1999) 'Comparative survival of *Salmonella* Typhimurium DT 104, *Listeria monocytogenes*, and *Escherichia coli* O157:H7 in preservative-free apple cider and simulated gastric fluid', *Int J Food Microbiol*, **46**, 263–269.

ROWBURY R J (1997) 'Regulatory components, including integration host factor, CysB and H-NS, that influence pH responses in *Escherichia coli*. A review', *Lett Appl Microbiol*, **24**, 319–328.

ROWBURY R J (2001) 'Cross-talk involving extracellular sensors and extracellular alarmones gives early warning to unstressed *Escherichia coli* O157:H7 of impending lethal chemical stress and leads to induction of tolerance responses', *J Appl Microbiol*, **90**, 677–695.

RYU J-H and BEUCHAT L R (1999) 'Changes in heat tolerance of *Escherichia coli* O157:H7 after exposure to acidic environments', *Food Microbiol*, **16**, 317–324.

RYU, J-H, DENG Y and BEUCHAT L R (1999) 'Behavior of acid-adapted and unadapted

Escherichia coli O157:H7 when exposed to reduced pH achieved with various organic acids', *J Food Prot*, **62**, 451–455.

SAMELIS J and METAXOPOULOS J (1999) 'Incidence and principal sources of *Listeria* spp. and *Listeria monocytogenes* contamination in processed meats and a meat processing plant', *Food Microbiol*, **16**, 465–477.

SAMELIS J and SOFOS J N (2003) 'Strategies to control stress-adapted pathogens', in Yousef A E and Juneja V K, *Microbial Stress Adaptation and Food Safety*, Boca Raton, FL, CRC Press, Chapter 9, 303–351.

SAMELIS J, METAXOPOULOS J, VLASSI M and PAPPA A (1998) 'Stability and safety of traditional Greek salami a microbiological ecology study', *Int J Food Microbiol*, **44**, 69–82.

SAMELIS J, SOFOS J N, KENDALL P A and SMITH G C (2001a) 'Fate of *Escherichia coli* O157:H7, *Salmonella* Typhimurium DT104 and *Listeria monocytogenes* in fresh beef decontamination fluids at 4 or 10 °C', *J Food Prot*, **64**, 950–957.

SAMELIS J, SOFOS J N, KENDALL P A and SMITH G C (2001b) 'Influence of the natural microbial flora on the acid tolerance response of *Listeria monocytogenes* in a model system of fresh meat decontamination fluids', *Appl Environ Microbiol*, **67**, 2410–2420.

SAMELIS J, SOFOS J N, KAIN M L, SCANGA J A, BELK K E and SMITH G C (2001c) 'Organic acids and their salts as dipping solutions to control *Listeria monocytogenes* inoculated following processing of sliced pork bologna stored at 4 °C in vacuum packages', *J Food Prot*, **64**, 1722–1729.

SAMELIS J, SOFOS J N, KENDALL P A and SMITH G C (2001d) 'Lactic acid sensitization of *Salmonella* Typhimurium DT 104 and *Listeria monocytogenes* in non-acid (water) meat decontamination fluids at 10 °C', in *88th Annual Meeting of International Association for Food Protection (IAFP)*, August 5–8, 2001, Minneapolis, Minnesota, Abst. P050, 59.

SAMELIS J, SOFOS J N, KENDALL P A and SMITH G C (2002a) 'Effect of acid adaptation on survival of *Escherichia coli* O157:H7 in meat decontamination washing fluids, and potential effects of organic acid interventions on the microbial ecology of the meat plant environment', *J Food Prot*, **65**, 33–40.

SAMELIS J, SOFOS J N, IKEDA J S, KENDALL P A and SMITH G C (2002b) 'Exposure to nonacid fresh meat decontamination washing fluids sensitizes *Escherichia coli* O157:H7 to organic acids', *Lett Appl Microbiol*, **34**, 7–12.

SAMELIS J, BEDIE G K, SOFOS J N, BELK K E, SCANGA J A and SMITH G C (2002c) 'Control of *Listeria monocytogenes* with combined antimicrobials after postprocess contamination and extended storage of frankfurters at 4 °C in vacuum packages', *J Food Prot*, **65**, 299–307.

SEMANCHECK J J and GOLDEN D A (1998) 'Influence of growth temperature on inactivation and injury of *Escherichia coli* O157:H7 by heat, acid, and freezing', *J Food Prot*, **61**, 395–401.

SHADBOLT C, ROSS T and MCMEEKIN T A (2001) 'Differentiation of the effects of lethal pH and water activity: food safety implications', *Lett Appl Microbiol*, **32**, 99–102.

SHELEF L A (1994) 'Antimicrobial effects of lactates: a review', *J Food Prot*, **57**, 445–450.

SHELEF L A and YANG Q (1991) 'Growth suppression of *Listeria monocytogenes* by lactates in broth, chicken and beef', *J Food Prot*, **54**, 283–287.

SHERIDAN J J and MCDOWELL D A (1998) 'Factors affecting the emergence of pathogens on foods', *Meat Sci*, **49**, S151–S167.

SILVA F M, GIBBS P, VIEIRA M C and SILVA C L M (1999) 'Thermal inactivation of *Alicyclobacillus* spores under different temperature, soluble solids and pH conditions for the design of fruit processes', *Int J Food Microbiol*, **51**, 95–103.

SMITTLE R B (2000) 'Microbiological safety of mayonnaise, salad dressings, and sauces produced in the United States: a review', *J Food Prot*, **63**, 1144–1153.

SOFOS J N (1989) *Sorbate Food Preservatives*, Boca Raton, FL, CRC Press.

SOFOS J N (2002) 'Microbial control in foods: needs and concerns', in Juneja V K and Sofos J N, *Control of Foodborne Microorganisms*, New York, Marcel Dekker, 1–11.

SOFOS J N and BUSTA F F (1993) 'Sorbic acid and sorbates', in Davidson P M and Branen A L, *Antimicrobials in Foods*, New York, Marcel Dekker, 49–94.

SOFOS J N and BUSTA F F (1999) 'Chemical food preservatives', in Russell A D, Hugo W B and Ayliffe G A J, *Principles and Practice of Disinfection, Preservation, and Sterilization*, London, Blackwell Science, 485–541.

SORRELLS K M, ENIGL D C and HATFIELD J R (1989) 'Effect of pH, acidulant, time, and temperature on the growth and survival of *Listeria monocytogenes', J Food Prot*, **52**, 571–573.

STEKELENBURG F K and KANT-MUERMANS M L T (2001) 'Effect of sodium lactate and other additives in a cooked ham product on sensory quality and development of a strain of *Lactobacillus curvatus* and *Listeria monocytogenes'*, *Int J Food Microbiol*, **66**, 197–203.

STOPFORTH J D, SAMELIS J, SOFOS J N, KENDALL P A and SMITH G C (2002) 'Biofilm formation by acid-adapted and nonadapted *Listeria monocytogenes* in fresh beef decontamination washings and its subsequent inactivation with sanitizers', *J Food Prot*, **65**, 1717–1727.

TAMBLYN K C and CONNER D E (1997), 'Bactericidal activity of organic acids against *Salmonella* Typhimurium attached to broiler chicken skin', *J Food Prot*, **60**, 629–633.

TEUBER M (2000) 'Fermented milk products', in Lund B M, Baird-Parker T C and Gould G W, *The Microbiological Safety and Quality of Food*, Maryland, Aspen Publishers, Vol. 1, Chapter 23, 535–589.

THRELFALL E J, WARD L R, FROST J A and WILLSHAW G A (2000) 'The emergence and spread of antibiotic resistance in food-borne bacteria', *Int J Food Microbiol*, **62**, 1–5.

TILDEN J J, YOUNG W, MCNAMARA A M, CUSTER C, BOESEL B, LAMBERT-FAIR M A, MAJKOWSKI J, VUGIA D, WERNER S B, HOLLINGSWORTH J and MORRIS J G (1996) 'A new route of transmission for *Escherichia coli* O157:H7: infection from dry fermented salami', *Am J Public Health*, **86**, 1142–1145.

TINNEY K S, MILLER M F, RAMSEY C B, THOMPSON L D and CARR M A (1997) 'Reduction of microorganisms on beef carcasses with electricity and acetic acid', *J Food Prot*, **60**, 625–628.

TOMICKA A, CHEN J, BARBUT S and GRIFFITHS M W (1997) 'Survival of bioluminescent *Escherichia coli* O157:H7 in a model system representing fermented sausage production', *J Food Prot*, **60**, 1487–1492.

TOMPKIN R B, SCOTT V N, BERNARD D T, SVEUM W H and GOMBAS K S (1999) 'Guidelines to prevent post-processing contamination from *Listeria monocytogenes*', *Dairy Food Environ. Sanit*, **19**, 551–562.

TSAI Y-A and INGHAM S C (1997) 'Survival of *Escherichia coli* O157:H7 and *Salmonella* spp. in acidic condiments', *J Food Prot*, **60**, 751–755.

ULJAS H E and INGHAM S C (1998) 'Survival of *Escherichia coli* O157:H7 in synthetic gastric fluid after cold and acid habituation in apple juice or trypticase soy broth acidified with hydrochloric acid or organic acids', *J Food Prot*, **61**, 939–947.

ULJAS H E and INGHAM S C (1999) 'Combination of intervention treatments resulting in 5–Log10–unit reductions in numbers of *Escherichia coli* O157:H7 and *Salmonella* Typhimurium DT104 organisms in apple cider', *Appl Environ Microbiol*, **65**, 1924–1929.

UNDA J R, MOLINS R A and WALKER H W (1991) '*Clostridium sporogenes* and *Listeria monocytogenes*: survival and inhibition in microwave-ready beef roasts containing selected antimicrobials', *J Food Sci*, **56**, 198–205, 219.

VAN NETTEN P, HUIS IN 'T VELD J H and MOSSEL D A A (1994a) 'The immediate bactericidal effect of lactic acid on meat-borne pathogens', *J Appl Bacteriol*, **77**, 490–496.

VAN NETTEN P, HUIS IN 'T VELD J H and MOSSEL D A A (1994b) 'The effect of lactic acid decontamination on the microflora on meat', *J Food Safety*, **14**, 243–257.

VAN NETTEN P, VALENTIJN A, MOSSEL D A A and HUIS IN'T VELD J H J (1997) 'Fate of low temperature and acid-adapted *Yersinia enterocolitica* and *Listeria monocytogenes* that contaminate lactic acid decontaminated meat during chill storage', *J Appl Microbiol*, **82**, 769–779.

VAN NETTEN P, VALENTIJN A, MOSSEL D A A and HUIS IN'T VELD J H J (1998) 'The survival and growth of acid-adapted mesophilic pathogens that contaminate meat after lactic acid decontamination', *J Appl Microbio*l, **84**, 559–567.

VOLD L, HOLCK A, WASTESON Y and NISSEN H (2000) 'High levels of background flora inhibits growth of *Escherichia coli* O157:H7 in ground beef', *Int J Food Microbiol*, **56**, 219–225.

WCAGANT S D, BRYANT J L and BARK D H (1994) 'Survival of *Escherichia coli* O157:H7 in mayonnaise and mayonnaise-based sauces at room temperatures', *J Food Prot*, **57**, 629–631.

WEAVER R A and SHELEF L A (1993) 'Antilisterial activity of sodium, potassium or calcium lactate in pork liver sausage', *J Food Saf*ety, **13**, 133–146.

WEDERQUIST H J, SOFOS J N and SCHMIDT G R (1994) '*Listeria monocytogenes* inhibition in refrigerated vacuum packaged turkey bologna by chemical additives', *J Food Sci*, **59**, 498–500, 516.

WEISSINGER W R and BEUCHAT L R (2000) 'Comparison of aqueous chemical treatments to eliminate *Salmonella* on alfalfa seeds', *J Food Prot*, **63**, 1475–1482.

WILDE S, JORGENSEN F, CAMPBELL A, ROWBURY R and HUMPHREY T (2000) 'Growth of *Salmonella* enterica Serovar Enteritidis PT4 in media containing glucose results in enhanced *RpoS*-independent heat and acid tolerance but does not affect the ability to survive air-drying on surfaces', *Food Microbiol*, **17**, 679–686.

WRIGHT J R, SUMNER S S, HACKNEY C R, PIERSON M D and ZOECKLEIN B W (2000) 'Reduction of *Escherichia coli* O157:H7 on apples using wash and chemical sanitizer treatments', *Dairy Food Environ Sanit*, **20**, 120–126.

WU F M, BEUCHAT L R, DOYLE M P, GARRETT V, WELLS J G and SWAMINATHAN B (2002) 'Fate of *Escherichia coli* O157:H7 in coleslaw during storage', *J Food Prot*, **65**, 845–847.

YANG Z, LI Y and SLAVIK M (1998), 'Use of antimicrobial spray applied with an inside-outside birdwasher to reduce bacterial contamination on prechilled chicken carcasses', *J Food Prot*, **61**, 829–832.

YOUNG K M and FOEGEDING P M (1993) 'Acetic, lactic and citric acids and pH inhibition of *Listeria monocytogenes* Scott A and the effect of intracellular pH', *J Appl Bacteriol*, **74**, 515–520.

ZHAO T and DOYLE M P (1994) 'Fate of enterohemorrhagic *Escherichia coli* O157:H7 in commercial mayonnaise', *J Food Prot*, **57**, 780–783.

ZHAO T, DOYLE M P and BESSER R E (1993) 'Fate of enterohemorrhagic *Escherichia coli* O157:H7 in apple cider with and without preservatives', *Appl Environ Microbiol*, **59**, 2526–2530.

ZHUANG R-Y, BEUCHAT L R and ANGULO F J (1995) 'Fate of *Salmonella montevideo* on and in raw tomatoes as affected by temperature and treatment with chlorine', *Appl Environ Microbiol*, **61**, 2127–2131.

7

Antimicrobials from animals

A. Satyanarayan Naidu, N-terminus Research Laboratory, USA

7.1 Introduction

Numerous antimicrobial agents exist in animals where they have evolved as host defense mechanisms. For example, the prophylactic and therapeutic benefits of milk and its bioactive components lactoferrin and lactoperoxidase are well known. As another example, the hen's egg efficiently resists microorganisms over a long period in the absence of systemic host defenses. The egg albumen hosts an array of antimicrobial compounds that together interfere with the invasion and proliferation of microorganisms. The compounds include: the bacteriolytic lysozyme that hydrolyzes peptidoglycans of the bacterial cell wall; ovotransferrin, the metal-binding protein that specifically chelates iron; avidin and ovoflavoprotein, the vitamin-binding proteins that chelate biotin and riboflavin; and the proteinase inhibitors ovoinhibitor, ovomucoid, ovomacroglobulin, and cystatin.

The emergence of muscle food pathogens such as the enterohemorrhagic *Escherichia coli* O157:H7, vancomycin-resistant *Enterococcus faecium*, multidrug resistant *Salmonella* Typhimurium DT104, and *Listeria monocytogenes* has focused much attention on animal antimicrobials. The new food safety paradigm is driven by recent advances in the elucidation of molecular mechanisms of bioactive compounds from animals, their broad-spectrum antimicrobial effects and their multifunctional properties. Futhermore, their natural abundance and new developments in large-scale isolation techniques have prompted several new applications not only in food but also in dietary supplements, pharmaceuticals, cosmetics and oral care products. The objective of this chapter is to describe the potential for food applications of three categories of animal antimicrobials: iron-chelators, enzymes, and immunoglobulins.

7.2 Iron-chelators

Iron occurs at an approximate level of 2 g in the body of a normal human adult and is essential for all living organisms. The ability of iron to alternate between two valence states, Fe^{2+} and Fe^{3+}, is its most important biological property, evident in many metabolic pathways, including the electron transport system that forms ATP by phosphorylation. Iron deprivation leads to conservation of bioenergy and inhibition of cellular multiplication. Accordingly, various iron-binding proteins have evolved in the animal physiological system to sequester iron from the milieu. Ovotransferrin (conalbumin) from the egg white was the first iron-binding protein to be purified (Osborne and Campbell, 1900). In 1939, Sörensen and Sörensen identified a red iron-binding protein in bovine milk. Based on structural and chemical homology with serum transferrin, Blanc and Isliker (1961) proposed the name 'lactoferrin' for this protein.

7.2.1 Lactoferrin (LF) and activated lactoferrin (ALF)

Lactoferrin (LF) is a major bioactive glycoprotein in milk although it has also been reported in saliva, tears, seminal fluid, etc. Its ability to bind two Fe^{3+} ions with high affinity, in cooperation with two HCO_3^- ions, contributes to its major functional properties, including antimicrobial activity. LF plays an important regulatory role in many physiological pathways and is considered a major component of the pre-immune innate defense in mammals (Naidu, 2000a).

Examples of early studies on the antimicrobial properties of LF, determined against individual target organisms *in vitro*, are given in Table 7.1. Despite wide citation, the biocidal effect of LF appears to be artifactual, caused by reactants other than LF in the milieu (Lassitter, 1990). Acid/pepsin hydrolysis of LF generates the cationic antimicrobial peptides known as the lactoferricins (LFcins) (Tomita *et al.*, 1991, 1994; Bellamy *et al.*, 1992, 1993). The broad biocidal spectrum of LFcins is non-specific and readily inhibited by salts in the milieu at physiological concentrations (Table 7.1).

The antimicrobial functionality of LF is dependent on its protein conformation (Naidu and Arnold, 1997). Bacteriostasis is enhanced when LF binds to the microbial cell surface (Dalamastri *et al.*, 1988; Naidu *et al.*, 1993). Specific LF-binding microbial targets exist in Gram-positive and Gram-negative bacteria (Naidu *et al.*, 1991a,b, 1992, 1993). The high-affinity interaction of LF with pore-forming outer-membrane proteins (OMPs) of Gram-negative enterics, including *E. coli*, is critical for the antimicrobial outcome of LF (Gado *et al.*, 1991; Tigyi *et al.*, 1992; Erdei *et al.*, 1994; Naidu and Arnold, 1997). LF-mediated outer-membrane (OM) damage in Gram-negative bacteria has been reported to potentiate antibiotic action (Naidu and Arnold, 1994).

The interaction of LF with microbial surfaces – OMPs of Gram-negative bacteria in particular – can inhibit microbial attachment to epithelial surfaces. Blocking of microbial attachment factors such as fimbriae by LF has been reported (Naidu and Bidlack, 1998). Furthermore, the immobilization of LF to

Table 7.1 *In vitro* antibacterial effects of bovine lactoferrin (LF) and lactoferricin B (LFcin) against foodborne pathogens and spoilage bacteria

Bacteria	Antimicrobial compound	Dose	Effect	Reference
Aeromonas hydrophila	LF	0.1%	Adhesion blocked 47%	Paulsson *et al.* (1993)
Bacillus cereus	LFcin	6 μM	Cidal (4 log CFU/ml, 100%)	Hoek *et al.* (1997)
Bacillus stearothermophilus	LF	1:20	Stasis	Reiter and Oram (1967)
Bacillus subtilis	LFcin	0.002%	Cidal (6 log CFU/ml, 100%)	Bellamy *et al.* (1992)
Clostridium innocuum	LF	0.1%	Agglutination	Tomita *et al.* (1994)
Clostridium perfringens	LFcin	0.024%	Cidal (6 log CFU/ml, 100%)	Bellamy *et al.* (1992)
Clostridium paraputrificum	LFcin	0.003%	As above	Bellamy *et al.* (1992)
Enterococcus faecalis	LFcin	0.06%	As above	Bellamy *et al.* (1992)
Escherichia coli	LF	0.1%	Stasis (24 h, 100%)	Naidu *et al.* (1993)
Escherichia coli	LF	0.1%	Adhesion blocked 50%	Paulsson *et al.* (1993)
Escherichia coli	LFcin	10 μM	Cidal (3 log CFU/ml)	Bellamy *et al.* (1992)
Escherichia coli	LF	20 μM	LPS release, OM damage	Yamauchi *et al.* (1993)
Klebsiella pneumoniae	LFcin	10 μM	Cidal (3 log CFU/ml)	Bellamy *et al.* (1992)
Listeria monocytogenes	LFcin	10 μM	Cidal (4 log CFU/ml)	Bellamy *et al.* (1992)
Listeria moncytogenes	LFcin	2 μM	Cidal (4 log CFU/ml, 100%)	Hoek *et al.* (1997)
Pseudomonas aeruginosa	LFcin	10 μM	Cidal (3 log CFU/ml)	Bellamy *et al.* (1992)
Pseudomonas fluorescens	LFcin	8 μM	Cidal (4 log CFU/ml, 100%)	Hoek *et al.* (1997)
Salmonella spp.	LF	0.1%	Stasis (24 h, 100%)	Naidu and Arnold (1994)
Salmonella Enteritidis	LFcin	0.012%	Cidal (6 log CFU/ml, 100%)	Bellamy *et al.* (1992)
Salmonella Montevideo	LF	20 μM	LPS release, OM damage	Yamauchi *et al.* (1993)
Salmonella Typhimurium	LF	0.5%	Stasis (64%)	Naidu *et al.* (1993)
Salmonella Typhimurium	LF	0.1%	Adhesion blocked 68%	Paulsson *et al.* (1993)
Salmonella Typhimurium	LF	20 μM	LPS release, OM damage	Yamauchi *et al.* (1993)
Shigella flexneri	LF	0.1%	Adhesion blocked 30%	Paulsson *et al.* (1993)
Staphylococcus albus	LF	0.5%	Stasis	Masson *et al.* (1966)
Staphylococcus aureus	LF	0.1%	Adhesion blocked 54%	Paulsson *et al.* (1993)
Staphylococcus aureus	LFcin	10 μM	Cidal (3 log CFU/ml)	Bellamy *et al.* (1992)
Staphylococcus spp.	LFcin	0.006%	Cidal (6 log CFU/ml, 100%)	Bellamy *et al.* (1992)
Streptococcus spp.	LFcin	0.003%	As above	Bellamy *et al.* (1992)

LPS = lipopolysaccharide
OM = outer membrane)
Source: adapted from Naidu (2000a).

mucosal surfaces containing sulfated glycans such as heparan sulfate were found to enhance the antimicrobial spectrum of LF by multifold.

Though credited with an impressive list of nutraceutical benefits, certain factors limit the functionality of LF after isolation from milk. Protein separation conditions pose the risk of denaturation or structural alteration of the LF molecule. The isolation process may generate cationic peptides as degradation products (Tomita *et al.*, 1994). The highly cationic N-terminus region of LF may facilitate charge-induced protein aggregation leading to inactivation. The isolated LF is highly susceptible to conformational changes, thermal uncoiling, and proteolysis. Milieu conditions such as pH, elevated calcium or phosphates, iron excess, and improper citrate/bicarbonate ratios can markedly diminish the antimicrobial activity of LF. Consequently, a new, patented technology has been developed to overcome these limitations.

Activated lactoferrin technology (Naidu, 2001) is based on the immobilization of milk LF via its N-terminus region on a food-grade galactose-rich polysaccharide (GRP) or carrageenan, neutralization of cationic peptides by salt, optimization of pH and the citrate/bicarbonate ratio, and the establishment of an equilibrium between bound (immobilized) and unbound LF. The result is a microbial blocking agent, which interferes with adhesion/colonization, detaches live or dead microorganisms from biological surfaces, inhibits microbial growth/multiplication, and neutralizes the activity of endotoxins.

Microbial attachment to epithelial mucosa is the initial step in the pathogenesis of many infections as well as spoilage of foods. Many enteric bacteria harbor fimbriae on their cell surface to promote bacterial adhesion and colonization on the mucosal surface (Parry and Rooke, 1985). For example, *E. coli* O157:H7 adheres tightly to epithelial surfaces (Louie *et al.*, 1993) and to collagens on beef tissue. Specific binding of ALF to OMPs of Gram-negative bacteria leads to inhibition of cellular functions and deregulation of adhesin/fimbrial synthesis on the bacterial surface. Furthermore, ALF–OMP interaction can also cure plasmids by diffusion, resulting in the loss of colonization-factor antigens and certain enterotoxins in *E. coli* (Erdei *et al.*, 1994; Naidu, 2002).

Binding to tissue-matrix components such as collagens is a pivotal anchor mechanism for many bacteria (Höök *et al.*, 1989). ALF binds to the same anchor sites on tissue surfaces with even greater affinity. In binding-displacement studies, ALF at low molar concentrations has been shown to detach tissue-bound bacteria, either viable or dead (Naidu, 2002). ALF also has broad-spectrum activity against DNA and RNA viruses by interacting with nucleic acids and binding with eukaryotic cells to prevent viral adhesion. Ongoing studies also indicate the effectiveness of ALF in the control of biofilms on nonbiological surfaces such as food processing equipment.

The growth-inhibiting activities, individual and synergistic, reported for LF, are also elicited by ALF but with a multifold potentiation over the original effect. Accordingly, the minimum inhibitory concentrations (MICs) for ALF are significantly lower than those for LF. In contrast to LF, ALF demonstrates a potent antimicrobial activity in iron-rich environments such as meats and biological fluids. ALF also shows syngergism with bacteriocins, colicins, and salivary histatins.

7.2.2 Ovotransferrin (OTF)

Ovotransferrin (OTF, also called conalbumin) is a monomeric glycoprotein that binds reversibly two Fe^{3+} ions per molecule concomitantly with two bicarbonate anions. OTF has a very high affinity for iron (~$10^{30}\,\mathrm{M}^{-1}$) and renders it unavailable for bacteria thereby inhibiting growth (Valenti *et al.*, 1983a; Tranter and Board, 1982). The antimicrobial activity of OTF is highly dependent on milieu conditions and the target organism. The antimicrobial potency of hen's egg albumin is enhanced at alkaline pH (9.5 in egg) and at elevated temperature ~40 °C (close to the physiological temperature of birds) (Tranter and Board, 1984). Supplementation of egg albumin with excess iron at 39.5 °C or conditions that promote iron-saturation of OTF do not affect antimicrobial acitivity. OTFs play a key role in the antimicrobial defense of the hen's egg and constitute 12 % of the total proteins of albumin (15 g/l of egg white).

Although many studies have demonstrated the inhibitory action of OTF, a biocidal effect (microbial killing independent of iron deprivation) against a wide spectrum of microorganisms (e.g. *Pseudomonas* spp., *E. coli*, *Streptococcus mutans*, *S. aureus*, *Proteus* spp., and *Klebsiella* spp.) has also been suggested (Valenti *et al.*, 1982, 1983b, 1984, 1985, 1987). The antibacterial activity of OTF complexed to several metals was studied *in vitro* and the OTF–zinc complex was found to be more potent than other metal complexes. This effect was not due to Zn^{2+} ions, nor iron deprivation, but to a specific activity of the OTF–Zn^{2+} complex (Valenti *et al.*, 1987). OTF is also antifungal against *Candida* spp., and this, too, seems to be independent of iron-deprivation (Valenti *et al.*, 1985). The bactericidal activity has been attributed to the presence of a peptide (OTAP-92), recognized as the bactericidal domain of OTF (Ibrahim, 2000). The purified peptide has been shown to have a strong bactericidal activity against both Gram-positive *Staph. aureus* (6 log CFU/ml kill) and Gram-negative *E. coli* (1 log CFU/ml kill) (Ibrahim, 1997; Ibrahim *et al.*, 1998).

7.3 Enzymes

7.3.1 Lactoperoxidase (LP)

Lactoperoxidase (LP) is the most abundant enzyme in bovine milk, constituting about 1% of the whey proteins or 10–30 μg/ml of milk (Reiter, 1985). The concentration of LP is low in the bovine colostrum and increases rapidly to reach a peak at four to five days of postpartum. Human LP concentration is high in colostrum and declines rapidly within one week (Kiermeier and Kuhlmann, 1972; Gothefors and Marklund, 1975). Involvement of hydrogen peroxide and thiocyanate in microbial inhibition by milk LP was first reported by Hogg and Jago (1970).

LP is an oxidoreductase and plays an important role in protecting the lactating mammary gland and the intestinal tract of the newborn infant against pathogenic microorganisms. LP catalyzes the oxidation of thiocyanate and iodide ions to generate highly reactive oxidizing agents. These products have a broad spectrum of antimicrobial activity against bacteria, fungi, and viruses (Table 7.2). The components of cells that are oxidized are the sulfhydryl groups, NADH, NADPH, and, under some conditions, aromatic amino acid residues. Oxidation of these components alters the functions of cellular systems. The cytoplasmic membrane, transport systems, and glycolytic enzymes may be damaged, resulting in cell death or growth inhibition. The major products responsible for these effects are the hypothiocyanite ion (OSCN−), hypothiocyanous acid (HOSCN), and iodine (I_2). However, minor products are also generated. These minor products are highly reactive and short-lived but have even more potent antimicrobial properties.

Microorganisms show great variability in their response to peroxidase systems. Killing of cells is usually greater at low pH, low temperature, and with I− as the electron donor. Gram-negative bacteria are more susceptible to killing and to cytoplasmic membrane damage than are Gram-positive species. However, the killing of both requires long incubation periods (in the order of an hour) and high concentrations ([OSCN−]) > 10 μM at pH 7.0) of the active products. Some bacteria can resist the antibacterial effects by generating substances that reduce OSCN− to SCN−. These reducing agents also enable cells to recover from inhibition by reversing the oxidation of sulfhydryl groups caused by peroxidases (Naidu, 2000b).

7.3.2 Lysozyme (LZ)

Lysozymes (LZ) are ubiquitous in both the animal and plant kingdoms and have an important role in natural defense mechanisms. LZ is attractive as a natural food preservative due to its endogenous occurrence in several foods, specific hydrolytic activity against bacterial cell walls, and non-toxicity to humans (Losso *et al.*, 2000). LZ is biocidal against Gram-positive bacteria by hydrolyzing the β-1,4 linkage between *N*-acetylmuramic acid (NAM) and *N*-acetyl-glucosamine (NAG) in the peptidoglycan of the cell wall. LZ is also known as *N*-acetylmuramideglycanohydrolase (EC 3.2.1.17). Egg-white LZ is the classic representative of the lysozyme family and the related enzymes are called *c* type (chicken- or conventional-type) lysozyme. One egg contains 0.3–0.4 g of LZ and is the easiest and most economical source for recovering this enzyme.

The rate of catalysis by LZ depends upon the pH of the medium. The pH profile is bell-shaped with a maximum at pH 5.0 and inflections at pH 3.8 and 6.7. The osmotic uptake of water leads to the expansion and eventual rupture of the cytoplasmic membrane, which, in turn, causes cell death. Gram-positive

Table 7.2 *In vitro* antimicrobial activity of the lactoperoxidase system against bacteria and fungi

Microorganism	Donor	H_2O_2 source	Inhibitory effect	Reference
Gram-positive bacteria				
Bacillus cereus	SCN^-	Reagent	Collagenase production	Tenovuo *et al.* (1983)
Bacillus megaterium	SCN^-	Reagent	Stasis	Klebanoff *et al.* (1966)
Lactobacillus casei	SCN^-	Bacteria	Stasis	Iwamoto *et al.* (1972)
Lactobacillus acidophilus	I^-	Bacteria	Lysine accumulation	Zeldow (1963)
Lactobacillus plantarum	SCN^-	Bacteria	Stasis	Hoogendron and Moorer (1973)
Lactobacillus bulgaricus	SCN^-	Bacteria	Acid production	Portman *et al.* (1962)
Sarcina lutea	SCN^-	Reagent	Stasis	Reiter (1978)
Staphylococcus albus	SCN^-	Reagent	Stasis	Klebanoff *et al.* (1966)
Staphylococcus aureus	I^- or SCN^-	*S. mitis*	Amino acid uptake	Hamon and Klebanoff (1973)
Streptococcus cremoris	Milk/ SCN^-	Bacteria	Glycolysis	Reiter (1985)
Streptococcus lactis	SCN^-	Reagent	Stasis	Marshall and Reiter (1980)
Streptococcus pyogenes	Milk	Bacteria	Cidal/acid production	Mickelson (1979)
Streptococcus agalactiae	Milk/ SCN^-	?	Lactic acid production	Brown and Mickelson (1979)
Streptococcus fecalis	SCN^-	Bacteria	Stasis	Klebanoff *et al.* (1966)
Gram-negative bacteria				
Escherichia coli	SCN^-	Reagent	Stasis	Klebanoff *et al.* (1966)
	I^-	Reagent	Stasis	Klebanoff (1967)
	Whey	Oxidase	Amino acid uptake	Hamon and Klebanoff (1973)
	Milk	Reagent	Stasis	Stephens *et al.* (1979)

Table 7.2 Continued

Microorganism	Donor	H_2O_2 source	Inhibitory effect	Reference
Legionella pneumophila	Cl^-	Reagent	Cidal	Lochner *et al.* (1983)
Salmonella typhimurium	SCN^-	Reagent	Cidal/stasis	Purdy *et al.* (1983)
Pseudomonas aeruginosa	SCN^-	Reagent	Cidal	Pruitt and Reiter (1985)
Pseudomonas fluorescens	Milk/ SCN^-	Oxidase	Cidal	Björck *et al.* (1975), Björck (1978)
Fungi				
Candida tropicalis	I^-	*S.mitis*	Cidal	Hamon and Klebanoff (1973)
Candida albicans	I^-	Reagent	Cidal	Lehrer (1969)
Rhodotorula rubra *Saccharomyces cerevisiae,* *Aspergillus niger* & *Byssochlamys fulva*	SCN^-	Oxidase	Stasis	Popper and Knorr (1997)

Source: adapted from Naidu (2000b).

Table 7.3 *In vitro* antimicrobial activity of lysozyme against Gram-positive bacteria

Family	Genus/species	Sensitivity
Bacillaceae	*Bacillus* spp.	+
	Bacillus cereus	+/–
	Bacillus megaterium	+
	Bacillus stearothermophilus	+
	Bacillus subtilis	+
Bifidobacterium	*Bifidobacterium* spp.	+
Clostridium	*Clostridium botulinum* A, B, E	m+
	Clostridium perfringens	–
	Clostridium sporogenes	+
	Clostridium tyrobutyricum	+
	Clostridium welchii	–
Corynebacterium	*Corynebacterium betae*	+
	Corynebacterium flaccumfaciens	+
	Corynebacterium poinsettae	+
	Corynebacterium tritici	–
Lactobacillaceae	*Lactobacillus acidophilus*	–
	Lactobacillus casei	–
	Lactobacillus fermentum	–
	Lactobacillus helveticus	+
	Lactobacillus lactis	+/–
	Lactobacillus plantarum	–
Leuconostoc	*Leuconostoc citrovorum*	–
Microccocaceae	*Micrococcus lysodeikticus*	+
	Sarcina lutea	+
	Staphylococcus aureus	+/–
Propionobacteriaceae	*Propionobacterium* spp.	–
Streptococcaceae	*Enterococcus faecium*	+
	Streptococcus bovis	–
	Streptococcus cremoris	+/–
	Streptococcus diacetilactis	–
	Streptococcus faecalis	–
	Streptococcus lactis	+/–
	Streptococcus liquefaciens	–
	Streptococcus thermophilus	+
	Streptococcus zymogenes	–

+, sensitive; m+, moderately sensitive; –, resistant.
Source: adapted from Losso *et al.* (2000).

bacteria are susceptible to LZ because their cell wall consists of about 90% peptidoglycan (Table 7.3). In some Gram-positive bacteria such as *B. cereus*, the absence of NAG residues confers resistance to LZ (Board, 1995). In Gram-negative bacteria, the peptidoglycan constitutes only 5–10% of the cell wall. The lipopolysaccharide layer of the outer membrane acts as a barrier against macromolecules and hydrophobic compounds. Hence Gram-negative bacteria are not sensitive to LZ. The antiviral activity of LZ is associated with its positive charge and not with its lytic activity (Cisani *et al.*, 1984).

Recently, Düring *et al.* (1999) unexpectedly discovered that heat-denatured LZ, in which the enzymatic activity was abolished, maintained its antimicrobial activity. The membrane-perturbing activity of the denatured LZ was demonstrated on bacterial, fungal, and plant cells. The amphiphatic C-terminal domains of the LZ seem to mediate the bactericidal and fungicidal activities. A synthetic peptide, with sequence homology to the amphiphatic C-terminal of T4 or hen's egg white LZ, showed bactericidal and fungicidal activity. These new data suggest that LZ could have both an enzymatic and non-enzymatic mode of biocidal action in its native and denatured state, respectively. Therefore, the antimicrobial activity of LZ may not be eliminated by heat treatment during food processing operations.

7.4 Immunoglobulins

Antibodies may be unique as food antimicrobials due to their action on both disease-causing organisms and virulence factors such as toxins. Since colostrum/milk and egg-derived antibodies are polyclonal, they present a multifaceted approach to pathogen control that transcends microbial ability to develop resistance to antimicrobials.

7.4.1 Lactoglobulins (IgG)

The benefit of colostrum, which contains high levels of Lacto-Ig (lactoglobulins), was well known more than one hundred years ago (Ehrlich, 1892). There are extensive reviews on colostrum and its ability to provide immunity to offspring (Yolken, 1985; Reddy *et al.*, 1988; Butler, 1994; Bostwick *et al.*, 2000).

Lactoglobulins have many advantages over synthetic antibiotics: they are low cost, polyclonal, limited in activity to the intestinal lumen and are less of a burden on the microbial gut ecology. Their entry into the market may be faster due to less stringent regulatory requirements. Lactoglobulin use in dietary supplements is under consideration. However, human applications are in their infancy due to uncertainties about dosage and timing of administration. Furthermore, lactoglobulins are heat-sensitive, subject to human digestive action and may present difficulties with respect to palatability or reduced shelf-life due to the action of proteolytic contaminating enzymes. Encapsulation of immunoglobulins with acid-resistant coatings is one approach under investigation. Clinical trials and sophisticated product development are needed.

7.4.2 Ovoglobulins (IgY)

Domestic avian species such as chickens, turkeys, and ducks produce antibodies in blood and eggs against avian diseases, as well as against other antigens. Immunoglobulin (Ig) in avian blood is transferred to the yolk of eggs to give acquired immunity to the newly hatched chick (Rose and Orlans, 1981). Egg albumen contains relatively low concentrations of immunoglobulins but the yolk contains a considerably higher concentration of 25 mg IgG/ml (Rose *et al.*, 1974). The antibody in egg yolk has been referred to as IgY (Leslie and Clem, 1969) because of its difference from mammalian IgG in structure and immunological properties. IgY is larger (Kobayashi and Hirai, 1980), slightly more acidic, and lower in molecular rigidity (Higgins, 1975) than mammalian IgG. Unlike mammalian IgG, IgY does not fix mammalian complement (Jensenius *et al.*, 1981; Akerstrom *et al.*, 1985). Since IgY is mainly composed of γ-livetin, which is a larger molecule than any other α-, β-livetin in egg yolk, it is relatively easy to separate from other proteins in the water-soluble fraction of egg yolk. Specific IgY antibodies isolated from hen eggs have been reported to elicit potent antimicrobial effects against a wide spectrum of bacterial and viral pathogens (Table 7.4).

A hen lays an average of 240 eggs a year. The annual production of IgY by a hen is about 24 g, which represents relatively high productivity compared with antibody production by mammals such as rabbits, mice and goats (Sim *et al.*, 2000). Chicken egg yolk has, therefore, received much attention as a good source of antibodies. As a food ingredient and/or reagent, IgY is a versatile molecule for prophylactic, therapeutic, and diagnostic purposes (Losch *et al.*, 1986; Shimizu *et al.*, 1988; Yolken *et al.*, 1988). The effectiveness of passive immunization by oral administration of IgY to prevent infection has been reported for rotavirus diarrhea in humans and animals (Ebina, 1996; Kuroki *et al.*, 1997), dental caries (Otake *et al.*, 1991; Hatta *et al.*, 1997), enteric colibacillosis (Yokoyama *et al.*, 1992; Imberechts *et al.*, 1997; Zuniga *et al.*, 1997), and salmonellosis (Peralta *et al.*, 1994; Sunwoo *et al.*, 1996; Yokoyama *et al.*, 1998b).

7.5 Applications in foods

7.5.1 Iron-chelators

Lactoferrin is available in ready-to-use form such as a liquid or spray-dried powder. In the past two decades, its metal-chelating property has been the primary claim for the application of LF in several infant food formulas in South-East Asian markets. The immuno-modulatory and antioxidant properties of LF have prompted its application in health foods but its use as a food antimicrobial has been cost-prohibitive since high doses of LF are required to obtain a preservative effect. The application of lactoferricins in foods has also been limited because the ionic environment in most foods can neutralize their antimicrobial activity (Venkitnarayan *et al.*, 1999).

Activation of LF by the patented immobilization process described in Section 7.2.1 (Naidu, 2001) has repositioned this iron-chelator as a novel microbial blocking agent for food applications. ALF is currently being explored as a decontaminant in beef packaging and poultry processing. Efficacy studies have

Table 7.4 Antimicrobial effects of ovoglobulins against bacterial and viral pathogens

Pathogen	Antimicrobial effect	Reference
Edwardsiella tarda	Prevention of edwardsiellosis of Japanese eels	Hatta *et al.* (1994)
Escherichia coli	Prevention of enterotoxigenic *E. coli* (ETEC) infection in neonatal piglets	Ikemorie *et al.* (1992)
	Protection of neonatal calves from fatal enteric colibacillosis	O'Farrelly *et al.* (1992)
	Prevention of diarrhea in rabbits challenged with ETEC	Zuniga *et al.* (1997)
	Reduction of gut colonization with ETEC in weaned pigs	Imberechts *et al.* (1997)
	Inhibition of pathogen shedding by infected pigs	Jin *et al.* (1998)
	Adhesion-blocking of ETEC to piglet intestinal mucus	Wiedemann *et al.* (1991)
	Alleviation of diarrhea in piglet infection model	Ozpinar *et al.* (1996)
	Reduction of diarrhea incidences in neonatal calves	Yokoyama *et al.* (1993)
	Protection of pig from ETEC-induced enterotoxemia	Yokoyama *et al.* (1992)
	Neutralization of ETEC heat-labile toxin	Akita *et al.* (1998)
Pseudomonas aeruginosa	Inhibition of growth-multiplication	Yoshiko *et al.* (1996)
Salmonella Dublin	Prevention of fatal salmonellosis in neonatal calves	Yokoyama *et al.* (1998b)
Salmonella Enteritidis	Prevention of gut colonization/organ invasion in chicks	Optiz *et al.* (1993)
Salmonella Enteritidis	Adhesion-blocking to human intestinal cells	Yoshiko *et al.* (1996)
Salmonella Typhimurium	Protection of mice from experimental salmonellosis	Yokoyama *et al.* (1998a)
Staphylococcus aureus	Inhibition of enterotoxin-A synthesis	Yoshiko *et al.* (1996)
Coronavirus	Protection of neonatal calves from diarrhea	Ikemori *et al.* (1997)
IBD-virus	Protection of chicks from infectious bursal disease	Eterradossi *et al.* (1997)
Rotavirus	Prevention of bovine rotavirus (BRV) induced diarrhea in rats	Kuroki *et al.* (1993)
	Protection of calves from BRV-induced diarrhea	Kuroki *et al.* (1994, 1997)
	Prevention of human rotavirus-induced gastroenteritis in mice	Ebina (1996)

Source: adapted from Sim *et al.* (2000).

been performed on beefsteaks sprayed with 1% LF or 1% ALF. A premarked area of 1 in^2 (2.54 cm^2) on a meat surface was inoculated with about 4 log CFU of *E. coli* O157:H7 and incubated at room temperature for 24 h. Spraying the meat with ALF-reduced *E. coli* O157:H7 by more than 2.5 log CFU compared with treatment with LF spray (Naidu, 2002).

Sensory studies have indicated that ALF treatment could extend the retail display life by 1.7 to 2.5 days for case-ready packaged steaks compared with non-treated steaks in conventional packages. ALF-treated and control samples showed no differences in tenderness, juiciness, or off-flavor of cooked beef, and the panelists preferred the flavor of the ALF-treated samples. Subprimals stored for 21 days following ALF application exhibited a five-fold reduction in total microbial plate counts compared with non-treated steaks (Naidu, 2002).

ALF is manufactured from bovine LF extracted from cheese whey or skim milk. Other ingredients in the formulation, i.e. sodium bicarbonate, citric acid, sodium chloride and carrageenan or protein, are commercially available chemicals of food-grade quality. The aqueous ALF formulation is constituted in de-ionized water at concentrations ranging from 1 to 4% depending on the application. ALF is delivered onto the meat surface as a fine mist using electrostatic or high-pressure liquid spray nozzles. The flow pattern and spray time are digitally monitored by a programmable logistic controller (PLC) for uniform coverage of ALF on the meat surface at optimum functional concentration (OFC). A solid-phase immunoassay based on a fluorescent-labeled LF-antibody has been developed to verify the coverage and quantify ALF-coating on the meat surface.

ALF has demonstrated microbial blocking activity against a variety of foodborne pathogens, including *E. coli.* O157:H7, *L. monocytogenes*, *Salmonella* spp., *Campylobacter* spp., *Vibrio* spp., *Aeromonas hydrophila*, and *Staph. aureus*, as well as the food spoilage microorganisms *Bacillus* spp., *Pseudomonas* spp., and *Klebsiella* spp. ALF also inhibits yeasts and molds as well as DNA and RNA viruses. These attributes make ALF a potential microbial intervention for poultry, pork, fish/seafood, and produce processing. ALF is also effective in detaching multidrug-resistant *Salmonella* Typhimurium DT104, vancomycin-resistant *Ent. faecium*, and methicillin-resistant *Staph. aureus* from biological surfaces. Studies are underway to evaluate the antimicrobial efficacy of ALF incorporated into edible food packaging materials (see also Chapter 12) and sausage casings.

7.5.2 Enzymes

The antimicrobial activity of lactoperoxidase has been exploited in the preservation of raw milk in warm climates, through the addition of hydrogen peroxide and thiocyanate. Also, LP itself has been used as an iodizing agent and is considered to be less disruptive to protein structure.

Lysozyme (LZ) is used in a number of food applications, most notably in cheese making. LZ is selective for certain Gram-positive spoilage bacteria but does not interfere with organisms that are considered beneficial for human health. Singly or in combination with other natural antimicrobial compounds, LZ has desirable antimicrobial properties for use in minimally processed foods. Japan leads the world for the variety of industrial uses of LZ. Food products containing LZ such as cheese, chewing gum and candies are on the market (Losso *et al.*, 2000).

Cheese, particularly Edam, Gouda and Italian cheeses, are classical examples of dairy products where LZ is used as a selective food antimicrobial against *Clostridium tyrobutyricum*. LZ is added at 20 to 35 mg/l to cheese vats in either a liquid or spray dried and granulated form, substituting for other, commonly used preservatives such as formaldehyde, nitrate, nisin, and hydrogen peroxide (Wasserfall and Teuber, 1979; Hughey *et al.*, 1989). The effectiveness of LZ is directly associated to its specificity. Thus, LZ inhibits the spoilage organism, while not interfering with the starter culture. For example, 2500 units/ml LZ did not inhibit the activity of four starter cultures in Gouda cheese while growth of coliforms was inhibited by 1000 units/ml (Bester and Lombard, 1990). Several patents claim the effectiveness of LZ at concentrations as low as 50 ppm to prevent the development of undesirable microorganisms in butter and cheese for more than 24 months. The cheese industry uses about 100 tonnes of LZ annually (Dell'Acqua *et al.*, 1989).

Until recently, sulfur dioxide has been used, worldwide, to control malolactic fermentation, inhibit spoilage bacteria and yeast and provide antioxidant properties to wines during storage. Unfortunately, the use of sulfur dioxide in wines is sometimes associated with side effects such as allergic reactions in asthmatic individuals. Addition of 500 mg/l of LZ to grape must has been shown to inhibit malolactic fermentation; however, addition of 125–250 mg/l LZ to red wines after malolactic fermentation has completely prevented acetic acid and biogenic amine formation (Gebraux *et al.*, 1997). LZ, because of the lack of antioxidant property, may not completely replace SO_2, but a combination of LZ with low levels of SO_2 may provide effective control of the bacterial flora (Villa, 1996). The role of LZ as an antimicrobial in unpasteurized beer containing 5% alcohol at pH 4.7 has also been studied (Makki and Durance, 1996). LZ in combination with β-glucopyranose aerodehydrogenase has been shown to preserve sake, the popular Japanese alcoholic beverage (Eisai, 1980).

LZ, at a concentration of 0.005–0.03% in combination with 0.25 to 0.35% NaCl was shown to preserve low-quality salmon or trout roe (Eisai, 1971). Furthermore, a combination of LZ and glycine has been shown to be an effective preservative of fish paste (Eisai, 1973). Two US patents (Dell'Acqua *et al.*, 1989; Johnson *et al.*, 1991) report on bacterial decontamination of vegetables using LZ. The synergistic combination of chelators with LZ seems to delay toxin production in potato suspensions (Cunningham *et al.*, 1991). Fresh vegetables, tofu, kimchi, japanese potato salad, sushi, chinese noodles, and creamed custards have been preserved using LZ or a combination of LZ with amino acids (Cunningham *et al.*, 1991). LZ when combined with glycine, lower fatty acid monoglyceride, phytic acid, and/or sodium acetate has been reported to preserve fruit and vegetables, fruit juice, uncooked noodles, bean jam, and custard cream (Eisai, 1971, 1972).

The well-known synergistic combination of LZ with nisin (see also Chapter 2) and EDTA is currently being investigated in biodegradable packaging films made from corn zein or soy protein isolate (Padgett *et al.*, 1998; see also Chapter 12). Payne *et al.* (1994) used a combination of LZ with LF and EDTA against *P. fluorescens, Salmonella* Typhimurium, *E. coli* 0157:H7, and *L. monocytogenes* in ultra-high-temperature pasteurized milk. Thus, LZ could have potential applications in the produce, meat, wine and animal feed industries.

7.5.3 Immunoglobulins

The market for immunoglobulins in animal feed has not grown because of the expense of producing bovine Ig concentrates. However, the use of antibiotics as growth promoters in feed has been subject to criticism and in some countries certain antibiotics are already banned due to the possible risk of developing multiresistant bacterial strains. Therefore, natural antimicrobials such as immunoglobulins, LF or LP may play a more prominent role in the future by replacing some of the current uses of antibiotics. Many other potential applications have been proposed including, for example, the use of immunoglobulins to control infantile diarrhea in developing countries and to impart passive immunity against oral pathogens such as streptococci and candida (Reddy *et al.*, 1988; Bogsted *et al.*, 1996; Weiner *et al.*, 1999). It has been calculated that low-cost antibodies could be produced at less than \$10/g (compared with more than \$20,000/g for mammalian IgG) by immunizing laying hens with antigens and collecting IgY from their egg yolks. Developments in IgY technology could therefore open up new market applications in medicine, public health, veterinary medicine, and food safety.

7.6 Toxicology

Bovine LF has been consumed by humans for thousands of years as a naturally occurring protein found in milk. Consumption of bovine LF through dairy sources has been estimated to be 73, 75, and 50 mg/day at the 90th percentile of intake for children, teens, and adults, respectively. Numerous clinical trials were performed to evaluate the effects of LF in both infants and adults on either iron absorption or modulation of microflora or prevention of infection. In studies with infants, dosage levels ranged from 1.4 mg/day (0.3 mg/kg/day) to 2.9 g/day (1.0 g/kg/day) and study durations were from 11 days to 5 months. In studies with adults, dosage levels ranged from 100 mg/day (1.7 mg/kg/day) to 3.6 g/day (60 mg/kg/day) and study durations varied from a single dose to a maximum of 8 weeks. In the above studies, though not designed specifically to test tolerance, no adverse health effects have been reported.

The safety of bovine LF was also assessed in an acute toxicity study (Nishimura, 1991), a 4 week and a 13 week oral toxicity study (Nishimura,

1997, 2000), and in an Ames assay (Kawai and Tanaka, 1997). The animal studies indicated no adverse effects related to LF consumption at levels up to 2000 mg/kg/day administered up to 13 weeks. The possibility that several-fold increases in exposure to LF in older children and adults might result in allergic sensitization is remote given the large existing background. However, individuals, especially children, who are allergic to cow's milk should be made aware of LF allergenicity.

Lysozyme, like many other low molecular proteins, is easily filtered through the kidney, reabsorbed by the proximal tubular cells and catabolized in the lysosomes. This property has opened the door to using LZ as anti-infective and carrier for renal delivery of various drugs including antibacterial agents (Haas *et al.*, 1997; Meijer *et al.*, 1996). Clinical trials of LZ have been reported with mixed results. Yamada *et al.* (1993) has reported the development of specific IgE antibody titers to hen's egg LZ in children allergic to egg. Several publications (Hoffman, 1982; Anet *et al.*, 1985; Pichler and Campi, 1992; Bernhisel-Broadbent *et al.*, 1994; Aabin *et al.*, 1996; Urisu *et al.*, 1997) reveal that commercially available egg white proteins, including LZ, act as major allergens. When individual egg white proteins were prepared to the highest purity available, LZ ranked last among egg white protein allergens. Ovomucoid, ovotransferrin, and ovalbumin were the major immunodominant proteins in egg white. Proteinase treatment of allergenic proteins is the most popular strategy for reducing allergenicity; however, this would also affect the functional properties of the protein.

7.7 Legislation and labeling

LF occurs naturally in milk, milk-derived ingredients and products, and, to a lesser extent, in beef tissue. Thus, persons who consume milk or beef already consume LF. Milk-derived ALF is considered GRAS (Generally Recognized As Safe) by the Food and Drug Administration [21 CFR.170.36(f)]. It is permitted at levels of 65.2 mg/kg of beef, according to the FDA Directive of October 23, 2001. The US Department of Agriculture approved the use of ALF on fresh beef in December 2001. For other countries, the status of ALF will need to be confirmed.

Currently, the EU does not have a specific regulation for LF but there is a Directive (83/417/EEC) for the use of protein derived from milk. LF could be included in this Directive, when it obtains a regulated status. LF is considered as a milk protein (provision 79/112/EEC allows all types of milk protein to be labeled as milk protein) and hence it would be allowed in foods.

In Japan, LF is permitted and specified in the List of Existing Food Additives. In South Korea, LF concentrates are also listed as authorized natural additives. In Taiwan, LF may be used in special nutritional foods under the condition 'only for supplementing foods with an insufficient nutritional content and may be used in appropriate amounts according to actual requirements'.

The FAO/WHO and several countries of the EU have acknowledged the non-toxicity of LZ and have therefore approved its use in some foods. LZ has been granted a tentative final ruling for GRAS status by the US FDA, and Canada is presently considering regulatory approval.

7.8 Future prospects

Most antimicrobials in use today kill microorganisms and leave debris as well as toxins on processed foods. In contrast, ALF is an extremely powerful antimicrobial intervention that prevents microbial proliferation and attachment on surfaces such as beef or poultry. The ability to detach the leftover microbial debris and to inactivate the surface-splattered endotoxins makes ALF a highly effective intervention all by itself or as an additional step in current multifactorial sanitizing systems.

Among animal enzymes, LZ from hen's egg white, used singly or in combination with other natural antimicrobial compounds, has desirable antimicrobial properties for use in minimally processed foods. Several patents claim the effectiveness of LZ, alone or in combination with other synergists, as a food preservative for fruits and vegetables, meat, and beverages. Until genetic engineering makes it possible to produce LZ with both Gram-positive and Gram-negative antimicrobial activity, the multifactorial approach will provide more opportunities for its use as a natural food preservative.

Immunoglobulins could potentially be used in foods as they have a wide spectrum of inhibitory activity and ability to neutralize microbial toxins. However, they are heat sensitive and this may present difficulties in certain food processing operations. Encapsulation of Ig with acid-resistant coatings may solve this problem in the future.

7.9 References

AABIN, B., POULSEN, L.K., EBBEHOJ, K., NORGAARD, A., FROKIOER, H., BINDSLEV-JENSEN, C., and BARKHOLT, V. (1996). Identification of IgE-binding egg white protein: Comparison of results obtained by different methods. *Int. Arch. Allergy. Immunol.* **109**:50–57.

AKERSTROM, G., BRODIN, T., REIS, K., and BORG, L. (1985). Protein G: A powerful tool for binding and detection of monoclonal and polyclonal antibodies. *J. Immunol.* **135**:2589–2592.

AKITA, E.M., LI-CHAN, E.C.Y., and NAKAI, S. (1998). Neutralization of enterotoxigenic *Escherichia coli* heat-labile toxin by chicken egg yolk immunoglobulin Y and its antigen-binding fragments. *Food Agri Immunol.* **10**:161–172.

ANET, J., BACK, J.F., BAKER, R.S., BARNETT, D., BURLEY, R.W., and HOWDEN, M.E.H. (1985). Allergens in white and yolk of hen's egg. A study of IgE binding by egg proteins. *Int. Archs. Allergy Appl. Immun.* **77**:364–371.

BELLAMY, W., TAKASE, M., YAMAUCHI, K., WAKABAYASHI, H., KAWASE, K., and TOMITA, M. (1992). Identification of the bactericidal domain of lactoferrin. *Biochim. Biophys.*

Acta **1121**:130–136.

BELLAMY, W., WAKABAYASHI, H., TAKASE, M., KAWASE, K., SHIMAMURA, S., and TOMITA, M. (1993). Killing of *Candida albicans* by lactoferricin B, a potent antimicrobial peptide derived from the N-terminal region of bovine lactoferrin. *Med. Microbiol. Immunol. (Berl)* **182**:97–105.

BERNHISEL-BROADBENT, J. DINTZIS, H.M., DINTZIS, R., and SAMPSON, H.A. (1994). Allergenicity and genicity of chicken egg ovomucoid (Gal d III) compared with ovalbumin (Gal d I) in children with egg allergy and in mice. *J. Allergy Clin. Immunol.* **93**:1047–1059.

BESTER, B.H., and LOMBARD, S.H. (1990). Influence of lysozyme on selected bacteria associated with Gouda cheese. *J. Food Prot.* **53**: 306–311.

BJÖRCK, L. (1978). Antibacterial effect of the lactoperoxidase system on psychrotrophic bacteria in milk. *J. Dairy Res.* **45**:109–118.

BJÖRCK, L., ROSEN, C., MARSHALL, V., and REITER, B. (1975). Antibacterial activity of the lactoperoxidase system in milk against pseudomonads and other gram-negative bacteria. *Appl. Microbiol.* **30**:199–204.

BLANC, B., and ISLIKER, H.C. (1961). Isolement et caracterisation de la proteine rouge siderophile dulait maternel: la lactoferrine. *Helv. Physiol. Pharmacol. Acta* **19**:C-13–14.

BOARD, R.G.. (1995). Natural antimicrobials from animals. In *New methods of food protection preservation*, ed. G.W. Gould, pp 41–57, London, Blackie Academic and Professional.

BOGSTEDT, A.K., JOHANSEN, K., HATTA, H., KIM, M., CASSWALL, T., SVENSSON, L. and HAMMERSTROM, L. (1996). Passive immunity against diarrhea. *Acta Paediatr.* **85**:125–128.

BOSTWICK, E.F., STEIJNS, J., and BRAUN, S. (2000). Lactoglobulins. In *Natural Food Antimicrobial Systems*, ed. A.S. Naidu, pp. 133–158. Boca Raton: CRC Press.

BROWN, R.W., and MICKELSON, M.N. (1979). Lactoperoxidase, thiocyanate, and free cystine in bovine mammary secretions in early dry period and at the start of lactation and their effect on Streptococcus agalactiae growth. *Am. J. Vet. Res.* **40**:250–255.

BUTLER, J. E. (1994). Passive immunity and Ig diversity. In: *Indigenous Antimicrobial Agents of Milk – Recent Developments*, Brussels, International Dairy Federation Press, 14–47.

CISANI, G., VARALDO, P.E., INGIANNI, A., POMPEI, R., and SATTA, G. (1984). Inhibition of herpes simplex virus-induced cytopathic effect by modified hen egg-white lysozymes. *Curr. Microbiol.* **10**: 35–40.

CUNNINGHAM, F.E., PROCTOR, V.A., and GRETSCH, S.J. (1991). Egg-white lysozyme as a food preservative: a review. *Poultr. Sci. J.* **47**: 141–163.

DALMASTRI, C., VALENTI, P., VISCA, P., VITTORIOSO, P., and ORSI, N. (1988). Enhanced antimicrobial activity of lactoferrin by binding to the bacterial surface. *Microbiologica* **11**: 225–230.

DELL'AQUA, E., BRUZZESE, T., and VAN DEN HEUVEL, H.H. (1989). *US Patent* 4,810,508.

DÜRING, K., PORSCH, P., MAHN, A., BRINKMANN, O., and GIEFFERS, W. (1999). The non-enzymatic microbicidal activity of lysozymes. *FEBS Lett* **449**: 93–100.

EBINA, T. (1996). Prophylaxis of rotavirus gastroenteritis using immunoglobulin. *Arch. Virol.* **12**:217–223.

EHRLICH, P. (1892). Der Immunitot durch Verebung und Zeugung, *Z. Hyg. Infekt Krankh* **12**:183–203.

EISAI KK (1971). Ikura and suzuko by lysozyme treatment of inferior roe. Japanese Patent

JP 7103732.

EISAI KK (1972). Preserving raw fishery product by saline lysozyme solution. Japanese Patent JP 7203732.

EISAI KK (1973). Preservation of foods-by adding a bacteriolytic enzyme and glycine. Japanese Patent JP 73016613.

EISAI KK. (1980). Storing Japanese sake with addition of lysozyme or its salt and β-glucopyranose aeodehydrogenase. Japanese Patent JP 80035105.

ERDEI, J., FORSGREN, A., and NAIDU, A.S. (1994). Lactoferrin binds to porins OmpF and OmpC in *Escherichia coli*. *Infect. Immun* **62**:1236–1240.

ETTERADOSSI, N., TOQUIN, D., ABBASSI, H., RIVALLAN, G., COTTE, J.P., and GUTTLET, M. (1997). Passive protection of specific pathogen free chicks against infectious bursal disease by in-ovo injection of semi-purified egg yolk antiviral immunoglobulins. *J. Vet. Med.* **B44**:371–383.

GADO, I., ERDEI, J., LASZLO, V.G., PASZTI, J., CZIROK, E., KONTROHR, T., TOTH, I., FORSGREN, A., and NAIDU, A.S. (1991). Correlation between human lactoferrin binding and colicin susceptibility in *Escherichia coli*. *Antimicrob. Agents Chemother.* **35**: 2538–2543.

GEBRAUX, V.V., MONAMY, C., and BERTRAND, A. (1997). Use of lysozyme to inhibit malolactic fermentation and to stabilize wine after malolactic fermentation. *Am. J. Enol. Vit.* **48**: 49–54.

GOTHEFORS, L., and MARKLUND, S. (1975). Lactoperoxidase activity in human milk and in saliva of newborn infants. *Infect. Immunol.* **11**:1210–1215.

HAAS, M., KLUPPEL, A.C.A., WARTNA, E.S., MOOLENAAR, F., MEIJER, D.K.F., DE JONG, P.E., and ZEUW, D. (1997). Drug-targeting to the kidney: renal delivery and degradation of a naproxen-lysozyme conjugate *in vivo*. *Kidney Intnl.* **52**: 1693–1699.

HAMON, C.B., and KLEBANOFF, S.J. (1973). A peroxidase-mediated, *Streptococcus mitis*-dependent antimicrobial system in saliva. *J. Exp. Med.* **137**:438–450.

HATTA, H., MABE, K., KIM, M., YAMAMOTO, T., GUITERREZ, M.A., and MIYAZAKI, T. (1994). Prevention of fish disease using egg yolk antibody. In Egg uses and Processing Technologies – New Developments, ed. Sim, J.S., Nakai, S., pp. 241. CAB International.

HATTA, H., TSUDA, K., OZEKI, M., KIM, M., YAMAMOTO, T., OTAKE, S., HIRASAWA, M., KATZ, J., KATZ, J., CHILDERS, N.K., and MICHALEK, S.M. (1997). Passive immunization against dental plaque formation in humans: Effect of a mouth rinse containing egg yolk antibodies (IgY) specific to *Streptococcus mutans*. *Caries Res.* **31**:268–274.

HIGGINS, D.A. (1975). Physical and chemical properties of fowl immunoglobulins. *Net. Bull.* **45**:139–154.

HOEK, K.S., MILNE, J.M., GRIEVE, P.A., DIONYSIUS, D.A., and SMITH, R. (1997). Antibacterial activity in bovine lactoferrin-derived peptides. *Antimicrob. Agents Chemother.* **41**:54–59.

HOFFMAN, D.R. (1982). Immunochemical identification of the allergens in egg white. *J. Allergy. Clin. Immunol.* **71**:481–486.

HOGG, D.M., and JAGO, G.R. (1970). The antibacterial action of lactoperoxidase: Nature of the bacterial inhibitor. *Biochem J.* **117**:779–790.

HOOGENDOORN, H., and MOORER, W.R. (1973). Lactoperoxidase in the prevention of plaque accumulation, gingivitis and dental caries. I. Effect on oral streptococci and lactobacilli. *Odontol. Revy* **24**:355–366.

HÖÖK, M., SWITALSKI, L., WADSTRÖM, T., and LINDBERG, M. (1989). Interactions of pathogenic microorganisms with fibronectin. In *Fibronectin*, ed. D.F. Mosher, pp. 295–308. New York: Academic Press.

HUGHEY, V.L., WILGER, P.A., and JOHNSON, E.A. (1989). Antibacterial activity of hen egg white lysozyme against *Listeria monocytogenes* Scott A in foods. *Appl. Envir. Microbiol.* **55**: 631–638.

IBRAHIM, H. R. (1997). Isolation and characterization of the bactericidal domain of ovotransferrin. *Nippon Nogeikagaku Kaishi* **71**:39–41.

IBRAHIM, H.R. (2000). Ovotransferrin. In *Natural Food Antimicrobial Systems*, ed. A.S. Naidu, pp. 211–226. Boca Raton: CRC Press.

IBRAHIM, H. R., IWAMORI, E., SUGIMOTO, Y. and AOKI, T. (1998). Identification of a distinct antibacterial domain within the N-lobe of ovotransferrin. *Biochim. Biophys. Acta*, **1401**:289–303.

IKEMORI, Y., KUROKI, M., PERALTA, R.C., YOKOYAMA, H., and KODAMA, Y. (1992). Protection of neonatal calves against fatal enteric colibacillosis by administration of egg yolk powder from hens immunized with K99–pilated enterotoxigenic *E.coli*. *Am. J. Vet. Res.* **53**:2005–2008.

IKEMORI, Y., OHTA, M., UMEDA, K., ICATLO, F.C., KUROKI, M., YOKOYAMA, H., and KODAMA, Y. (1997). Passive protection of neonatal calves against bovine coronavirus-induced diarrhea by administration of egg yolk or colostrum antibody powder. *Vet. Microbiol.* **58**:105–111.

IMBERECHTS, H., DEPREZ, P., VAN DRIESSCHE, E., and POHL, P. (1997). Chicken egg yolk antibodies against F18 fimbriae of *Escherichia coli* inhibit shedding of F18 positive *E.coli* by experimentally infected pigs. *Vet. Microbiol.* **54**:329–341.

IWAMOTO, Y., NAKAMURA, R., WATANABE, T., and TSUNEMITSU, A. (1972). Heterogeneity of peroxidase related to antibacterial activity in human parotid saliva. *J. Dent. Res.* **51**:503–508.

JENSENIUS, J.C., ANDERSEN, I., HAU, J., CRONE, M., and KOCH, C. (1981). Eggs: conveniently packaged antibodies. Methods for purification of yolk IgG. *J. Immunol. Methods* **46**:63–68.

JIN, L.Z., BAIDOO, S.K., MARQUARDT, R.R., and FROHLICH, A.A. (1998). In vitro inhibition of adhesion of enterotoxigenic *Escherichia coli* K88 to piglet intestinal mucus by egg-yolk antibodies. *FEMS Immunol. Med. Microbiol.* **21**:313–321.

JOHNSON, E.A., DELL'ACQUA, E., and FERRARI, L. (1991). Process for bacterial decontamination of vegetable foods. US Patent 5,019,411.

KAWAI, A., and TANAKA, K. (1997). *Reverse mutation test of Monl-01 using bacteria*. Gotemba Laboratory. Bozo Research Center Inc., Setagaya-ku, Tokyo, Japan.

KIERMEIER, F., and KUHLMANN, H. (1972). Lactoperoxidase activity in human and in cows' milk. Compartive studies. *Munch. Med. Wochenshr.* **114**:2144–2146.

KLEBANOFF, S. J. (1967). Iodination of bacteria: a bactericidal mechanism. *J. Exp. Med.* **126**:1063–1078.

KLEBANOFF, S.J., CLEM, W. H., and LUEBKE, R. G. (1966). The peroxidase-thiocyanate-hydrogen peroxide antimicrobial system. *Biochim. Biophys. Acta.* **117**:63–72.

KOBAYASHI, K., and HIRAI, J. (1980). Studies on subunit components of chicken polymeric Immunoglobulins. *J. Immunol.* **124**:1695–1704.

KUROKI, M., IKEMORI, Y., YOKOYAMA, H., PERALTA, R.C., ICATLO, F.C., and KODAMA, Y. (1993). Passive protection against bovine rotavirus-induced diarrhea in murine model by specific immunoglobulins from chicken egg yolk. *Vet. Microbiol.* **37**:135–146.

KUROKI, M., OHTA, M., IKEMORI, Y., PERALTA, R.C., YOKOYAMA, H., and KODAMA, Y. (1994). Passive protection against bovine rotavirus in calves by specific immunoglobulins from chicken egg yolk. *Arch. Virol.* **138**:143–148.

KUROKI, M., OHTA, M., IKEMORI, Y., ICATLO, F.C., KOBAYASHI, C., YOKOYAMA, H., and KODAMA,

Y. (1997). Field evaluation of chicken egg yolk immunoglobulins specific for bovine rotavirus in neonatal calves. *Arch. Virol.* **142**:843–851.

LASSITER, M.O. (1990). Ethylenediaminetetraacetate (EDTA): Influence on the antimicrobial activity of human lactoferrin against *Escherichia coli* and *Streptococcus mutans*. Doctoral Thesis. Department of Pathology, School of Medicine. Atlanta, Georgia: Emory University.

LEHRER, R.I. (1969). Antifungal effects of peroxidase systems. *J. Bacteriol.* **99**:361–365.

LESLIE, G.A., and CLEM, L.W. (1969). Phylogeny of immunoglobulin structure and function. III. Immunoglobulins of the chicken. *J. Exp. Med.* **130**:1337–1352.

LOCHNER, J. E., FRIEDMAN, R. L., BIGLEY, R. H., and IGLEWSKI, B. H., (1983). Effect of oxygen-dependent antimicrobial systems on *Legionella pneumophila*. *Infect. Immun.* **39**:487–489.

LOSCH. Y., SCHRANNER, I., WANKE, R., and JURGENS, L. (1986). The chicken egg, an antibody source. *J. Vet. Med.* **B33**:609–619.

LOSSO, J.N., NAKAI, S., and CHARTER, E.A. (2000). Lysozyme. In *Natural Food Antimicrobial Systems*, ed. A.S. Naidu, pp. 17–102. Boca Raton: CRC Press.

LOUIE, M., DE AZAVEDO, J.C,, HANDELSMAN, M.Y., CLARK, C.G., ALLY, B., DYTOC, M., SHERMAN, P., and BRUNTON, J. (1993). Expression and characterization of the *eaeA* gene product of *Escherichia coli* serotype O157:H7. *Infect. Immunol.* **61**:4085–4092.

MAKKI, F., and DURANCE, T.D. (1996). Thermal inactivation of lysozyme as influenced by pH, sucrose, and sodium chloride and inactivation and preservative effect in beer. *Food Res. Int.* **29**: 635–645.

MARSHALL, V.M., and REITER, B. (1980). Comparison of the antibacterial activity of the hypothiocyanite anion towards *Streptococcus lactis* and *Escherichia coli. J Gen. Microbiol.* **120**:513–516

MASSON, P.L., HEREMANS, J.F., PRIGNOT, J.J., and WAUTERS, G. (1966). Immunohistochemical localization and bacteriostatic properties of an iron-binding protein from bronchial mucus. *Thorax* **21**:538–544.

MEIJER, D.K.F. , MOLEMA, G., MOOLENAAR, F., ZEEUW, D., and SWART, P.J. (1996). (Glyco)-protein drug carriers with an intrinsic therapeutic activity: The concept of dual targeting. *J. Control. Release* **39**:163–172.

MICKELSON, M.N. (1979). Antibacterial action of lactoperoxidase-thiocyanate-hydrogen peroxide on Streptococcus agalactiae. *Appl. Environ. Microbiol.* **38**:821–826.

NAIDU, A.S. (2000a). Lactoferrin. In *Natural Food Antimicrobial Systems*, ed. A.S. Naidu, pp. 17–102. Boca Raton: CRC Press.

NAIDU, A.S. (2000b). Lactoperoxidase. In *Natural Food Antimicrobial Systems*, ed. A.S. Naidu, pp. 17–102. Boca Raton: CRC Press.

NAIDU, A.S. (2001). Immobilized lactoferrin antimicrobial agents and the use. US Patent 6,172,040.B1.

NAIDU, A.S. (2002). Activated lactoferrin – a new approach to meat safety. *Food Technol.* 56:40–45.

NAIDU, A.S., ANDERSSON, M., and FORSGREN, A. (1992). Identification of a human lactoferrin-binding protein in *Staphylococcus aureus*. *J. Med. Microbiol.* **36**:177–183.

NAIDU, A.S., and ARNOLD, R.R. (1994). Lactoferrin interaction with salmonellae potentiates antibiotic susceptibility *in vitro*. *Diagn. Microbiol. Infect. Dis.* **20**:69–75.

NAIDU, A.S., and ARNOLD, R.R. (1997). Influence of lactoferrin on host-microbe interactions. In. *Lactoferrin – Interactions and Biological Functions*, ed. T.W. Hutchens, and B. Lonnerdal, pp. 259–275. Totowa, NJ: Humana Press.

NAIDU, A.S., and BIDLACK, W.R. (1998). Milk lactoferrin – Natural microbial blocking agent (MBA) for food safety. *Environ. Nutr. Interact.* **2**:35–50.

NAIDU, A.S., MIEDZOBRODZKI, J., MUSSER, J.M., ROSDAHL, V.T., HEDSTROM, S.A., and FORSGREN, A. (1991a). Human lactoferrin binding in clinical isolates of *Staphylococcus aureus*. *J. Med. Microbiol.* **34**:323–328.

NAIDU, S.S., ERDEI, J., CZIROK, E., KALFAS, S., GADO, I., THOREN, A., FORSGREN, A., and NAIDU, A.S. (1991b). Specific binding of lactoferrin to *Escherichia coli* isolated from human intestinal infections. *APMIS.* **99**:1142–1150.

NAIDU, S.S., SVENSSON, U., KISHORE, A.R., and NAIDU, A.S. (1993). Relationship between antibacterial activity and porin binding of lactoferrin in *Escherichia coli* and *Salmonella* Typhimurium. *Antimicrob. Agents Chemother.* **37**:240–245.

NISHIMURA, N. (1991). *Single dose oral toxicity study of monl-01 and monl-02 in rats.* Gotemba Laboratory, Bozo Research Center Inc., Setagaya-ku, Tokyo, Japan.

NISHIMURA, N. (1997). *Four-week oral repeated dose toxicity study of monl-01 in rats.* Gotemba Laboratory. Bozo Research Center Inc., Setagaya-ku, Tokyo, Japan.

NISHIMURA, N. (2000). Thirteen-week oral repeated dose toxicity study of monl-01 in rats. *Food Chem. Tox.* **38**:489–498.

O'FARRELLY, C., BRANTON, D., and WANKE, C.A. (1992). Oral ingestion of egg yolk immunoglobulin from hens immunized with an enterotoxigenic *Escherichia coli* strain prevents diarrhea in rabbits challenged with the same strain. *Infect. Immun.* **60**:2593–2597.

OPTIZ, H.M., EL-BEGEARMI, M., FLEGG, P., and BEANE, D. (1993). Effectiveness of five feed additives in chicks infected with *Salmonella enteritidis* phage type 13A. *J. Appl. Poultry Res.* **2**:147–153.

OSBORNE, T.B., and CAMPBELL, G.F. (1900). The protein constituents of egg white. *J. Am. Chem. Soc.* **22**:422–426.

OTAKE, S., NISHIHARA, Y., MAKIMURA, M., HATTA, H., KIM, M., YAMAMOTO, T., and HIRASAWA, M. (1991). Protection of rats against caries by passive immunization with hen-egg-yolk antibody (IgY). *J. Dent. Res.* **70**:162–166.

OZPINAR, H., ERHARD, M.H., AYTUG, N., OZPINAR, A., BAKLACI, C., KARAMUPTUOGLU, S., HOFMANN, A., and LOSCH, U. (1996). Dose-dependant effects of specific egg-yolk antibodies on diarrhea of newborn calves. *Preventive Vet. Med.* **27**:67–73.

PADGETT, T., HAN, I.Y., and DAWSON, P.I. (1998). Incorporation of food-grade antimicrobial compounds into biodegradable packaging films. *J. Food Protection* **61**:1330–1335.

PARRY, S.H., and ROOKE, D.M. (1985). Adhesins and colonization factors of *Escherichia coli*. In *The virulence of* Escherichia coli – *Reviews and Methods*, ed. M. Sussman, pp. 79–155. London: Academic Press.

PAULSSON, M.A., SVENSSON, U., KISHORE, A.R., and NAIDU, A.S. (1993). Thermal behavior of bovine lactoferrin in water and its relation to bacterial interaction and antibacterial activity. *J. Dairy Sci.* **76**:3711–3720.

PAYNE, K.D., OLIVER, S.P., and DAVIDSON, P.M. (1994). Comparison of EDTA and apo-lactoferrin with lysozyme on the growth of foodborne pathogenic and spoilage bacteria. *J. Food Prot.* **57**:62–65.

PERALTA, R.C., YOKOYAMA, H., IKEMORI, Y., KUROKI, M., and KODAMA, Y. (1994). Passive immunization against experimental salmonellosis in mice by orally administered hen egg yolk antibodies specific for 14–kDa fimbriae of *Salmonella* Enteritidis. *J. Med. Microbiol.* **41**:29–35.

PICHLER, W.J., and CAMPI, P. (1992). Allergy to lysozyme/egg white-containing vaginal suppositories. *Annal. Allergy* **69**:521–525.

POPPER, L., and KNORR, D. (1997). Inactivation of yeast and filamentous fungi by the lactoperoxidase-hydrogen peroxide-thiocyanate-system. *Nahrung* **41**:29–33.

PORTMAN, A., GATE, Y., and AUCLAIR, J. (1962). *Sixteenth International Dairy Congress.* Vol. B. Copenhagen, Denmark.

PRUITT, K.M., and REITER, B. (1985). Biochemistry of peroxidase system – antimicrobial effects. In *The Peroxidase System – Chemistry and Biological Significance*, ed. K.M. Pruitt, and J.O. Tenuvuo, pp. 143–177. New York: Marcel Dekker.

PURDY, M.A., TENOVUO, J., PRUITT, K.M., and WHITE, W.E. JR. (1983). Effect of growth phase and cell envelope structure on susceptibility of *Salmonella typhimurium* to the lactoperoxidase-thiocyanate-hydrogen peroxide system. *Infect. Immun.* **39**:1187–1195

REDDY, N.R., ROTH, S.M., EIGEL, W.N., and PIERSON, M.D. (1988), Foods and food ingredients for prevention of diarrheal disease in children in developing countries, *J. Food Prot.* **51**:66–75.

REITER, B. (1978). Review of nonspecific antimicrobial factors in colostrum. *Ann. Rech. Vet.* **9**:205–224.

REITER, B. (1985). The lactoperoxidase system of bovine milk. In *The Peroxidase System – Chemistry and Biological Significance*, ed. K.M. Pruitt, and J.O. Tenuvuo, pp. 123–141. New York: Marcel Dekker.

REITER, B., and ORAM, J.D. (1967). Bacterial inhibitors in milk and other biological fluids. *Nature* **216**:328–330.

ROSE, M.E., and ORLANS, E. (1981). Immunoglobulin in the egg, embryo and young chick. *Devel. Comp. Immunol.* **5**:15–20, 371–375.

ROSE, M.E., ORLANS, E., and BUTTRESS, N. (1974). Immunoglobulin classes in the hen's egg: their segregation in yolk and white. *Eur. J. Immunol.* **4**:521–523.

SHIMIZU, M., FITZSIMMONS, R.C., and NAKAI, S. (1988). Anti-*E.coli* Immunoglobulin Y isolated from egg yolk of immunized chickens as potential food ingredient. *J. Food Sci.* **53**:1360–1366.

SIM, J.S., SUNWOO, H.H., and LEE, E.N. (2000). Ovoglobulin IgY. In *Natural Food Antimicrobial Systems*, ed. A.S. Naidu, pp. 227–252. Boca Raton: CRC Press.

SØRENSEN, M., and SØRENSEN. S. P. L.. (1939). The proteins in whey. *CR. Trav. Lab. Carlsberg.* **23**:55–99.

STEPHENS, S., HARKNESS, R.A., and COCKLE, S.M. (1979). Lactoperoxidase activity in guinea-pig milk and saliva: correlation in milk of lactoperoxidase with bactericidal activity against *Escherichia coli*. *Br. J. Exp. Pathol.* **60**:252–258.

SUNWOO, H.H., NAKANO, T., DIXON, W.T., and SIM, J.S. (1996). Immune responses in chickens against lipopolysaccharide of *Escherichia coli* and *Salmonella* Typhimurium. *Poultry Sci.* **75**:342–345.

TENOVUO, J., MAKINEN, K. K., and SIEVERS, G. (1983). Antibacterial effect of lactoperoxidase and myeloperoxidase against *Bacillus cereus*. *Antimicrob. Agents Chemother.* **27**:96–101.

TIGYI, Z., KISHORE, A.R., MAELAND, J.A., FORSGREN, A., and NAIDU, A.S. (1992). Lactoferrin-binding proteins in *Shigella flexneri*. *Infect Immunol.* **60**:2619–2626.

TOMITA, M., BELLAMY, W., TAKASE, M., YAMAUCHI, K., WAKABAYASHI, H., and KAWASE, K. (1991). Potent antibacterial peptides generated by pepsin digestion of bovine lactoferrin. *J. Dairy Sci.* **74**:4137–4142.

TOMITA, M., TAKASE, M., WAKABAYASHI, H., and BELLAMY, W. (1994). Antimicrobial peptides of lactoferrin. *Adv. Exp. Med. Biol.* **357**: 209–218.

TRANTER, H.S., and BOARD, R.G. (1982). The antimicrobial defense of avian eggs:

Biological perspective and chemical basis. *J. Appl. Biochem.* **4**:295–338.

TRANTER, H. S., and BOARD, R. G. (1984). The influence of incubation temperature and pH on the antimicrobial properties of hen egg albumen. *J. Appl. Bacteriol.* **56**:53–61.

URISU, A., ANDO, H., MORITA, Y., WADA, E., YASAKI, T., YAMADA, K., KOMADA, K., TORI, S., GOTO, M., and WAKAMATSU, T. (1997). Allergenic activity of heated and ovomucoid-depleted egg white. *J. Allergy Clin. Immunol.* **100**:171–176.

VALENTI, P., ANTONINI, G., FANELLI, M.R., ORSI, N., and ANTONINI, E. (1982). Antibacterial activity of matrix-bound ovotransferrin. *Antimicrob. Agents Chemother.* **21**:840–841.

VALENTI, P., ANTONINI, G., VON-HUNOLSTEIN, C., VISCA, P., ORSI, N., and ANTONINI, E. (1983a). Studies of the antimicrobial activity of ovotransferrin. *Int. J. Tissue React.* **5**:97–105.

VALENTI, P., VISCA, P., VON-HUNOLSTEIN, C., ANTONINI, G., CREO, C., and ORSI, N. (1983b). Importance of the presence of metals in the antibacterial activity of ovotransferrin. *Ann. Ist Super Sanita* **18**:471–472.

VALENTI, P., ANTONINI, G., VON-HUNOLSTEIN, C., VISCA, P., ORSI, N., and ANTONINI, E. (1984). Studies of the antimicrobial activity of ovotransferrin. *J. Appl. Bacteriol.* **56**:53–61.

VALENTI, P., VISCA, P., ANTONINI, G., and ORSI, N. (1985). Antifungal activity of ovotransferrin towards genus Candida. *Mycopathologia* **89**:169–175.

VALENTI, P., ANTONINI, G., VON-HUNOLSTEIN, C., VISCA, P., ORSI, N., and ANTONINI, E. (1987). Studies of the antimicrobial activity of ovotransferrin. *Int. J. Tissue React.* **5**:97–105.

VENKITNARAYANAN, K.S., ZHAO, T., and DOYLE, M.P. (1999). Antibacterial effect of lactoferricin B on *Escherichia coli* O157:H7 in ground beef. *J. Food Prot.* **62**:747–750.

VILLA, A. (1996). Controlling malolactic fermentation (MLF) with lysozyme: applications and results. Wine Spoilage Microbiology Conference. California State University, Fresno, CA., 8 March 1996.

WASSERFALL, F.E., and TEUBER, M. (1979). Action of egg-white lysozyme on *Clostridium tyrobutyricum*. *Appl. Environ. Microbiol.* **38**:197–199.

WEINER, C., PAN, Q., HURTIF, M., BOREN, T., BOSTWICK, E., and HAMMERSTROM, L. (1999). Passive immunity against human pathogens using bovine antibodies. *Clin. Exp. Immunol.* **116**:193–205.

WIEDEMANN, V., LINCKH, E., KUHLMANN, R., SCHMIDT, P., and LOSCH, U. (1991). Chicken egg antibodies for prophylaxis and therapy of infectious intestinal diseases. *J. Vet. Med.* **B38**:283–291.

YAMADA, T., YAMADA, M., SASAMOTO, K., NAKAMURA, H., MISHIMA, T., YASUEDA, H., SHIDA, T., and IKIKURA, Y. (1993). Specific IgE antibody titers to hen's egg white lysozyme in allergic children to egg. *Japanese J. Allerg.* **42**:136–141.

YAMAUCHI, K., TOMITA, M., GIEHL, T.J., and ELLISON, R.T. (1993). Antibacterial activity of lactoferrin and a pepsin-derived lactoferrin peptide fragment. *Infect. Immun.* **61**:719–728.

YOKOYAMA, H., PERALTA, R.C., DIAZ, R., SENDO, S., IKEMORI, Y., and KODAMA, Y. (1992). Passive protective effect of chicken egg yolk Immunoglobulins against experimental enterotoxigenic *Escherichia coli*. *Infect. Immunol.* **60**:998–1007.

YOKOYAMA, H., PERALTA, R.C., SENDO, S., IKEMORI, Y., and KODAMA, Y. (1993). Detection of passage and absorption of chicken egg yolk immunoglobulins in the gastrointestinal tract of pigs by use of enzyme-linked immunosorbent assay and fluorescent antibody testing. *Am. J. Vet. Res.* **54**:867–872.

YOKOYAMA, H., UMEDA, K., PERALTA, R.C., HASHI, T., ICATLO, F.C., KUROKI, M., IKEMORI, Y., and KODAMA, Y. (1998a). Oral passive immunization against experimental salmonellosis in mice using chicken egg yolk antibodies specific for *Salmonella enteritidis* and *S. typhimurium*. *Vaccine* **16**:388–393.

YOKOYAMA, H., PERALTA, R.C., UMEDA, K., HASHI, T., ICATLO, F.C., KUROKI, M., IKEMORI, Y., and KODAMA, Y. (1998b). Prevention of fatal salmonellosis in neonatal calves, using orally administered chicken egg yolk salmonella-specific antibodies. *Am. J. Vet. Res.* **59**:416–420.

YOLKEN, R.H., LOSONSKY, G.A., VONDERFECT, S., LEISTER, F., and WEE, S-B. (1985). Antibody to human rotavirus in cow's milk. *N. Engl. J. Med.* **312**:605–610.

YOLKEN, R.H., LEISTER, F., WEE, S.B., MISKUFF, R., and VONDERFECHT, S. (1988). Antibodies to rotaviruses in chickens' eggs: a potential source of antiviral immunoglobulins suitable for human consumption. *Pediatrics.* **81**:291–295.

YOSHIKO, S.K., SHIBATA, K., YUN, S.S., YUKIKO, H.K., YAMAGUCHI, K., and KUMAGAI, S. (1996). Immune functions of immunoglobulin Y isolated from egg yolk of hens immunized with various infectious bacteria. *Biosci. Biotech. Biochem.* **60**:886–888.

ZELDOW, B.J. (1963). Studies on the antibacterial action of human saliva. III. Cofactor requirements of a lactobacillus bactericidin. *J. Immunol.* **90**:12–16.

ZUNIGA, A., YOKIAYAMA, H., ALBICKER-RIPPINGER, P., EGGENBERGER, E., and BERTSCHINGER, H.U. (1997). Reduced intestinal colonization with F18-positive enterotoxigenic *Escherichia coli* in weaned pigs fed chicken egg antibody against the fimbriae. *FEMS Immunol. Med. Microbiol.* **18**:153–161.

8

Chitosan: new food preservative or laboratory curiosity?

S. Roller, Thames Valley University, UK

8.1 Introduction

Chitosan consists of polymeric 1,4-linked 2-amino-2-deoxy-β-D-glucose and is made commercially by alkaline deacetylation of chitin. The latter is an abundant constituent of crustacean shells and fungi (including edible mushrooms). The physical, chemical and biological properties of chitin, chitosan and their derivatives have been studied extensively and hundreds of documents per year are published on them. Numerous applications have been proposed for chitosan including heavy metal chelation in wastewater treatment, as a hypocholesterolemic agent, and as an adjunct in obesity treatment. For the reader interested in learning more about the general properties and applications of chitin and chitosan, the reviews by Winterowd and Sandford (1995) and Shahidi *et al.* (1999) provide a useful starting point.

The inhibitory and biocidal action of chitosan has been reported widely in the scientific literature mainly on the basis of *in vitro* trials against individual microorganisms. Such studies have led to suggestions that chitosan could be used as a novel food preservative. However, most foods consist of mixtures of different compounds (carbohydrates, proteins, fats, minerals, vitamins, salts, organic acids, etc.), many of which interact with chitosan leading to loss or enhancement of antimicrobial activity. Furthermore, all foods harbour complex microbial ecosystems and a selective agent such as chitosan could have profound implications for both the shelf-life and safety of those foods. Other factors such as solubility, stability under conventional food processing operations, regulatory status and cost also play a role in the commercialisation of any novel food preservative, including chitosan. In this chapter, some of the reported antimicrobial properties of chitosan are critically reviewed and discussed in relation to the applicability of chitosan as a natural preservative in real food systems. Recent data on the uses of chitosan in combination with reduced concentrations of traditional food preservatives such as benzoate and sulfite are also presented.

8.2 The antimicrobial properties of chitosan *in vitro*

As shown in Table 8.1, minimum inhibitory concentrations (MICs) ranging from 0.0004% to 1% of chitosan against moulds in laboratory media have been reported, depending on the type of chitosan used (e.g. soluble or particulate), the conditions of testing (e.g. pH, temperature, medium) and the target organism. Some phytopathogenic fungi such as *Fusarium solani* appear remarkably sensitive to the antimicrobial action of chitosan with MICs as low as 0.0004–0.0018% in liquid media (Kendra and Hadwiger, 1984). By contrast, several common spoilage agents of processed foods (*Aspergillus flavus*, *Cladosporium cladosporioides* and *Penicillium aurantiogriseum*) have been reported as resistant to chitosan at levels of 1% and above in solid laboratory media (Roller and Covill, 1999).

Many spoilage yeasts are also inhibited and/or inactivated by chitosan, as shown in Table 8.1. MICs of 0.01–0.05% for six chitosan salts (glutamate, malate, lactate, chloride, formate and acetate) against 17 out of 18 yeasts isolated from spoiled fruit-based beverages have been reported (Sagoo, 2003). However, *Torulaspora delbrueckii*, a spoilage yeast isolated from fruit juice concentrate, was resistant to chitosan at levels as high as 1%.

A typical inactivation curve for 0.025 and 0.05% chitosan against a foodborne yeast at ambient temperatures is shown in Fig. 8.1 using *Saccharomycodes ludwigii*, a strain isolated from spoiled cider, as an example. In the first 2–3 min of exposure to chitosan, the reduction in microbial numbers obeys first-order kinetics of the kind commonly associated with thermal processing. However, this is typically followed by a much slower rate of decline in numbers. The shapes of the survivor curves shown in Fig. 8.1 are very similar in the presence of the two concentrations of chitosan (0.025 and 0.05%) tested, although the absolute number of surviving organisms is greater at the lower concentration, suggesting some concentration dependency. The 'tailing' in the survivor curves shown in Fig. 8.1 has been reported for numerous yeasts and bacteria exposed to chitosan (Papineau *et al.*, 1991; Roller, 2002; Sagoo *et al.*, 2002a,b; Sagoo, 2003). Non-linear semi-logarithmic survival curves with upward concavity are not unusual in non-physical preservation processes (Peleg, 2000). Clearly, a surviving population of microorganisms following treatment with an antimicrobial compound or process has implications for both the shelf-life and safety of foods as it poses a risk of re-growth during storage.

Several reports in the literature suggest that chitosan inhibits growth of foodborne bacteria including pathogens at concentrations as low as 0.01% (summarised in Table 8.2). However, contradictory results showing resistance of

Table 8.1 Inhibition of mould and yeast growth by chitosan from crab in laboratory media (at 22–25 °C, unless otherwise indicated)

Source/form of chitosan	Microorganisms and media	Growth inhibition reported and concentration of chitosan	Reference
Water-washed, particles ≤5 μm	46 phytopathogenic fungi in liquid media in microtitre plates	MICs 0.01–0.025% for 10 strains MICs 0.05–0.1% for 22 strains No inhibition of 14 strains incl. *Botrytis* sp. & *Mucor* sp. at 0.1%	Allan and Hadwiger (1979)
Water-washed, particles ≤5 μm	*Saccharomyces* sp. in liquid media in microtitre plates	MIC 0.05%	Allan and Hadwiger (1979)
Snow crab	2 strains of *F. solani* in liquid media in microtitre plates	MICs: 0.0004–0.0018%	Kendra and Hadwiger (1984)
66% deacetylated, particles ≤50 μm	7 phytopathogenic fungi including *F. solani* on agar media	Radial growth reduced by 20–93% (depending on strain, medium and degree of deacetylation) at 1% chitosan	Stossel and Leuba (1984)
MW>300 000, DP>1500	*F. oxysporum* on agar media	Radial growth reduced by 33% at 0.25% chitosan and by 60% at 0.5% chitosan	Leuba and Stossel (1986)
Chitosan lactate, MW>400 000 95% deacetylated	18 phytopathogenic fungi on agar media	Radial growth of 16 strains reduced by 16-51% at 0.1% chitosan No inhibition of 2 strains	Hirano and Nagao (1989)
72% deacetylated	*Asp. niger* in liquid media	41% less mycelium at 0.5% chitosan	Fang *et al.* (1994)
Chitosan glutamate, 75–85% deacetylated	7 foodborne moulds on agar media:	MICs (%)	Roller and Covill (1999)
	M. racemosus	0.2–0.5 (Radial growth reduced by 73% at 0.2% chitosan on agar at pH 4.5, 18 °C)	
	3 *Byssochlamys* spp.	0.5	
	Asp. flavus, Clad. cladosporioides & *Pen. aurantiogriseum*	>1	
Chitosan glutamate, 75–85% deacetylated	*Z. bailii* in liquid media	No inhibition at 0.5% chitosan	Roller and Covill (2000)
6 chitosan salts	17 out of 18 foodborne yeasts, including strains of *Z. bailii*, on agar media	MIC 0.01–0.05%	Sagoo (2003)
	T. delbrueckii on agar media	MIC > 1%	

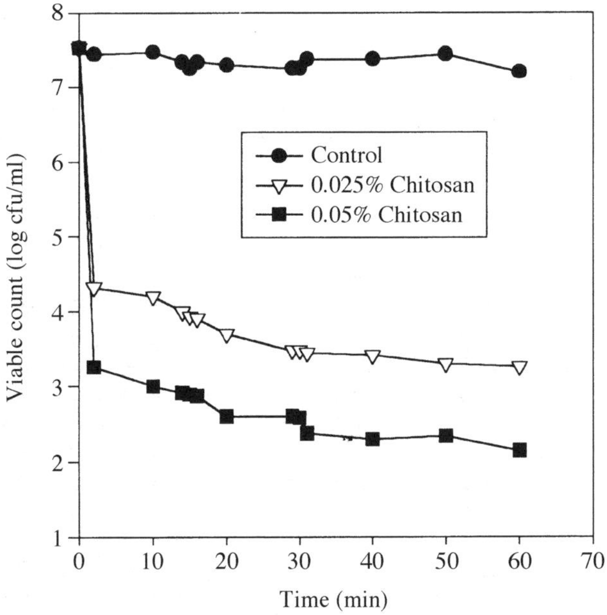

Fig. 8.1 Survival of *Saccharomycodes ludwigii* exposed to chitosan glutamate in saline at room temperature, pH 6.2. Adapted from Sagoo *et al.* (2002b).

some foodborne pathogens such as *Salmonella* Enteritidis to the inhibitory action of chitosan at concentrations as high as 1% in laboratory media have also been reported (Roller and Covill, 2000). Inconsistencies in the extent of biocidal activity of chitosan against both bacteria and yeasts in water, buffers and laboratory media have also been reported and these are summarised in Table 8.3. Some of the inconsistencies in results are related to the use of selective media to enumerate microorganisms, which can lead to overestimates of chitosan efficacy. For example, it has been demonstrated that the recovery of *Escherichia coli* O157:H7 treated with 0.5% chitosan glutamate for 10 min was reduced by up to 4 log CFU/ml when plated out on a selective medium (VRBG Agar containing Crystal Violet, Neutral Red, bile salts and glucose) compared with a general-purpose complex medium such as plate count agar (PCA) containing no added inhibitors (Helander *et al.*, 2001; Roller, 2002). The large variability in the raw material (chitosan) used in different studies is probably also responsible for the large variations in results (Strand *et al.*, 2001).

In spite of contradictory reports in the literature, some general conclusions can be drawn from *in vitro* studies. Firstly, concentrations of chitosan required to inactivate or inhibit growth of bacteria appear to be at least one order of magnitude higher than those needed to show an effect on yeasts and moulds (Tables 8.1–8.3). Secondly, Gram-negative bacteria are generally more resistant than Gram-positive organisms. Thirdly, the antimicrobial activity of chitosan is more pronounced at pH values below its pK_a of 6.3 and at temperatures above 20 °C (Tables 8.1–8.3). And finally, the molecular weight, degree of

Table 8.2 Growth inhibition of bacteria by chitosan in laboratory media

Source/form of chitosan	Temp. (ºC)	Microorganisms and media	Growth inhibition reported and concentration of chitosan	Reference
Unspecified	37	*E. coli*, *L. monocytogenes*, *S.* Typhimurium, *Staph. aureus* and *Y. enterocolitica* in nutrient broth, pH 6.5 for up to 8 days, survivors plated on selective media	0.5–2.5% chitosan inhibited growth initially but by Days 2–8, numbers were the same as in control (At pH 5.5, bacteria were inactivated, see Table 8.3)	Wang (1992)
Crab, chitosan extracted in-house, adjusted to pH 5.6 with 0.1 M acetic acid	37	*S.* Typhimurium, *Staph. aureus*, *E. coli*, *P. fluorescens, Pr. vulgaris* and *B. cereus*	0.005–0.02% chitosan inhibited growth of most bacteria for 5 days	Simpson *et al.* (1997)
Shrimp, 69% deacetylated	37	11 bacteria in nutrient broth, pH 6 for 2 days	MICs 0.01–0.02% for 7 bacteria including *E. coli*, *P. aeruginosa*, *Staph. aureus* and *L. monocytogenes* MICs 0.1–0.2% for *B. cereus* and *A. hydrophila* MIC > 0.2% for *S.* Typhimurium	Chen *et al.* (1998)
Crab, chitosan glutamate, 75–85% deacetylated	25	*S.* Enteritidis and *Lac. fructivorans* in liquid media for 48 h	*S.* Enteritidis not inhibited at 0.5%; *Lac fructivorans* inhibited at 0.01% for 48 h	Roller and Covill (2000)
Unspecified, 89% deacetylated	37	*E. coli* in liquid media for 32 h	0.5% chitosan inhibited growth for 32 h	Jeon and Kim (2000)
Crab	28	*A hydrophila* in liquid media for 24 h	Growth inhibited at 0.02–0.04% for 24 h	Taha and Swailam (2002)

Table 8.3 The biocidal properties of chitosan in water, buffers and liquid laboratory media against bacteria and yeasts.

Source/form of chitosan	Conditions	Microorganisms	Conc. (%)	Reduction in viable count (log CFU/ml)	Reference
Crab, chitosan lactate, boiled and dialysed against water	Distilled water, pH approx 5.6, 37 °C for 60 min	*Staph. aureus*	0.02 0.1	4 6	Papineau *et al.* (1991)
		E. coli	0.02 0.1 0.5	4 3–3.5 1.5	
		Sac. cerevisiae	0.1	7	
Crab, chitosan lactate and glutamate, MW ~1 million	Sodium phosphate buffer (100 mM), pH 5.8, 32 °C for 60 min	9 bacteria	0.2	*E. coli* & *Staph. aureus*: 1 *S.* Typhimurium: 2 Other bacteria: 4–5	Sudarshan *et al.* (1992)
Crab, chitosan glutamate, MW ~ 1 million	Sodium phosphate buffer (100 mM), pH 5.8, 32 °C for 60 min	*Staph. aureus*	0.01 0.2–0.5	2 0.5	Sudarshan *et al.* (1992)
Unspecified	Nutrient broth, pH 5.5 adjusted with acetic acid, 30 °C for up to 8 days, survivors plated on selective media	*Staph. aureus* *E. coli* *Y. enterocolitica* *L. monocytogenes* *S.* Typhimurium	0.5–1.5	3.5–5.4 log in 1–3 days (At pH 6.5, no inactivation but growth was inhibited, see Table 8.2)	Wang (1992)
Shrimp, prepared in-house from chitin, 98 % deacetylated	Phosphate buffer (5 mM), pH 6, 37 °C for up to 24 h	*E. coli*	0.015	2–3.5 in 5 h	Tsai and Su (1999)
				3.5–4.5 in 5 h @ 4 and 15 °C >6.5 in 1–5 h @ 25 and 37 °C No inactivation at pH 9 2, 4, 5 and 6 log in 7 h @ pH 8, 7, 6 and 5, respectively	

Table 8.3 Continued

Source/form of chitosan	Conditions	Microorganisms	Conc. (%)	Reduction in viable count (log CFU/ml)	Reference
Crab, chitosan chloride, 83% deacetylated	Saline, pH 6.2, room temperature	*S.* Tyhpimurium (log phase cells)	2.0	None in first h; 3 log reduction after 18 h (stationary phase cells resistant)	Helander *et al.* (2001)
Crab, chitosan glutamate, 79% deacetylated	Saline, pH 7.2, 20 °C	*L. monocytogenes, S.* Typhimurium and *Staph. aureus*	2.0	0.8–1.7 (planktonic) 1.8–2.4 (dried films)	Knowles and Roller (2001)
Crab, chitosan glutamate, 79% deacetylated	Saline, pH 6.2, room temperature	*Sac. exiguus* *Scd. ludwigii* *T. delbrueckii*	0.05	3.5–5.0 in 1–2 h 2.5 in 1–2 h 0 in 2 h	Sagoo *et al.* (2002a)
Crab, chitosan glutamate, 79% deacetylated	Saline, pH 6.2, room temperature	*Scd. ludwigii*	0.025 0.05	4.0 5 +	Sagoo *et al.* (2002b)
		L. innocua	0.25 0.5	2.5 5.5	
		Lac. viridescens	0.25 0.5	2.5 4.0	
Crab	Liquid media, pH 6 and 7, exposed to chitosan for 24 h at 28 °C	*A. hydrophila*	0.04 0.10	3 (pH 6), 1 (pH 7) 5 (pH 6), 2 (pH 7)	Taha and Swailam (2002)

deacetylation and ionic strength of the medium also play a role in the flocculation and hence antimicrobial efficacy of chitosan (Helander *et al.*, 2001; Rhoades and Roller, 2000; Shahidi *et al.*, 1999; Strand *et al.*, 2001).

8.3 The antimicrobial properties of chitosan in foods and beverages

The loss of antimicrobial efficacy of natural antimicrobial compounds in complex food systems (compared to *in vitro* systems) is a recurring theme in this book and chitosan is no different (Knowles and Roller, 2001; Roller and Covill, 1999, 2000; Sagoo *et al.*, 2002a). For example, the concentrations of chitosan required to achieve the same reduction in viable numbers of *Scd. ludwigii* as those seen in Fig. 8.1 (in saline solution) were up to one order of magnitude higher in apple juice (Roller and Covill, 1999). Similarly, levels of chitosan sufficient to inactivate 7 log CFU/ml of *Sac. cerevisiae* in distilled water achieved a reduction of only 2 log CFU/ml in apple juice (Boguslawski *et al.*, 1990; Papineau *et al.*, 1991). Nevertheless, some successful applications of chitosan in foods have been reported.

Chitosan dips, sprays and films have been shown to prevent post-harvest spoilage of many fruits and vegetables by *Botrytis cinerea* and other spoilage fungi and are being investigated as replacements for synthetic fungicides. For a detailed review of chitosan applications in post-harvest storage, the reader is referred to Chapter 10.

The applications of chitosan in minimally processed fruits, acidified sauces, dips, fruit juices, salads and meat products are summarised in Table 8.4. Notably, Rhoades and Roller (2000) reported that much higher concentrations of chitosan were required to inhibit microbial growth in houmous (0.5%) than in apple and elderflower juice (0.03%) in spite of the lower initial microbial load in houmous (approx 2 log CFU/g) than in juice (approx 4 log CFU/ml). It is possible that the more particulate nature of the dip compared with the juice restricted mass transfer of the relatively large, polymeric chitosan molecules, thereby reducing the chances of contact with microorganisms in the houmous. In addition, other factors such as protein content, pH (4.2 in houmous and 3.3 in juice), fat content (25.6% in houmous and 0% in juice) and type of acid present (citric in houmous and malic in juice) may have contributed to the differences in antimicrobial activity.

The influence of food ingredients on the antimicrobial action of chitosan in complex food systems has not been studied systematically. Some reports suggest that salt and protein may interact with chitosan in a way that makes it unavailable to exert its antimicrobial action against microorganisms. For example, the presence of 2% salt or 10% whey protein isolate (at pH 6) in liquid laboratory media prevented growth inhibition of *Candida lambica* by 0.005 and 0.01% chitosan at 7 °C but the presence of 10% sunflower oil emulsified with Tween 80 had no effect on its inhibitory properties (Devlieghere *et al.*, 2002).

Table 8.4 The antimicrobial properties of chitosan in foods and beverages

Food or beverage	Source/form of chitosan	Conc. (%)	Temp (C)	Microorganism/s	Result	Reference
Apple juice	Chitosan glutamate	0.01	21	*Sac. cerevisiae* *Lac. plantarum*	Yeasts reduced from 6 to 4 log CFU/ml; bacteria reduced from 7 to 2 log CFU/ml	Boguslawski *et al.* (1990)
Apple juice, pH 3.4	Crab, chitosan glutamate, 75–85% deacetylated	0.01–0.5	25	8 spoilage yeasts	Initial reduction up to 3 log CFU/ml, extended lag phase, yeast-free conditions maintained for 14–32 days, depending on strain and chitosan concentration	Roller and Covill (1999)
Apple and elderflower juice, pasteurised	Chitosan hydrochloride, 89% deacetylated	0.03	7	Total viable counts, yeasts and lactic acid bacteria	Yeasts eliminated; growth of total and lactic acid bacteria slower than in control for 13 days	Rhoades and Roller (2000)
Beef patties	Unspecified	1.0	4	Total viable counts, lactic acid bacteria etc.	After 10 days counts reduced by 1–2 log CFU/ml; no inhibition at 30 °C or at 0.2 and 0.5% chitosan at 4 or 30 °C	Darmadji and Izumimoto (1994)
Houmous (chickpea dip)	Chitosan hydrochloride, 89% deacetylated	0.03 0.1 0.5	7	Total viable counts, yeasts and lactic acid bacteria	Lower concentrations had no effect; at 0.5% chitosan, bacteria but not yeast inhibited	Rhoades and Roller (2000)
Kumquat, candied	Crab, 72% deacetylated	0.6	Ambient	*Asp. niger*	Sugar content was reduced from 65 to 62 Brix and mould-free shelf-life mainained at 65 days	Fang *et al.* (1994)
Mayonnaise, pH adjusted to 4.5 with acetic acid or lemon juice	Crab, chitosan glutamate, 75–85% deacetylated	0.3	5 and 25	*S.* Enteritidis, *Z. bailii*, *Lac. fructivorans*	Reduction of 5 log CFU/g of *Lac. fructivorans*, 1–2 log CFU/g of *Z. bailii* and no effect on *S.* Enteritidis in mayo acidified with acetic acid. No effects in mayo acidified with lemon juice	Roller and Covill (2000)

Mayonnaise-based shrimp salad	Crab, chitosan glutamate, 75–85% deacetylated, used to coat shrimps	Approx. 9 mg/g shrimp	5 and 25	Total viable counts, yeasts and lactic acid bacteria	At 5 °C, chitosan reduced flora from 8 to 4 log CFU/g for 4 weeks. No effect at 25 °C	Roller and Covill (2000)
Oysters	Shrimp, 69% deacetylated	0.2	5	Total counts, coliforms, *Vibrio* spp.	Total and coliform counts 1–2.5 log CFU/g lower in chitosan-treated than in control oysters for up to 14 days; detection of *Vibrio* spp. delayed from 8 to 10 days	Chen *et al.* (1998)
Pork sausages, raw	Crab, chitosan glutamate	0.6	4	Total viable counts, lactic acid bacteria, yeasts, enterics	No effect on its own but in combination with 170 ppm sulfite, all counts up to 3 log lower than in controls	Roller *et al.* (2002)
Pork, raw minced	Crab, chitosan glutamate	0.3 0.6	4	Total viable counts, yeasts and lactic acid bacteria	Counts consistently 2–3 logs lower than in control for up to 18 days	Sagoo *et al.* (2002a)
Sausage meat	MW 30 000 and 120 000	0.35 & 0.5	30	Total viable count	Initial inactivation of 2 log CFU/g followed by rapid growth within 1 day of storage	Youn *et al.* (1999)
Shrimp, raw	Crab, prepared in-house, adjusted to pH 5.6 with Na acetate	1 & 2	4–7	Total viable counts	Up to 1 log CFU/g fewer organisms in shrimp treated by dipping in 2% chitosan solutions	Simpson *et al.* (1997)

Jumaa *et al.* (2002) have suggested that lipids such as castor oil and medium chain triglycerides were less detrimental to chitosan action in emulsions than in aqueous formulations for pharmaceutical applications. However, work undertaken in dairy emulsions made up of fats, proteins and sucrose (designed to mimic chilled dairy desserts) has shown that the presence of 9, 26 and 32% fat prevented the inhibitory action of 0.25% chitosan against *Listeria innocua* (inoculated at 5–6 log CFU/g); the organisms in the chitosan-treated models grew at the same rate as those in the untreated controls at 4, 20 and 30 °C. By contrast, in the presence of 1% fat and 0.25% chitosan, *L. innocua* died off over a period of 15 days at all three temperatures studied (O'Mahony *et al.*, 2001; Roller, unpublished results). These contradictory results demonstrate that more research is needed on the ingredient interactions that may occur with chitosan in a food system before reliable predictions about preservative efficacy can be made.

There is considerable experimental evidence that chitosan may be useful as a preservative for raw minced meat products. Darmadji and Izumimoto (1994) have reported that the total microbial counts in beef patties containing 1.0% chitosan were 1–2 log CFU/g lower than in the controls after 10 days of storage at 4 °C. However, at lower chitosan concentrations (0.5 and 0.2%) and higher storage temperature (30 °C), no inhibition of microbial growth was observed. The number of viable organisms present in the meat used by Darmadji and Izumimoto (1994) at the start of their experiment was high (>10^7 CFU/g). This level of initial microbial contamination would not be acceptable for minced meat sold as such under the European Community Hygiene Directive (EC Council Directive, 1994). This specifies that the aerobic mesophilic count in minced meat is unacceptable if it exceeds 5×10^6 CFU/g in one or more samples out of every five taken. It is possible that chitosan addition could have been more effective in Darmadji and Izumimoto's study had lower initial populations been present in the meat. More recently, Youn *et al.* (1999) reported inactivation of up to 2 log CFU/g of the total microbial flora in sausage meat immediately following addition of 0.35 or 0.5% chitosan; however, after storage at 30 °C for just one day, total counts in the chitosan-treated sausage were the same as in the control. It is possible that better control of microbial proliferation could have been achieved at lower storage temperatures.

In raw minced pork, chitosan at 0.3% and 0.6% inhibited growth of the total microflora by up to 3 log CFU/g for 18 days at 4 °C, compared with the untreated control, as shown in Fig. 8.2. Similar inhibition was also reported for yeasts and lactic acid bacteria in minced pork (Sagoo *et al.*, 2002b).

Total viable counts, lactic acid bacteria, yeasts and moulds were reduced by approximately 1–3 log CFU/g in both skinless and standard raw pork sausages dipped in chitosan solution (1%) and stored at 7 °C for 18 days. The total counts of sausages treated with 1% chitosan were maintained below the maximum acceptable levels (10^7 CFU/g) specified by microbiological criteria for fresh meat products (Anon., 1999) for up to 15 days, thereby doubling the shelf-life of the product at chill temperatures. Notably, when both types of sausages were

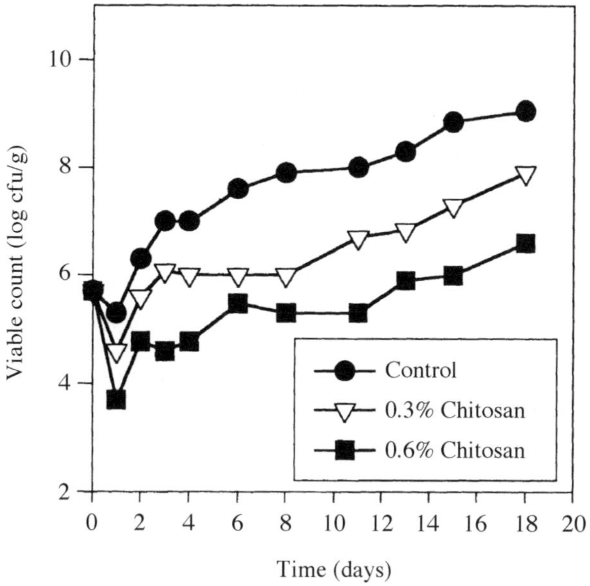

Fig. 8.2 Total viable count in raw minced pork with and without chitosan at 4 °C. Adapted from Sagoo *et al.* (2002b).

stored at 15 and 25 °C, no differences in counts between the treated and untreated samples were observed, indicating that chitosan films would not protect the product from microbial growth at abuse temperatures (Sagoo *et al.*, 2002b). Chitosan has also been reported as an effective adjunct to sulfite in raw sausages, as described in more detail in Section 8.4 (Roller *et al.*, 2002).

In contrast to the above studies, no inhibition of *Serratia liquefaciens* or *Lac. sake* was reported on cooked ham coated with chitosan films and stored at 4 °C under vacuum for 21 days, unless the chitosan film was impregnated with acetic acid, propionic acid and/or cinnamaldehyde (Ouattara *et al.*, 2000). It is possible that some of the components of ham may have interacted with chitosan to render it inactive against microorganisms. For example, there is evidence from *in vitro* studies that salt can have a detrimental effect on chitosan action (Tsai and Su, 1999; Strand *et al.*, 2001).

8.4 Chitosan in combination with traditional preservatives

Antimicrobial compounds in nature rarely function in isolation; combination systems are far more common (Dillon and Board, 1994). Similarly, it is unlikely that a single food preservative, whether old or new, synthetic or natural, will provide the 'magic bullet' for the prevention of food spoilage and poisoning. Furthermore, recent work using a range of membrane-probing techniques, electron microscopy and lipopolysaccharide analysis has shown that chitosan acts on Gram-negative bacteria by disrupting the barrier properties of the outer membrane (Helander *et al.*, 2001). Therefore, like many other membrane-permeabilising agents such as ethylenediaminetetraacetic acid (EDTA), chitosan may be useful as an adjunct to other food preservatives.

The use of chitosan in combination with traditional acidifying agents, including acetic acid in mayonnaise and mayonnaise-based shrimp salads, has been investigated. Roller and Covill (2000) have shown that 0.3% chitosan glutamate was biocidal to *Lactobacillus fructivorans*, a typical spoilage agent of mayonnaise, when the latter was acidified with acetic acid but not with lemon juice (citric acid).

Common food preservatives such as nitrite, sodium benzoate and sodium metabisulfite have a long history of safe use (Russell and Gould, 2003) but reports of occasional allergic reactions in sensitive individuals, the formation of potentially carcinogenic by-products (e.g. nitrosamines from nitrite) and increasingly vociferous consumer demands to eliminate synthetic food additives from foods have created a need for alternative solutions. Reductions in the permitted levels of use of many traditional preservatives are considered to be the likely outcome of the legislation reviews that are in progress in many countries (Stratford *et al.*, 2000). Resistance of common spoilage organisms such as *Zygosaccharomyces bailii* and *Saccharomycodes ludwigii* to levels of benzoate that are well above those permitted legally or acceptable organoleptically has added urgency to the search for alternatives to many preservatives. In the following paragraphs, the potential for using chitosan in combination with reduced concentrations of nitrite, benzoate and/or sulfite is reviewed (Youn *et al.*, 1999; Roller, 2002; Sagoo *et al.*, 2002a).

8.4.1 Chitosan and nitrite

Youn *et al.* (1999) investigated the possible substitution of 50% of the nitrite in cured Korean sausages with 0.2% chitosan. The total bacterial counts in sausages containing 150 ppm nitrite alone were just over 2 log CFU/g for four days at 30 °C and then increased to 3 log CFU/g by Day 7 of storage. By contrast, the viable numbers in sausages containing the chitosan + nitrite combination were below the detection limit of the plate counting method from Day 0 to Day 4 and then increased to 3 log CFU/g by Day 7 at 30 °C. The bacterial counts in sausages containing chitosan only (0.2, 0.35 and 0.5%) were reduced immediately upon preparation of the product but rose rapidly to equal numbers in the controls within the first one or two days of storage at 30 °C. The red colour, emulsion stability and hardness of the sausages were maintained with the combination treatment provided that the chitosan had a molecular weight of 30 000 and not 120 000. The use of the higher molecular weight chitosan led to detrimental sensorial changes.

8.4.2 Chitosan and benzoate

The combination of benzoate with chitosan has been studied *in vitro* (Sagoo *et al.*, 2002b). Three spoilage yeasts, *Sac. exiguus*, *Scd. ludwigii* and *T. delbrueckii*, were suspended in 0.05 and 0.005% chitosan glutamate and 0.025% sodium benzoate, alone or in combination, in 0.9% saline solutions at pH 6.2 and 4.5. Survivor curves were constructed from viable counts determined periodically for up to 120 min. Chitosan at 0.005% almost doubled the extent of death caused by 0.025% benzoate alone from about 1–2 log to about 2–4 log CFU/ml, depending on pH and target organism. It was concluded that chitosan (0.005%) and 0.025% sodium benzoate acted synergistically against spoilage yeasts in saline solutions.

The potentiation of the lethal effect of antimicrobial agents by exposure of organisms to combinations of several sub-lethal stresses has been reported previously. For example, mild heat treatment enhanced the inhibitory properties of 0.1% benzoate against *Bacillus stearothermophilus*, while alkali or enzyme-based cleaning agents increased the sporicidal activity of peroxygen disinfectants against *Bacillus cereus* (Lopez *et al.*, 1998; Langsrud *et al.*, 2000).

The mode of interaction between chitosan and benzoate is unclear and warrants further study. As a polymeric macromolecule, chitosan is unable to pass through the cell wall of yeasts or bacteria and, unlike benzoate, to interact directly with intracellular components. Below its pK_a of 6.3, chitosan is highly polycationic and can be expected to interact with the predominantly anionic components of the microbial surface, which may, in turn, facilitate the energy-dissipating action of benzoate. The results obtained *in vitro* need to be confirmed in selected foods before a realistic assessment could be made about the utility of the benzoate + chitosan combination.

8.4.3 Chitosan and sulfite

Sulfite cannot be completely eliminated from fresh pork sausages (when sold raw) not only because of its preservative efficacy against Gram-negative bacteria but also because it gives the sausages a pink colour, an essential sensory feature of this product. Therefore, a study was undertaken in which the effects of low levels of sulfite used singly or in combination with chitosan on the total microbial flora, yeasts, lactic acid bacteria, enteric bacteria, pH, appearance and odour were investigated. An example of the results is illustrated in Fig. 8.3. The results show that fresh pork sausages containing a combination of chitosan (0.6%) and low level of sulfite (170 ppm) had the lowest total viable count for the longest period of storage at 4 °C. In addition, these sausages had fewer unfavourable odours and better visual acceptability as judged by a trained sensory panel. As shown in Fig. 8.3, the maximum count (as recommended by microbiological criteria; Anon., 1999) was exceeded in all the sausages preserved using single compounds (chitosan or sulfite) by Day 6 of storage at 4 °C. Notably, these criteria were not exceeded in the sausages containing the chitosan + sulfite combination until Day 18. It was calculated that the shelf-life

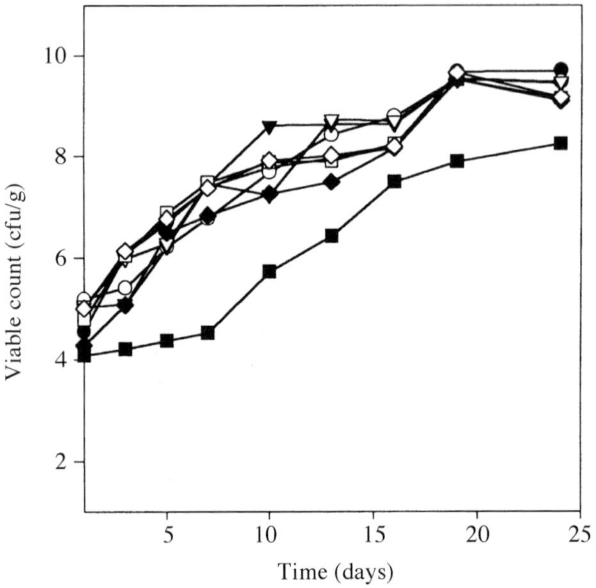

Fig. 8.3 Total viable count in fresh pork sausages preserved with a combination of 0.6% chitosan and 170 ppm sulfite (■) and a range of chitosan and sulfite concentrations used individually at 4 °C. Adapted from Roller *et al.* (2002).

of the sausages at 4 °C was extended nearly three times by the presence of the chitosan + sulfite combination (Roller *et al.*, 2002).

The mechanism of preservation by the chitosan + sulfite combination may be two-fold. Firstly, chitosan may have acted as a 'slow-release' agent for sulfite, thereby preventing its premature degradation or irreversible binding with other ingredients in the sausage. With a pK_a of 6.3/6.5, approximately one-half of the chitosan present in the sausage on the day of manufacture would have been highly positively charged and could have interacted readily but reversibly with the negatively charged sulfite moieties. Secondly, any unbound chitosan may have selectively inhibited the yeast flora, thereby preventing acetaldehyde production and the consequent inactivation of sulfite by binding. The efficacy of the chitosan + sulfite combination could therefore be explained on the basis of protection from sulfite breakdown by chitosan (Roller *et al.*, 2002).

8.5 Conclusions and future prospects

Chitosan is not currently on the list of permitted food preservatives in Europe. However, in the USA, it has recently been self-affirmed as Generally Recognised As Safe (GRAS; Anon., 2001). The widespread commercial use of chitosan in foods is currently confined to Japan (Hirano, 1989, 1997). Although dietary studies on test animals suggest that the toxicity of chitosan is low when ingested at a level of <5% of the total diet (Winterowd and Sandford, 1995; Koide, 1998), the potential for adverse consequences during long-term feeding in humans has not been investigated thoroughly. There is a need to ensure that long-term intake of low levels of chitosan would not selectively inactivate beneficial species of the gut flora and/or impede the absorption of fat-soluble vitamins and minerals. It is worth noting that the concentrations of chitosan needed to obtain antimicrobial activity in foods are several orders of magnitude lower than the doses currently used in some weight- and cholesterol-reducing programmes in Europe (Krotkiewski *et al.*, 2001). Nevertheless, the regulatory framework for food additives is becoming increasingly restrictive (see Chapter 15) and it is likely that toxicological studies would be required before chitosan or any other novel preservative (natural or synthetic) would be permitted for use in foods.

Chitosan efficacy as a preservative in foods is not readily predicted on the basis of *in vitro* work alone. Extensive applications work in real food systems is essential for full exploitation of the antimicrobial properties of this interesting polymer. Combinations of chitosan with reduced concentrations of traditional food preservatives hold particular promise for the future.

8.6 References

ALLAN, C.R. and HADWIGER, L.A. (1979) The fungicidal effect of chitosan on fungi of varying cell wall composition. *Exp. Mycol.* **3**, 285–287.

ANON. (1999) *Development and Use of Microbiological Criteria for Foods*. Institute of Food Science and Technology, London: IFST.

ANON. (2001) US Food and Drug Administration, Center for Food Safety and Applied Nutrition. Office of Premarket Approval. GRAS Notices Received in 2001. http:/vm.cfsan.fda.gov

BOGUSLAWSKI, S., BUNZEIT, M. and KNORR, D. (1990) Effects of chitosan treatment on clarity and microbial counts of apple juice. *Zeit. Lebensmitt. Tech.* **41**, 42–44.

CHEN, C-S., LIAU, W-Y. and TSAI, G-J. (1998) Antibacterial effects of *N*-sulfonated and N-sulfobenzoyl chitosan and application in oyster preservation. *J. Food Prot.* **61**, 1124–1128.

DARMADJI, P. and IZUMIMOTO, M. (1994) Effect of chitosan in meat preservation. *Meat Sci.* **38**, 243–254.

DEVLIEGHERE, F., VERMEULEN, A. and DEBEVERE, J. (2002) Chitosan as a food preservative: possibilities and limitations. *Proceedings of Food Micro 2002, 18th Intl. ICFMH Symp.*, Lillehammer, Norway. 18–23 August 2002. pp. 83–86.

DILLON, V.M. and BOARD, R.G. (1994) *Natural Antimicrobial Systems and Food Preservation*. CAB International, Wallingford.

EC COUNCIL DIRECTIVE 94/65/EC (1994) Requirements for the production and placing on the market of minced meat and meat preparations. *Official J. Eur. Communities* 31.12.94, No. **L368**, 10–31.

FANG, S.W., LI, C.F. and SHIH, D.Y.C. (1994) Antifungal activity of chitosan and its preservative effect on low-sugar candied kumquat. *J. Food Prot.* **56**, 136–140.

HELANDER, I.M., NURMIAHO-LASSILA, E.-L., AHVENAINEN, R., RHOADES, J. and ROLLER, S.

(2001) Chitosan disrupts the barrier properties of the outer membrane of Gram-negative bacteria. *Intl. J. Food Microbiol.* **71**, 235–244.

HIRANO, S. (1989) Production and application of chitin and chitosan in Japan. In: *Chitin and Chitosan: Sources, Chemistry, Biochemistry, Physical Properties and Applications*, Eds. G. Skjak-Braek, T. Anthonsen and P. Sandford, Elsevier Applied Science, New York. pp. 37–40.

HIRANO, S. (1997) Applications of chitin and chitosan in the ecological and environmental fields. In: *Applications of Chitin and Chitosan*. Ed. M.F.A. Goosen. Technomic Publishing Co., Inc., Lancaster, PA, USA, pp. 31–54.

HIRANO, S. and NAGAO, N. (1989) Effects of chitosan, pectic acid, lysozyme and chitinase on the growth of several phytopathogens. *Agric. Biol. Chem.* **53**, 3065–3066.

JEON, Y.-J. and KIM, S-K. (2000) Production of chitooligosaccharides using an ultrafiltration membrane reactor and their antibacterial activity. *Carb. Polymers* **41**, 133–141.

JUMAA, M., FURKERT, F.H. and MULLER, B.W. (2002) A new lipid emulsion formulation with high antimicrobial efficacy using chitosan. *Eur. J. Pharma. Biopharma* **53**, 115–123.

KENDRA, D.F. and HADWIGER, L.A. (1984) Characterisation of the smallest chitosan oligomer that is maximally antifungal to *Fusarium solani* and elicits pisatin formation in *Pisum sativum*. *Exp. Mycol.* **8**, 276–281.

KNOWLES, J.R. and ROLLER, S. (2001) Efficacy of chitosan, carvacrol and a hydrogen-peroxide-based biocide against foodborne microorganisms in suspension and adhered to stainless steel. *J. Food. Prot.* **64**, 1542–1548.

KOIDE, S.S. (1998) Chitin-chitosan: properties, benefits and risks. *Nutrition Res.* **18**, 1091–1101.

KROTKIEWSKI, M., ZAHORSKA-MARKIEWICZ, B., OLSZANECKA-GLINIANOWICZ, M. and ZURAKOWSKI, A. (2001) Chitosan as an adjunct to dietary treatment of obesity. *4th Conference of the European Chitin Society*, 6–10 May 2001, Senigallia, Italy. Abstracts Booklet, p. S24.

LANGSRUD, S., BAARDSEN, B. and SUNDHEIM, G. (2000) Potentiation of the lethal effect of peroxygen on *Bacillus cereus* spores by alkali and enzyme wash. *Int. J. Food Microbiol.* **56**, 81–86.

LEUBA, J.L. and STOSSEL, P. (1986) Chitosan and other poyamines: Antifungal activity and interaction with biological membranes. In: *Chitin in Nature and Technology*, Eds. R. Muzzarelli, C. Jeuniaux and G. W. Gooday, Plenum Press, New York, pp. 215–222.

LOPEZ, M., MARTINEZ, S., GONZALES, J., MARTIN, R. and BERNARDO, A. (1998) Sensitization of thermally injured spores of *Bacillus stearothermophilus* to sodium benzoate and potassium sorbate. *Letts. Appl. Microbiol.* **27**, 331–335.

O'MOHONY, T., REKHIF, N., CAVADINI, C. and FITZGERALD, G.F. (2001) The application of a fermented food ingredient containing variacin, a novel antimicrobial produced by *Kocuria varians*, to control the growth of *Bacillus cereus* in chilled dairy products. *J. Appl. Microbiol.* **90**, 106–114.

OUATTARA, B., SIMARD, R.E., PIETTE, G., BÉGIN, A. and HOLLEY, R. (2000) Inhibition of surface spoilage bacteria in processed meats by application of antimicrobial films prepared with chitosan. *Int. J. Food Microbiol.* **62**, 139–148.

PAPINEAU, A.M., HOOVER, D.G., KNORR, D. and FARKAS, D.F. (1991) Antimicrobial effect of water-soluble chitosans with high hydrostatic pressure. *Food Biotechnol.* **5**, 45–57.

PELEG, M. (2000) Microbial survival curves – the reality of flat 'shoulders' and absolute thermal death times. *Food Res. Int.* **33**, 531–538.

RHOADES, J. and ROLLER, S. (2000) Antimicrobial actions of degraded and native chitosan against spoilage organisms in laboratory media and foods. *Appl. Env. Microbiol.* **66**, 80–86.

ROLLER, S. (2002) Chitosan: A novel food preservative? In: *Chitosan in Pharmacy and Chemistry*, Eds. R.A.A. Muzzarelli and C. Muzzarelli, Atec Edizioni, Grottammare, Italy, pp. 177–182.

ROLLER, S. and COVILL, N. (1999) The antifungal properties of chitosan in laboratory media and apple juice. *Int. J. Food Microbiol.* **47**, 67–77.

ROLLER, S. and COVILL, N. (2000) The antimicrobial properties of chitosan in mayonnaise and mayonnaise-based shrimp salads. *J. Food Prot.* **63**(2), 202–209.

ROLLER, S., SAGOO, S., BOARD, R., O'MOHONY, T., CAPLICE, E., FITZGERALD, G., FOGDEN, M., OWEN, M. and FLETCHER, H. (2002) Novel combinations of chitosan, carnocin and sulphite for the preservation of chilled pork sausages. *Meat Sci.* **62**, 165–177.

RUSSELL, N.J. and GOULD, G.W. (2003) *Food Preservatives*, 2nd Edn, Kluwer Academic/ Plenum Publishers, New York.

SAGOO, S. (2003) The antimicrobial action of chitosan. PhD Thesis, South Bank University, UK.

SAGOO, S., BOARD, R. and ROLLER, S. (2002a) Chitosan potentiates the antimicrobial action of sodium benzoate on spoilage yeasts. *Lett in Appl. Microbiol.* **34**, 168–172.

SAGOO, S., BOARD, R. and ROLLER, S. (2002b) Chitosan inhibits growth of spoilage microorganisms in chilled pork products. *Food Microbiol.* **19**, 175–182.

SHAHIDI, F. and ARACHCHI, J.K.V. and JEON, Y-J (1999) Food applications of chitin and chitosan. *Trends in Food Science and Technology* **10**, 37–51.

SIMPSON, B.K., GAGNÉ, N., ASHIE, I.N.A. and NOROOZI, E. (1997) Utilization of chitosan for preservation of raw shrimp. *Food Biotechnol.* **11**, 25–44.

STOSSEL, P. and LEUBA, J.L. (1984) Effect of chitosan, chitin and some amino-sugars on growth of various soilborne phytopathogenic fungi. *Phytopath. Z.* **111**, 82–86.

STRAND, S.P., VANDVIK, M.S., VÅRUM, K.M. and ØSTGAARD, K. (2001) Screening of chitosans and conditions for bacterial flocculation. *Biomacromolecules* **2**, 126–133.

STRATFORD, M., HOFMAN, P.D. and COLE, M.B. (2000) Fruit juices, fruit drinks and soft drinks. In *The Microbiological Safety and Quality of Food*, Vol. I, Eds. B.M. Lund, T.C. Baird-Parker and G.W. Gould, Aspen Publishers Inc., Gaithersburg, pp. 836–869.

SUDARSHAN, N.R., HOOVER, D.G. and KNORR, D.G. (1992) Antibacterial action of chitosan. *Food Biotechnol.* **6** (3), 257–272.

TAHA, S.M.A. and SWAILAM, M.H. (2002) Antibacterial activity of chitosan against *Aeromonas hydrophila. Nahrung/Food* **46**, 337–340.

TSAI, G-J. and SU, W-H. (1999) Antibacterial activity of shrimp chitosan against *Escherichia coli. J. Food Prot.* **62**, 239–243.

WANG, G.-H. (1992) Inhibition and inactivation of five species of food-borne pathogens by chitosan. *J. Food Prot.* **55**, 916–919.

WINTEROWD, J.G. and SANDFORD, P. (1995) Chitin and chitosan. In: *Food Polysaccharides and Their Applications*, Ed. A.M. Stephen, Marcel Dekker, Inc., New York. pp. 441–462.

YOUN, S.-K., PARK, S.-M., KIM, Y.-J. and AHN, D.-H. (1999) Effect on storage property and quality in meat sausage by added chitosan. *J. Chitin Chitosan* **4**, 189–194.

9

Antimicrobials from herbs and spices

G.-J. E. Nychas and P. N. Skandamis, Agricultural University of Athens, Greece and C. C. Tassou, National Agricultural Research Foundation, Greece

9.1 Introduction

Herbs and spices are used widely in the food industry as flavours and fragrances. However, they also exhibit useful antimicrobial and antioxidant properties. Many plant-derived antimicrobial compounds have a wide spectrum of activity against bacteria, fungi and mycobacteria and this has led to suggestions that they could be used as natural preservatives in foods (Farag *et al.*, 1989; Ramadan *et al.*, 1972; Conner and Beuchat, 1984a,b; Galli *et al.*, 1985). Although more than 1300 plants have been reported as potential sources of antimicrobial agents (Wilkins and Board, 1989), such alternative compounds have not been sufficiently exploited in foods to date.

In this chapter, the antimicrobial compounds from herbs and spices are reviewed and the barriers to the adoption of these substances as food preservatives are discussed. The mode of action of essential oils and the potential for development of resistance are also discussed. The focus is primarily on bacteria and fungi in prepared foods, while in the following chapter (Chapter 10), greater attention is given to the control of postharvest pathogens (mainly fungi) of edible crops.

9.2 Barriers to the adoption of flavouring substances as antimicrobials in foods

Since ancient times, spices and herbs have not been consciously added to foods as preservatives but mainly as seasoning additives due to their aromatic properties. Although the majority of essential oils from herbs and spices are classified as Generally Recognized As Safe (GRAS) (Kabara, 1991), their use in foods as preservatives is limited because of flavour considerations, since effective antimicrobial doses may exceed organoleptically acceptable levels. This problem could possibly be overcome if answers could be given to the following questions:

- Can the inhibitory effect of an essential oil (a mixture of many compounds) be attributed to one or several key constituents?
- Does the essential oil provide a synergy of activity, which simple mixtures of components cannot deliver?
- What is the minimum inhibitory concentration (MIC) of the active compound(s) of the essential oil?
- How is the behaviour of the antimicrobial substance(s) affected by the homogeneous (liquid, semi-solid) or heterogeneous (emulsions, mixtures of solids and semisolids) structure of foodstuffs?
- Could efficacy be enhanced by combinations with traditional (salting, heating, acidification) and modern (vacuum packing, VP, modified atmosphere packing, MAP) methods of food preservation?

An in-depth understanding of the antimicrobial properties of these compounds is needed to answer these questions but such understanding has been lacking, despite the burgeoning literature on the subject. Methodological limitations (discussed in more detail below) in the evaluation of antimicrobial activity *in vitro* have led to many contradictory results. Moreover, there have been too few studies in real foods (these are considered laborious and often lead to negative outcomes). There is also a need to investigate the appropriate mode of application of an essential oil in a foodstuff. For instance, immersion, mixing, encapsulation, surface-spraying, and evaporating onto active packaging are some promising methods of adding these compounds to foods that have not been extensively investigated.

9.3 Methodological issues

The antimicrobial activity of plant-derived compounds against many different microorganisms, tested individually and *in vitro*, is well documented in the literature (Tables 9.1 and 9.2; Chapter 10, Tables 10.1–10.3). However, the results reported in different studies are difficult to compare directly. Indeed, contradictory data have been reported by different authors for the same antimicrobial compound (Mann and Markham, 1998; Manou *et al.*, 1998; Skandamis, 2001; Skandamis *et al.*, 2001). Also, it is not always apparent whether the methods cited measure bacteriostatic or bactericidal activities, or a combination of both.

Antimicrobial assays described in the literature include measurement of:

Table 9.1 Plant essential oils tested for antibacterial properties

Achiote[14], Allspice[16], Almond[1] (bitter, sweet), Aloe Vera[14], Anethole[11], Angelica[1], Anise[1,5,6], Asafoetida[14] (*Ferula* sp.)
Basil[1,10,31], Bay[1,20,28,31], Bergamot[1], Birch[14]
Cajeput[32], Calmus[1], Camomile-German[10], Cananga, Caraway[1,3], Cardamon[1], Carrot seed[39], Cedarwood[39], Celery[39], Chilli[39], Cinnamon casia[1,19,16,18], Cinnamon[28,33] (bark leaf), Cinnamon[28], Citronella[1], Clove[1,3,8,10,1,12,15,16,18,19,40], Coriander[1,5,8], Cornmint[5], Cortuk[17], Cumin[3,5,10], Cymbopogon[38]
Dill[1,5]
Elecampane, Estragon[10], Eucalyptus[24,35,38], Evening primrose[39]
Frankincense[39], Fennel[1,5,10,23]
Gale (sweet), Gardenia[39], Garlic[10,16,18,22], Geranium[1], Ginger[1,10], Grapefruit[6]
Horseradish, Hassaku Fruit Peel[27]
Jasmine[14,32]
Laurel[1,5,10], Lavender[1], Lemon[1,5,6,10], Lime[1,6], Linden flower[2], Liquorice, Lovage[1], Lemongrass[24,31,36]
Mace[20], Mandarin[1,6,10], Marigold *Tagetes*[39], Marjoram[1,10,31], Mastich gum tree[14] (*Pistachia lentiscus* var. *chia*) , Melissa[1], Mint[1,29,30] (apple), mugwort[39], Musky bugle, Mustard[16], Mountain tea[14] (*Sideritis* app.)
Neroly[10], Nutmeg[1,8,10,20]
Onion[10,16,18,22], Orange[1,5,6,10,21], Oregano[4,9,10,16,18,31], Ocicum[38]
Palmarosa[24], Paprika[16], Parsley[1,5,10], Patchouli[39], Pennyroyal, Pepper, Peppermint[1,10,24], Pettigrain[10], Pimento[1,10,18],
Ravensara[39], Rose[1], Rosemary[1,3,7,10,16], Rosewood[39]
Saffron[10], Sage[1,3,5,7,10,16], Sagebrush[13], Savoury[5], Sassafras[1], Sideritis[37], Senecio[34] (chachacoma), Spike[1], Spearmint[1], Star Anise[1], St John?s Wort[1]
Tangarine[39] Tarragon[4], Tea Thuja[1], Thyme[1,3,4,5,9,10,18,40], Tuberose[39], Turmeric[16], Tea-tree[25,26]
Valerian[1], Verbena[1], Vanilla[10]
Wintergreen[39], Wormwood[39]

Data from: [1]Deans and Ritchie (1987); [2]Aktug and Karapinar (1987); [3]Farag *et al.* (1989); [4]Paster *et al.* (1990); [5]Akgul and Kivanc (1989); [6]Dabbah *et al.* (1970); [7]Shelef *et al.* (1980); [8]Stecchini *et al.* (1993); [9]Salmeron *et al.* (1990); [10]Aureli *et al.* (1992); [11]Kubo and Himejima (1991); [12]Briozzo *et al.* (1989); [13]Nagy and Tengerdy (1967); [14]Nychas and Tassou (2000); [15]Al-Khayat and Blank (1985); [16]Azzouz and Bullerman (1982); [17]Kivanc and Akgul (1990); [18]Ismaiel and Pierson (1990); [19]Blank *et al.* (1987); [20]Hall and Maurer (1986); [21]Sankaran (1976); [22]Elnima *et al.* (1983); [23]Davidson and Branen (1993); [24]Pattnaik *et al.* (1995a,b,c); [25]Mann *et al.* (2000); [26]Nelson (2000); [27]Takahashi *et al.* (2002); [28]Smith-Palmer *et al.* (2001); [29]Iscan *et al.* (2002); [30]Tassou *et al.* (2000); [31]Mejlholm and Dalgaard (2002); [32]Skandamis (2001); [33]Chang *et al.* (2001); [34]Perez *et al* (1999); [35]Oyedeji *et al.* (1999); [36]Carlson *et al.* (2001); [37]Ozcan *et al.* (2001); [38]Cimanga *et al.* (2002); [39]Nychas unpublished; [40]Smith-Palmer *et al.* (1998).

Table 9.2 Naturally occurring antimicrobial compounds in plants

Apigenin-7-glucose, aureptan
Benzoic acid, berbamine, berberine, borneol
Caffeine, caffeic acid, 3-*o*-caffeylquinic acid, 4-*o*-caffeylquinic acid, 5-*o*-caffeylquinic acid, camphene camphor, carnosol, carnosic acid, carvacrol*, caryophelene, catechin, 1,8 cineole, cinnamaldehyde*, cinnamic acid, citral, chlorogenic acid, chicorin, columbamine, coumarine, *p*-coumaric acid, *o*-coumaric, *p*-cymene, cynarine
Dihydrocaffeic acid, dimethyloleuropein,
Esculin, eugenol*
Ferulic acid
Gallic acid, geraniol, gingerols,
Humulone, hydroxytyrosol, 4-hydroxybenzoic acid, 4-hydroxycinnamic acid
Isovanillic, isoborneol
Linalool, lupulone, luteoline-5-glucoside, ligustroside, *S*-limonene
Myricetin, 3-methoxybenzoic acid, menthol, menthofurane
Oleuropein
Paradols, protocatechuic acid, *o*-pyrocatechic, *a*-pinene, *b*-pinene, pulegone
Quercetin
Rutin, resocrylic
Salicylaldehyde, sesamol, shogoals, syringic acid, sinapic
Tannins, thymol*, tyrosol, 3,4,5-trimethoxybenzoic acid, 3,4,5-thihydroxyphenylacetic acid
Verbascoside, vanillin, vanillic acid

* Chemical structures are shown in Chapter 2, Fig. 2.1.

Data from: Nychas and Tassou (2000); Iscan *et al.* (2002); Flamini *et al.* (2002); Mourey and Canillac (2002); Takahashi *et al.* (2002); Amakura *et al.* (2002); Gounaris *et al.* (2002), Hayes and Markovic (2002); Chang *et al.* (2001); Perez *et al.* (1999); Oyedeji *et al.* (1999); Carlson-Castelan *et al.* (2001); Ozcan *et al.* (2001); Cimanga *et al.* (2002)

- the radius or diameter of the zone of inhibition of bacterial growth around paper disks impregnated with (or wells containing) an antimicrobial compound on agar media;
- the inhibition of bacterial growth on an agar medium with the antimicrobial compound diffused in the agar;
- the minimum inhibitory concentration (MIC) of the antimicrobial compound in liquid media;
- the changes in optical density or impedance in a liquid growth medium containing the antimicrobial compound.

Three main factors may influence the outcome of the above methods when used with essential oils of plants: (i) the composition of the sample tested (type of plant, geographical location and time of the year), (ii) the microorganism (strain, conditions of growth, inoculum size, etc.), and (iii) the method used for growing and enumerating the surviving bacteria. Many studies have been based on subjective assessment of growth inhibition, as in the disc diffusion method, or on rapid techniques such as optical density (turbidimetry) without accounting for the limitations inherent in such methods. In the disc method, the inhibition area depends on the ability of the essential oil to diffuse uniformly through the agar as well as on the released oil vapours. Other factors that may influence results involve the presence of multiple active components. These active compounds at low concentrations may interact antagonistically, additively or synergistically with each other. Some of the differences in the antimicrobial activity of oils observed in complex foods compared with their activity when used alone in laboratory media could be due to the partitioning of active components between lipid and aqueous phases in foods (Stechini *et al.*, 1993, 1998).

Turbidimetry is a rapid, non-destructive and inexpensive method that is easily automated but has low sensitivity. Turbidimetry detects only the upper part of growth curves, and requires calibration in order to correlate the results with viable counts obtained on agar media (Koch, 1981; Bloomfield, 1991; Cuppers and Smelt, 1993; McClure *et al.*, 1993; Dalgaard and Koutsoumanis, 2001; Skandamis *et al.*, 2001b). The changes in absorbance are only evident when population levels reach 10^6–10^7 CFU/ml, and are influenced by the size of the bacterial cells at different growth stages. The physiological state of the cells (injured or healthy), the state of oxidation of the essential oil as well as inadequate dissolution of the compound tested may also affect absorbance measurements in growth media.

Unlike the plate counting technique, impedance-based methods can be used to monitor microbial metabolism in real time mode. The impedimetric method is recognized as an alternative rapid method not only for screening the biocide activity of novel antimicrobial agents but also for estimation of growth kinetics in mathematical modelling (Ayres *et al.*, 1993, 1998; Tranter *et al.*, 1993; Tassou *et al.*, 1995, 1997; Johansen *et al.*, 1995; Tassou and Nychas, 1995a,b,c; Koutsoumanis *et al.*, 1997, 1998; MacRae *et al.*, 1997; Lachowicz *et al.*, 1998). The technique depends on using a medium that offers a sharp detectable impedimetric change as the bacterial population grows and converts the low conductivity nutrients into highly charged products. As with turbidometry, calibration of impedimetric data with plate counts is necessary (Dumont and Slabyj, 1993; Koutsoumanis *et al.*, 1998).

Although time-consuming and laborious, the traditional microbiological method of determining viable numbers by plate counting remains the gold standard in antimicrobial studies. The latter method has a major advantage of requiring little capital investment; however, it is material-intensive, requires a long elapse time and may have poor reproducibility.

MICs are measured by serial dilution of the tested agents in broth media followed by growth determination by either absorbance reading or plate-counting (Carson *et al.*, 1995). The MIC technique has been miniaturized and automated using the bioscreen microbiological growth analyzer (Lambert and Pearson, 2000) to allow a high throughput of compounds and microorganisms (Lambert *et al.*, 2001). The advantage of this method is the simultaneous examination of multiple concentrations of one or more preservatives and subsequent determination of MIC based on mathematical processing.

9.4 Studies *in vitro*

Almost all essential oils from spices and herbs inhibit microbial growth as well as toxin production. The antimicrobial effect is concentration-dependent and may become strongly bactericidal at high concentrations. Gram-positive bacteria (spore- and non-spore-formers), Gram-negative bacteria, yeasts (Tables 9.1–9.3) and moulds (Chapter 10, Tables 10.1–10.3) are all affected by a wide range of essential oils. Well-known examples include the essential oils from allspice, almond, bay, black pepper, caraway, cinnamon, clove, coriander, cumin, garlic, grapefruit, lemon, mace, mandarin, onion, orange, oregano, rosemary, sage and thyme. The active compounds of some of these essential oils are shown in Tables 9.2 and 9.4 and illustrated in Chapter 2, Fig. 2.1. The antimicrobial activity of these compounds is influenced by the culture medium, the temperature of incubation and the inoculum size. In addition, a strong synergism with some membrane chelators acting as permeabilising agents (e.g. ethylenediaminetetraacetic acid, EDTA) against Gram-negative bacteria has been reported (Tassou, 1993; Ayers *et al.*, 1998; Brul and Coote, 1999; Skandamis, 2001; Skandamis *et al.*, 2001b).

9.5 Applications in foods

There have been relatively few studies of the antimicrobial action of essential oils in model food systems and in real foods (Table 9.5). The efficacy of essential oils *in vitro* is often much greater than *in vivo* or *in situ*, i.e. in foods (Nychas and Tassou, 2000; Davidson, 1997; Skandamis *et al.*, 1999b). For example, the essential oil of mint (*Mentha piperita*) has been shown to inhibit the growth of *Salmonella* Enteritidis and *Listeria monocytogenes* in culture media for 2 days at 30 °C. However, the effect of mint essential oil in the traditional Greek appetizers tzatziki (pH 4.5) and taramasalata (pH 5.0) and in paté (pH 6.8) at 4 °C and 10 °C was variable. *Salmonella* Enteritidis died off in the appetizers under all conditions examined but not when inoculated in paté and maintained at 10 °C. Similarly, *L. monocytogenes* numbers declined in the appetizers but increased in paté (Tassou *et al.*, 1995a,b, 2000).

Growth of *Escherichia coli, Salmonella* spp., *L. monocytogenes* and *Staphylococcus aureus* was inhibited by oregano essential oil (EO) in broth cultures. However, the antimicrobial action of this EO in an emulsion or pseudo-emulsion type of food such as aubergine salad, taramasalata and mayonnaise depended on environmental factors such as pH, temperature and oil (vegetable or olive) used. Homemade aubergine salad and taramasalata were inoculated with *E. coli* O157:H7 and *Salmonella* Enteritidis, respectively. The pH of these products was adjusted to 4–5.3. A range of concentrations (0–2.1%) of oregano essential oil was added and the foods were incubated at temperatures from 0 to 20 °C. The survival curves for *E. coli* O157:H7 in aubergine salad at 0 and 15 °C, modelled according to Baranyi, are shown in Figs 9.1 and 9.2. A reduction in

Table 9.3 Some examples of microorganisms sensitive to the antimicrobial action of essential oils from herbs and spices

Gram-positive bacteria	Gram-negative bacteria	Yeasts/fungi*
Arthobacter sp.	*Acetobacter* spp.	*Aspergillus niger*
Bacillus sp.	*Acinetobacter* sp.	*As. parasiticus*
B. subtilis	*A.calcoaceticus*	*As. flavus*
B. cereus	*Aeromonas hydrophila*	*As. ochraceus*
B. megaterium	*Alcaligenes* sp.	*Candida albicans*
Brevibacterium	*A.faecalis*	*Candida tropicalis*
ammoniagenes	*Campylobacter jejuni*	*Dekkera bruxellensis*
Brev. linens	*Citrobacter* sp.	*Fusarium oxysporum*
Brochothrix	*C. freundii*	*F. culmorum*
thermosphacta	*Edwardsiella* sp.	*Mucor* sp.
Clostridium botulinum	*Enterobacter* sp.	*Pichia anomala*
Cl. perfrigenes	*En. aerogenes*	*Penicillium* sp.
Cl. sporogenes	*Escherichia coli*	*Pen. chrysogenum*
Corynebacterium sp.	*E. coli* O157:H7	*Pen. patulum,*
Enterococcus feacalis	*Erwinia carotovora*	*Pen. roquefortii*
Lactobacillus sp.	*Flavobacterium* sp.	*Pen. citrinum*
Lac. plantarum,	*Fl. suaveolens*	*Rhizopus* sp.
Lac. minor	*Klebsiella* sp.	*Saccharomyces cerevisiae*
Leuconostoc sp.	*K. pneumoniae*	*Trichophyton*
Leuc. cremoris	*Moraxella* sp.	*mentagrophytes*
Listeria monocytogenes	*Neisseria* sp.	*Torulopsis holmii*
L. inocua	*N. sicca*	*Pityrosporum ovale*
Micrococcus sp.	*Mycobacterium smegmatis*	
M. luteus	*Pseudomonas* spp.	
M. roseus	*P. aeruginosa, fluorescens,*	
Pediococcus spp.	*fragi* and *clavigerum*	
Photobacterium	*Proteus* spp.	
phosphoreum	*Pr. vulgaris*	
Propionibacterium acnes	*Salmonella* spp.	
Sarcina spp.	*Salmonella* Enteritidis,	
Staphylococcus spp.	Senftenberg, Typhimurium,	
Staph. aureus,	Flexneri, Pullorum	
Staph. epidermidis	*Serratia* sp.	
Streptococcus faecalis	*S. marcecens*	
	Vibrio sp.	
	V. parahaemolyticus	
	Yersinia enterocolitica	

* Additional antimicrobial activities against fungi important in postharvest diseases are shown in Chapter 10, Tables 10.1–10.3.

Based on: Nychas (1995); Mejlholm and Dalgaard (2002); Thangadural *et al.* (2002); Mangena and Muyima (1999); Karaman *et al.* (2001); Hayes and Markovic (2002); Chang *et al.* (2001); Cimagna *et al.* (2002).

Table 9.4 Examples of essential oils commonly used for food preservation and their main active constituents

Herb/spice	Active compound	Herb/spice	Active compound
Allspice	eugenol* methyl eugenol	Mint	α-, β-pinene limonene 1,8-cineole
Caraway	carvone	Onion	*d-n*-propyl disulfide methyl-*n*-propyl disulfide
Cinnamon	cinnamaldehyde* eugenol*	Oregano	thymol* carvacrol*
Cloves	eugenol* eugenol acetate	Pepper	monoterpenes
Coriander	*d*-linalool *d*-α-pinene β-pinene	Rosemary	borneol 1,8-cineole camphor bornyl acetate
Cumin	cuminaldehyde	Sage	thujone 1,8-cineol borneol
Garlic	diallyl disulfide diallyl trisulfide allyl propyl disulfide	Thyme	thymol* carvacrol* menthol, menthone

* Chemical structures are shown in Chapter 2, Fig. 2.1.

Data from: Skandamis (2001); Iscan *et al.* (2002); Flamini *et al.* (2002); Karaman *et al.* (2001); Gounaris *et al.* (2002), Oyedeji *et al.* (1999); Ozcan *et al.* (2001); Cimanga *et al.* (2002).

viable counts for both pathogens in both foods tested was observed and their death rate depended on the pH, the storage temperature and the essential oil concentration (Koutsoumanis *et al.*, 1999; Skandamis and Nychas, 2000; Skandamis *et al.*, 1999a, 2002b) (see Fig. 9.3).

The type of oil or fat present in a food can affect the antimicrobial efficacy of essential oils. This was evident when the efficiency of four plant essential oils (bay, clove, cinnamon and thyme) was assessed in low-fat and full-fat soft cheese against *L. monocytogenes* and *Salmonella* Enteritidis at 4 and 10 °C, respectively, over a 14 day period. In the low-fat cheese, all four oils at 1% reduced *L. monocytogenes* to below the detection limit of the plating method. In contrast, in the full-fat cheese, the oil of clove was the only substance to achieve such reduction. The oil of thyme was ineffective against *Salmonella* Enteritidis in the full-fat cheese despite the fact that this organism was completely inhibited in broth culture (Skandamis, 2001). Thyme oil was as effective as the other three oils in the low fat cheese, reducing *Salmonella* Enteritidis to less than 1 log CFU/g from day 4 onwards (Smith-Palmer *et al.*, 2001).

Table 9.5 Applications of essential oils in foods

Food	Microorganisms	Essential oil	References
Milk (fresh, skimmed)	*Staph. aureus* *Salmonella* Enteritidis *P. fragi*	Mastic gum	Tassou and Nychas (1995c)
Dairy products: soft cheese, mozzarella	*L. monocytogenes* *Salmonella* Enteritidis	Clove, cinnamon, thyme	Smith-Palmer *et al.* (2001); Menon and Garg (2001)
Fresh meat: block or minced	*Salmonella* Typhimurium and Enteritidis *Staph. aureus* *P. fragi* *L. monocytogenes* Lactic acid bacteria *Br. thermosphacta* Enterobacteriaceae Yeasts & indigenous flora	Oregano, clove, basil, sage	Tassou and Nychas (1995b); Menon and Garg (2001); Skandamis and Nychas (2001, 2002a,b); Tsigarida *et al.* (2000); Skandamis *et al.* (2002a,b); Stecchini *et al.* (1993)
Meat products: paté	*L. monocytogenes* *Salmonella* Enteritidis Indigenous flora	Mint	Tassou *et al.* (1995)
sausage	*Br. thermosphacta,* *E. coli*	Mustard oil	Lemay *et al.* (2002)

Fish: Gilt-head bream	*Salmonella* Enteritidis *Staph. aureus* Resident flora	Oregano	Tassou *et al*. (1996)
Cod fillets, salmon	*Photobacterium phosphoreum*	Basil, bay, cinnamon, clove, lemongrass, marjoram, oregano, sage, thyme	Mejlholm and Dalgaard (2002)
Salads and dressings: tuna, potato augergine (egg plant), taramasalata, mayonnaise, tzatziki	*Staph. aureus* *Salmonella* Enteritidis	Carob	Tassou *et al*. (1997)
	P. fragi *L. monocytogenes* *Sh. putrefaciens* *Br. thermosphacta* *E. coli* Indigenous flora	Mint, oregano, basil, sage	Tassou and Nychas (1995c); Tassou *et al*. (1995); Koutsoumanis *et al.* (1999); Skandamis and Nychas (2000); Skandamis *et al*. (1999a,b, 2001a, 2002c)
Sauces: meat gravy	*Salmonella* Enteritidis and Typhimurium *Staph. aureus* *P. fragi*	Basil, sage	Tassou and Nychas (1995c)

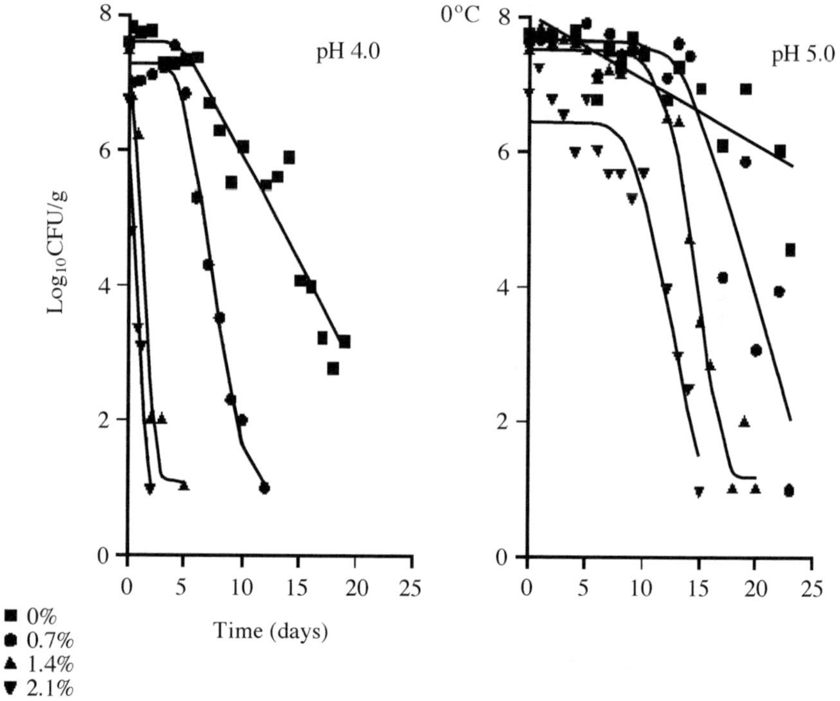

Fig. 9.1 Survival curves for *E. coli* O157:H7 in aubergine (egg plant) salad at 0 °C, pH 4.0 and 5.0, in the presence of 0, 0.7, 1.4 and 2.1% oregano essential oil. Data from Skandamis and Nychas (2000).

Table 9.5 summarizes some of the studies on the inhibitory action of essential oils in solid foods (e.g fish and meat) stored under various packaging conditions (VP, MAP). For example, *L. monocytogenes* and *Salmonella* Typhimurium were inhibited in meat treated with clove and oregano essential oil, respectively (Menon and Garg, 2001; Tsigarida *et al.*, 2000; Skandamis *et al.*, 2002a). *Salmonella* Typhimurium survived in untreated meat, while the addition of oregano essential oil at a concentration of 0.8% v/w reduced viable numbers by 1–2 log CFU/g. The same level of oregano essential oil reduced the counts of *L. monocytogenes* by 2–3 log CFU/g on meat. A marked reduction of *Aeromonas hydrophila* was also reported in cooked, non-cured pork treated with clove or coriander oils and packaged either under vacuum or air and stored at 2 and 10 °C. The lethal effect of these two oils was more pronounced under vacuum than in aerobic conditions (Stecchini *et al.*, 1993).

The availability of oxygen can affect the antimicrobial efficacy of essential oils. Paster *et al.* (1990, 1995) observed that the antimicrobial activity of the oregano essential oil on *Staph. aureus* and *Salmonella* Enteritidis was enhanced when these organisms were incubated under microaerobic or anaerobic conditions. Under conditions of low oxygen tension, there are fewer oxidative

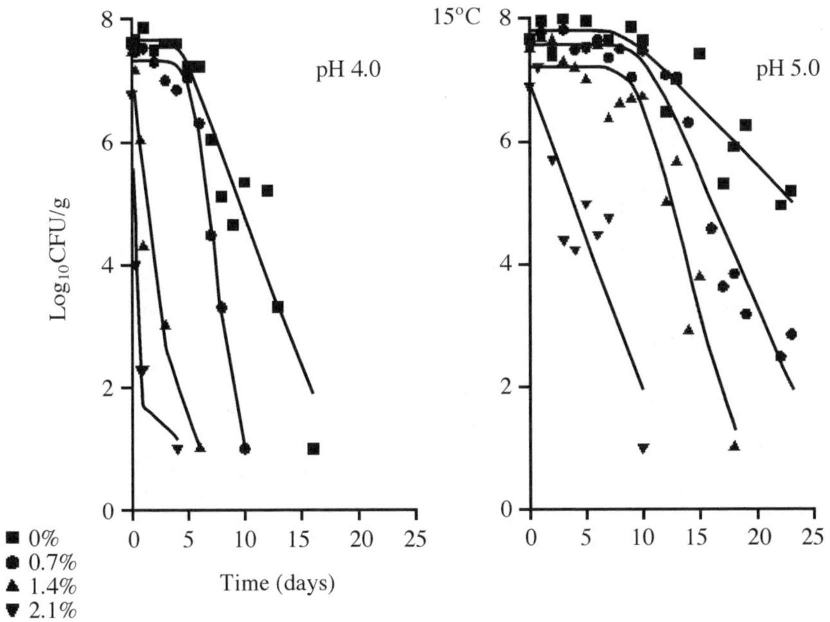

Fig. 9.2 Survival curves for *E. coli* O157:H7 in aubergine (egg plant) salad at 15 °C, pH 4.0 and 5.0, in the presence of 0, 0.7, 1.4 and 2.1% oregano essential oil. Data from Skandamis and Nychas (2000).

changes in the essential oil (Paster *et al.*, 1990, 1995). Moreover, oregano essential oil was more effective under vacuum and a 40% CO_2 : 30% O_2 : 30% N_2 atmosphere when an impermeable film was used compared to aerobic incubation or packaging in bags that allowed O_2 to permeate the package (Tsigarida *et al.*, 2000; Skandamis *et al.*, 2002a).

Oregano EO has both bacteriostatic and bacteriocidal effects on raw fish (*Sparus aurata*) inoculated with *Staph. aureus* and *Salmonella* Enteritidis and stored under MAP (40% CO_2, 30% O_2 and 30% N_2) or in air at 1 °C. Growth of spoilage organisms such as *Shewanella putrefaciens* and *Photobacterium phosphereum* is also inhibited on gilt head seabream and cod treated with oregano EO (Tassou *et al.*, 1996; Mejlholm and Dalgaard, 2002). Similar reductions were also reported for many other meat and fish organisms, as shown in Table 9.5 (Greer *et al.*, 2000; Mejlholm and Dalgaard, 2002; Skandamis and Nychas, 2001, 2002a,b).

The studies reviewed above all show that antimicrobial activity demonstrated *in vitro* is not necessarily a good indication of practical value in food preservation. The active compounds of essential oils are often bound with food components (e.g. proteins, fats, sugars, salts). Therefore, only a proportion of the total dose of EO added to a food remains free to exert antibacterial activity. Extrinsic factors such as temperature also limit the antimicrobial action of essential oils (Davidson, 1997). Moreover, the spatial distribution of the

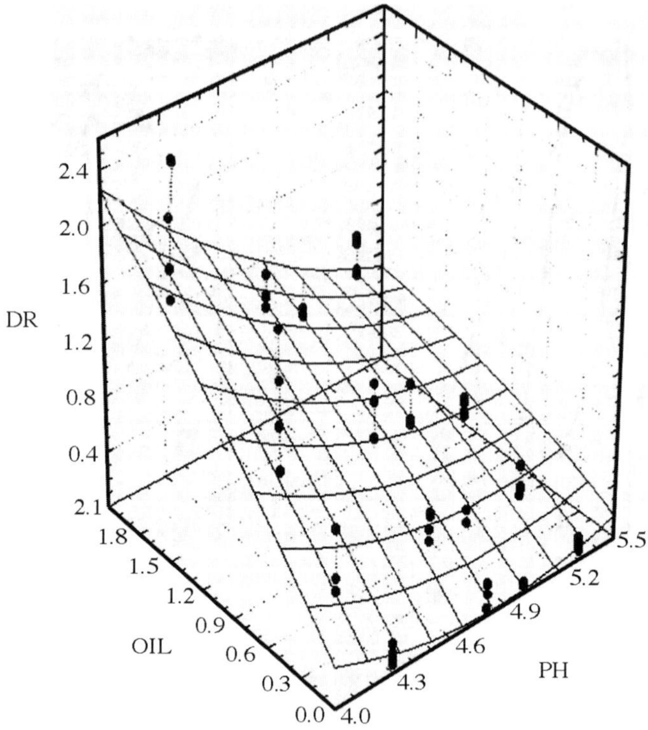

Fig. 9.3 Quadratic response surfaces predicting the death rate (DR) of *Salmonella* Typhimurium in taramasalata as a function of pH and oregano essential oil (OIL%). Data from Koutsoumanis *et al.* (1999).

different phases (solid/liquid) in a food and the lack of homogeneity of pH and water can also play a role in efficacy. Interactions between the different components in the food may create pH gradients in the final product as well as different bulk concentrations of the antimicrobial in the different phases. The local buffering capacity of the food ingredients determines the pH within specific regions of complex foods. Since the spatial distribution of microorganisms is not homogeneous, the antimicrobial activity could also depend on their population density, on the food structure *per se*, and on carbon source availability governed by diffusion factors. The microbial ecology of specific foodstuffs, buffering capacity, local pH and food structure should all be taken into account during the evaluation of an antimicrobial compound.

The growth of bacteria in liquids occurs planktonically, in contrast with the discrete colonies formed either on or within a solid matrix (Robins *et al.*, 1994; Wilson *et al.*, 2002). In the latter case, the cells are immobilized and localised in high densities in the food matrix (Skandamis *et al.*, 2000; Wilson *et al.*, 2002). Challenge tests have revealed that the physiological attributes of bacteria grown in model food matrices were significantly different from those of cells growing freely in liquid cultures (Brocklehurst *et al.*, 1997; Skandamis *et al.*, 2000; Wilson *et al.*, 2002). These differences can be accounted for by: (i) the population density *per se*, (ii) diffusivity and thus availability of major nutrients, (iii) oxygen availability, and (iv) accumulation of end products (Stecchini *et al.*, 1993, 1998; Thomas *et al.*, 1997; Skandamis *et al.*, 2000). Bacteria within solid matrices grow as submerged 'nests' (Thomas *et al.*, 1997). While the diffusivity of low molecular weight nutrients such as glucose may be very similar in liquids and gel matrices, that of antimicrobial agents may be very different and may strongly influence the efficacy of the agents in a solid matrix (Diaz *et al.*, 1993; Stecchini *et al.*, 1998). Oily substances within emulsions form droplets with diameters of 10–18μm (Wilson *et al.*, 2002). The diffusion of such large droplets is very likely to be affected by the density, viscosity, tortuosity and other structure-related properties of the medium. Thus, the higher mobility of essential oil droplets in liquid media may be the most important factor enhancing inhibition of target bacteria.

9.6 Mode of action and development of resistance

In general, the mode of action of essential oils is concentration-dependent (Prindle and Wright, 1977). Low concentrations inhibit enzymes associated with energy production while higher amounts may precipitate proteins. However, it is uncertain whether membrane damage is quantitatively related to the amount of active antimicrobial compound to which the cell is exposed, or the effect is such that, once small injuries are caused, the breakdown of the cell follows (Judis, 1963).

Essential oils damage the structural and functional properties of membranes and this is reflected in the dissipation of the two components of the proton motive force: the pH gradient (ΔpH) and the electrical potential ($\Delta\psi$) (Sikkema *et al.*, 1995, Davidson, 1997; Ultee *et al.*, 1999, 2000, 2002). Carvacrol, an active component of many essential oils, has been shown to destabilize the cytoplasmic and outer membranes and act as a 'proton exhanger', resulting in a reduction of the pH gradient across the cytoplasmic membrane (Helander *et al.*, 1998; Lambert *et al.*, 2001; Ultee *et al.*, 2002). The collapse of the proton motive force and depletion of the ATP pool eventually led to cell death (Ultee *et al.*, 2002). Like other many preservatives, the essential oils cause leakage of ions, ATP, nucleic acids and amino acids (Tranter *et al.*, 1993; Gonzalez *et al.*, 1996; Tahara *et al.*, 1996; Helander *et al.*, 1998; Cox *et al.*, 1998; Ultee *et al.*, 1999; Tassou *et al.*, 2000). Like carvacrol, the essential oils from tea and mint cause leakage of cellular material including potassium ions and 260 nm-absorbing substances (Cox *et al.*, 1998; Gustafson *et al.*, 1998; Ultee *et al.*, 1999). Nutrient uptake, nucleic acid synthesis and ATPase activity may also be affected, leading to further damage to the cell. Several reports have demonstrated that most essential oils (at approx. 100 mg/l) impair the respiratory activity of bacteria and yeasts (e.g. *Saccharomyces cerevisiae*) (Conner *et al.*, 1984a,b; Denyer and Hugo, 1991; Tassou *et al.*, 2000).

Unlike many antibiotics, essential oils are capable of gaining access to the periplasm of Gram-negative bacteria through the porin proteins of the outer membrane (Helander *et al.*, 1998). The permeability of cell membranes is dependent on their composition and the hydrophobicity of the solutes that cross it (Sikkema *et al.*, 1995; Helander *et al.*, 1998; Ultee *et al.*, 2002). Low temperatures decrease the solubility of essential oils and hamper penetration of the lipid phase of the membrane (Wanda *et al.*, 1976). The partition coefficient of essential oils in cell membranes is a crucial determinant of antimicrobial efficacy.

The solubility of essential trace elements such as iron is negatively affected by essential oils. Consequently, reduced availability of iron could inhibit bacterial growth. Additionally, the reaction of ferrous ion with phenolic compounds can indirectly damage cells by causing oxidative stress (Friedman and Smith, 1984; Nagaraj, 2001). The highly reactive aldehyde groups of some plant-derived antimicrobial compounds (e.g. citral, salicyldehyde) form Schiff's bases with membrane proteins and so prevent cell wall biosynthesis (Friedman, 1996, 1999; Patte, 1996). Phenolics, essential oils and phytoalexins generally cause static rather than outright toxic effects (Tokutake *et al.*, 1992); cell membranes that leak or function poorly would not necessarily lead to cell death but would most probably cause a deceleration of metabolic processes such as cell division (Darvill and Albersheim, 1984; Kubo *et al.*, 1985).

Antibiotics and related drugs have substantially reduced the threat posed by infectious diseases in the last century. However, the emergence and spread of antibiotic-resistant bacteria has, more recently, become a major concern. This concern has widened to include all microorganisms exposed to antimicrobial agents, including the so-called 'natural' compounds. However, there is relatively little information on the resistance mechanisms of microorganisms against plant-derived antimicrobial compounds.

Deans and Ritchie (1987), who studied the effect of 50 plant essential oils against 25 genera of bacteria, concluded that Gram-positive and Gram-negative organisms were equally susceptible to the antimicrobial action of essential oils. However, this conclusion is now under dispute. In general, Gram-positive bacteria are more sensitive than Gram-negative organisms to the antimicrobial compounds in spices (Dabbah *et al.*, 1970; Farag *et al.*, 1989; Shelef, 1983; Tassou *et al.*, 1995b,c, 1999). However, variation in the rate or extent of inhibition is also evident among the Gram-negative bacteria. For example, *E. coli* was less resistant than *Pseudomonas fluorescens* or *Serratia marcescens* to essential oils from sage, rosemary, cumin, caraway, clove and thyme (Farag *et al.*, 1989). *Salmonella* Enteritidis and Typhimurium were less sensitive than *P. fragi* to sage and mastic gum oils (Tassou and Nychas, unpublished) whereas *Salmonella* Typhimurium was more sensitive than *P. aeruginosa* to the essential oils from oregano and thyme (Paster *et al.*, 1990). *Pseudomonas putida* and *P. aeruginosa* have been reported as relatively tolerant of essential oils and efflux pumps in the outer membrane have been suggested as possible mechanisms for this resistance (Pattnaik *et al.*, 1995a,b; Isken and de Bond, 1998; Mann *et al.*,

2000). Mutants of *E. coli* and sub-populations of *Staph. aureus* resistant to pine and tea-tree oil, respectively, have also been reported (Moken *et al.*, 1997; Nelson, 2000).

9.7 Legislation

Many essential oils from herbs and spices are used widely in the food, health and personal care industries and are classified as GRAS substances or are permitted food additives (Kabara, 1991). A large number of these compounds have been the subject of extensive toxicological scrutiny and an example of the data available is shown in Table 9.6. However, their principal function is to impart desirable flavours and aromas and not necessarily to act as antimicrobial agents. Therefore, it is possible that additional safety and toxicological data would be required before regulatory approval for their use as novel food preservatives would be granted. For a more detailed review of legislation and how it impinges on natural antimicrobials, the reader is referred to Chapter 15.

9.8 Future prospects and multifactorial preservation

Given the high flavour and aroma impact of plant essential oils, the future for using these compounds as food preservatives lies in the careful selection and

Table 9.6 Lethal dose (LD_{50}) of some essential oils determined in rats

Plant/herb	LD_{50} (g/kg)	Plant/herb	LD_{50} (g/kg)
Prunus amygdalus	<1.0	*Juniperus communis*	>5
Angelica archangelica	2–>5	*Laurus nobilis*	2–5
Pimpinella anisum	2–5	*Lavandula angustifolia*	2–>5
Ocinum basilicum	1–2	*Citrus limonum*	>5
Pimenta racemosa	1–2	*Origanum marjorana*	2–5
Citrus bergamia	>5	*Pistacia lentiscus*	>5
Cinnamomum camphora	2–5	*Citrus aurantium*	>5
Anethum graveolens	2–5	*Origanum vulgare*	1–2
Allium sativum	>5	*Petroselinum sativum*	1–5
Anthemis nobilis	>5	*Piper nigrum*	>5
Cinnamomum zeylanicum	2–5	*Rosemarinus officinalis*	>5
Daucus carota	>5	*Menta viridis*	2–5
Cinnamomum cassia	2–5	*Salvia officinalis*	2–5
Syzygium aromaticum	1–5	*Thymus vulgaris*	2–5
Eucalyptys globules	2–5	*Citrus reticulata*	>5
Foeniculum vulgare	2–5	*Coriandrum sativum*	2–5
Zingiber officinale	>5	*Camellia sinensis*	2–5

Data modified from: Skandamis (2001).

evaluation of their efficacy at low concentrations but in combination with other chemical preservatives or preservation processes. Synergistic combinations have been identified between garlic extract and nisin, carvacrol and nisin (Chapter 2, Section 2.9.1 and Fig. 2.1), vanillin or citral and sorbate (Chapter 11, Section 11.2 and Figs. 11.1–11.3), thyme oil and/or cinnamaldehyde in an edible coating (Chapter 12, Table 12.7), and low-dose gamma irradiation and extracts of rosemary or thyme (Chapter 13, Section 13.3.2 and Table 13.1).

9.9 References

AKGUL, A. and KIVANC, M. (1989) Sensitivity of four foodborne moulds to essential oils from Turkish spices, herbs and citrus peel. *Journal of Sciences & Food Agriculture* **47**, 129–132.

AKTUG, S.E. and KARAPINAR, M. (1987) Inhibition of foodborne pathogens by thymol, eugenol, menthol and anethole. *International Journal of Food Microbiology* **4**, 161–166.

AL-KHAYAT, M.A. and BLANK, G. (1985) Phenolic spice components sporostatic to *Bacillus subtilis*. *Journal of Food Science* **50**, 971–980.

AMAKURA, Y., UMINO, Y., TSUJI, S., ITO, H., HATANO, T., YOSHIDA, T. and TONOGAI, Y. (2002) Constituents and their antioxidative effects in eucalyptus leaf extract used as natural food additive. *Food Chemistry* **77**, 47–56.

AURELI, P., CONSTANTINI, A. and ZOLEA, S. (1992) Antimicrobial activity of some plant essential oils against *Listeria monocytogenes*. *Journal of Food Protection* **55**, 344–348.

AYRES, H.M., FURR, J.R. and RUSSELL, A.D. (1993) A rapid method of evaluating permeabilizing activity against *Pseudomonas aeruginosa*. *Letters in Applied Microbiology* **17**, 149–151.

AYRES H.M., PAYNE, D.N., FURR, J.R. and RUSSELL, A.D. (1998) Use of the Malthus-AT system to assess the efficacy of permeabilizing agents on the activity of antibacterial agents against *Pseudomonas aeruginosa*. *Letters in Applied Microbiology* **26**, 422.

AZZOUZ, M.A. and BULLERMAN, L.B. (1982) Comparative antimycotic effects of selected herbs, spices plant components and commercial antifungal agents. *Journal of Food Protection* **45**, 1298–1301.

BLANK, G., AL-KHAYAT, M. and ISMOND, M.A.H. (1987) Germination and heat resistance of *Bacillus subtilis* spores produced on clove and eugenol based media. *Food Microbiology* **4**, 35–42.

BLOOMFIELD, S.F. (1991) Methods for assessing antimicrobial activity. In *Mechanisms of Action of Chemical Biocides; Their Study and Exploitation*, Denyer, S.P. and Hugo, W.B. (eds), Society for Applied Bacteriology, Technical Series No. 27, Blackwell Scientific Publications, Oxford pp. 1–22.

BRIOZZO, J., NUNEZ, L., CHIRIFE, J. and HERSZAGE, L. (1989) Antimicrobial activity of clove oil dispresed in a concentrated sugar solution. *Journal of Applied Bacteriology* **66**, 69–75.

BROCKLEHURST, T.F, MITCHELL, G.A. and SMITH, A.C. (1997) A model experimental gel surface for the growth of bacteria on foods. *Food Microbiology* **14**, 303–311.

BRUL, S. and COOTE, P. (1999) Preservative agents in foods. Mode of action and microbial resistance mechanisms. *International Journal of Food Microbiology* **50**, 1–17.

CARLSON-CASTELAN, LH., MACHADO, R.A.F., SPRICIGO, C.B., PEREIRA, L. K. and BOLZAN, A. (2001) Extraction of lemongrass essential oil with dense carbon dioxide. *Journal of Supercritical Fluids* **21**, 33–39.

CARSON, C.F., HAMMER, K.A. and RILEY, T.V. (1995) Broth microdilution method for determining the susceptibility of *Escherichia coli* and *Stapylococcus aureus* to the essential oil of *Melaleuca alternifolia* (tea tree oil). *Microbios* **82**, 181–185.

CHANG, S-T., CHEN,P-F. and CHANG, S-C. (2001) Antibacterial activity of leaf essential oils and their constituents from *Cinamomum osmophloeum. Journal of Ethnopharmacology* **77**, 123–127.

CIMANGA, K. KAMBU, K., TONA., L., APERS, S., DE BRUYEN, T., HERMANS, N., TOTTE, J., PIETERS, L. and VLIENTINCK, A.J. (2002) Correlation between chemical composition and antibacterial activity of essential oisl of some aromatic medicinal plants growing in the Democratic Republic of Congo. *Journal of Ethnopharmacology* **79**, 213–220.

CONNER, D.E. and BEUCHAT, L.R. (1984a) Sensitivity of heat-stressed yeasts to essential oils of plants. *Applied Environmental Microbiology* **47**, 229–233.

CONNER, D.E. and BEUCHAT, L.R. (1984b) Effects of essential oils from plants on growth of food spoilage yeasts. *Journal of Food Science* **49**, 429–434.

COX, S.D., GUSTAFSON, J.E., MANN, C.M., MARKHAM, J.L., LIEW, Y.C., HARTLAND, R.P., BELL, H.C., WARMINGTON, J.R. and WYLLIE, S.G. (1998) Tea tree oil causes K^+ leakage and inhibits respiration in *Escherichia coli. Letters in Applied Microbiology* **26**, 355–358.

CUPPERS, H.G.A.M. and SMELT, J.P.P.M. (1993) Time to turbidity measurement as a tool for modeling spoilage by *Lactobacillus. Journal of Industrial Microbiolology* **12**, 168–171.

DABBAH, R., EDWARDS, V.M. and MOATS, W.A. (1970) Antimicrobial action of some citrus fruit oils on selected food-borne bacteria. *Applied Microbiology* **19**, 27—31.

DALGAARD, P. and KOUTSOUMANIS, K. (2001) Comparison of maximum specific growth rates and lag times estimated from absorbance and viable count data by different mathematical models. *Journal of Microbiological Methods* **43**, 183–196.

DARVILL, A.G. and ALBERSHEIM, P. (1984) Phytoalexins and their elicitors: a defence against microbial infection in plants. *Annual Review of Plant Physiology* **35**, 243–275.

DAVIDSON, P.M. (1997) Chemical Preservatives and Natural antimicrobial compounds. In *Food Microbiology Fundamentals and Frontiers*, Doyle, M.P., Beuchat, L.R., Montville, T.J. (eds), pp. 520–556, NY: ASM Press.

DAVIDSON, P.M. and BRANEN, A.L. (1993) *Antimicrobials in Foods*. Marcel Dekker, NY.

DEANS, S.G AND RITCHIE, G. (1987) Antibacterial properties of plant essential oils. *International Journal of Food Microbiol.* **5**, 165–180.

DENYER, S.P. and HUGO, W.B. (1991) Biocide induced damage to the bacterial cytoplasmic membrane. In *Mechanisms of Action of Chemical Biocides; Their Study and Exploitation*, Denyer, S.P. and Hugo, W.B. (eds), The Society for Applied Bacteriology, Technical Series No. 27. Blackwell Scientific Publications, Oxford.

DIAZ, G., WOLF, W., KOSTAROPOULOS, A.E. and SPIESS, W.E.L. (1993) Diffusion of low-molecular weight compounds in food model system. *Journal of Food Processing and Preservation* **17**, 437–454.

DUMONT L.E. and SLABYJ B.M. (1993) Impedimetric estimation of bacterial load in commercial carrageenan. *Food Microbiol*ogy **11**, 375–383.

ELNIMA, N.I., SYED, A.A., MEKKAWI, A.G. and MOSSA, J.S. (1983) Antimicrobial activity of garlic and onion extracts. *Pharmazie* **38**, 743–748.

FARAG, R.S., DAW, Z.Y, HEWEDI, F.M. and EL-BAROTY, G.S.A (1989) Antimicrobial activity of some egyptian spice essential oils. *Journal of Food Protection* **52**, 665–667.

FLAMINI, G., CIONI, P.L., MORELLI, I., MACCHIA, M. and CECCARINI, L. (2002) Main agronomic-productive characteristics of two ecotypes of *Rosmarinus officinalis* L. and chemical composition of their essential oils. *Journal of Agricultural and Food Chemistry* **50**, 3512–3517.

FRIEDMAN, M. (1996) Food browning and its prevention. *Journal of Agricultural and Food Chemistry* **47**, 1523–1540.

FRIEDMAN, M. (1999) Chemistry nutrition and microbiology of D-amino acids. *Journal of Agricultural and Food Chemistry* 47, 3457–3479.

FRIEDMAN, M. and SMITH, G.A. (1984) Inactivation of quercetin mutagenicity. *Food Chemistry & Toxicology* **22**, 535–539.

GALLI, A., FRANZETTI, L. and BRIGUGLIO, D. (1985) Antimicrobial properties *in vitro* of essential oils and extract of spices used for food. *Industrial Alimentaries*, **24**, 463–466.

GONZALEZ, B., GLAASKER, E., KUNJI, E.R.S., DRIESSEN, A.J.M., SUAREZ, J.E. and KONINGS, W.N. (1996) Bactericidal mode of action of plantaricin C. *Applied Environmental Microbiology* **62**, 2701–2709.

GOUNARIS, Y., SKOULA, M., FOURNARAKI, C., DRAKAKAKI, G. and MAKRIS, A. (2002) Comparison of essential oils and genetic relationship of Origanum X intercedens to its parental taxa in the island of Crete. *Biochemical Systematics and Ecology* **30**, 249–258.

GREER, G.G. PAQUET, A. and DILTS, B.D. (2000) Inhibition of *Brochothrix thermosphacta* in broths and pork by myristoyl-L-methionine. *Food Microbiology* **17**, 177–183.

GUSTAFSON, J.E., LIEW, Y.C., CHEW, S., MARKHAM, J., BELL, H.C., WYLLIE, S.G. and WARMINGTON, J.R. (1998) Effects of tea tree oil on *Escherichia coli*. *Letters in Applied Microbiology* **26**, 194–198.

HALL, M.A. and MAURER, A.J. (1986) Spice extracts and propylene Glycols as inhibitors of *Clostridium botulinum* in turkey frankfurter slurries. *Poultry Science* **65**, 1167–1171.

HAYES, A.J. and MARKOVIC, B. (2002) Toxicity of Australian essential oil *Backhousia citriodora* (Lemon myrtle) Part 1. Antimicrobial activity and in vitro cytoxicity. *Food And Chemical Toxicology* **40**, 535–543.

HELANDER, I.K., ALAKOMI, H.L., LATVA-KALA, K., MATTILA-SANDHOLM, T., POL, I., SMID, E.J. and VON WRIGHT, A. (1998) Characterization of the action of selected essential oil components on Gram negative bacteria. *Journal of Agricultural Chemistry* **46**, 3590–3595.

JOHANSEN, C., GILL, T. and GRAM, L. (1995) Antibacterial effect of protamine assayed by impedimetry. *Journal of Applied Bacteriology* **78**, 97–303.

JUDIS, J. (1963) Studies on the mechanism of action of phenolic disinfectants II. *Journal of Pharmacological Science* **52**(2), 126–131.

ISCAN, G., KIRIMER, N., KURKCUOGLU, M., CAN BASER, K.H. and DEMIRCI, F. (2002) Antimicrobial screening of *Menta piperita* Essential oils. *Journal of Agricultural Food Chemistry* **50**, 3943–3946.

ISKEN, S. and DE BOND, J.A.M (1998) Bacterial tolerance to organic solvents. *Extremophiles* **2**, 29–238.

ISMAIEL, A.A. and PIERSON, M.D (1990) Inhibition of germination, outgrowth and vegetative growth of *Clostridium botulinum* 67B by spice oils. *Journal of Food Protection* **53**, 755–758.

KABARA, J.J. (1991) Phenols and chelators. In *Food Preservatives*, Russell, N.J. and Gould, G.W. (eds), pp. 200–214. Blackie, Glasgow & London.

KARAMAN, S., DIGRAK, M., RAVID, U. and ILCIM, A. (2001) Antibacterial and antifugal activity of the essential oils of *Thymus revolutus* Celak from Turkey. *Journal of Ethnopharmacology* **76**, 183–186.

KIVANC, M. and AKGUL, A. (1990) Mould growth on black table olives and prevention by sorbic acid, methyl-eugenol and spice essential oil. *Die Nahrung* **34**, 369–373.

KOCH, A.L. (1981) Growth measurement,. In *Manual of Methods for General Bacteriology*, Gerhardt, P., Murray, R.G.E and Costilow R.N., (eds), pp. 179–207. American Society for Microbiology, Washington DC.

KOUTSOUMANIS, K., TAOUKIS, P.S., TASSOU, C.C. and NYCHAS G.-J.E. (1997) Predictive modelling of the growth of *Salmonella enteritidis*; the effect of temperature, initial pH and oleuropein concentration. *Proceedings of the International Conference on Predictive Microbiology Applied to Chilled Food Preservation*, 16–18 June, 1997 Quimper, France, pp. 113–119.

KOUTSOUMANIS, K., TASSOU, C.C., TAOUKIS, P.S. and NYCHAS, G.-J.E. (1998) Modelling the effectiveness of a natural antimicrobial on *Salmonella* Enteritidis as a function of concentration, temperature and pH, using conductance measurements. *Journal of Applied Microbiology* **84**, 981–987.

KOUTSOUMANIS, K., LAMBROPOULOU, K. and NYCHAS, G.J.E. (1999) A predictive model for the non-thermal inactivation of *Salmonella* Enteritidis in a food model system supplemented with a natural antimicrobial. *International Journal of Food Microbiology* **49**, 67–74.

KUBO, I. and HIMEJIMA, M. (1991) Anethole, a synergist of polygodial against filamentous microorganisms. *Journal of Agricultural Food Chemistry* **39**, 2290–2292.

KUBO,I., MATSUMOTO, A. and TAKASE, I. (1985) A multichemical defense mechanism of bitter olive *Olea europaea* (Oleaceae). Is oleuropein a phytoalexin precursor? *Journal of Chemical Ecology* **11**(2), 251–263.

LACHOWICZ, K.J., JONES, G.P., BRIGGS, D.R., BIENVENU, F.E., WAN, J., WILCOCK, A. and COVENTRY, M.J. (1998) The synergistic preservative effects of the essential oils of sweet basil (*Ocimum basillicum* L.) against acid-tolerant food microflora. *Letters in Applied Microbiology* **26**, 209–214.

LAMBERT, R.J.W. and PEARSON, J. (2000) Susceptibility testing: accurate and reproducible minimum inhibitory concentration (MIC) and non-inhibitory concentration (NIC) values. *Journal of Applied Microbiology* **88**, 784–790.

LAMBERT, R.J.W., SKANDAMIS, P., COOTE, P.J. and NYCHAS, G.-J.E. (2001) A study of the minimum inhibitory concentration and mode of action or oregano essential oil, thymol and carvacrol. *Journal of Applied Microbiology* **91**, 453–462.

LEMAY, M-J., CHOQUETTE, J., DELAQUIS, P.J., GARIEPY, C., RODRIGUE, N. and SAUCIER L. (2002) Antimicrobial effect of natural preservatives in a cooked and acidified chicken meat model. *International Journal of Food Microbiology* **78**, 217–226.

MACRAE, M., REBATE, T., JOHNSTON, M. and OGDEN, I.D. (1997) The sensitivity of *Escherichia coli* O157 to some antimicrobials by conventional and conductance assays. *Letters in Applied Microbiology* **25**, 135–137.

MANGENA, T. and MUYIMA, N.Y.O. (1999) Comparative evaluation of the antimicrobial activities of essential oils of *Artemisis afra, Pteronia incana* and *Rosmarinus ofiicinalis* on selected bacteria and yeasts strains. *Letters in Applied Microbiology* **28,** 291–296.

MANN, C.M. and MARKHAM, J.L. (1998) A new method for determining the minimum inhibtion concentration of essential oils. *Journal of Applied Microbiology* **84**, 538–544.

MANN, C.M, COX, S.D. and MARKHAM, J.L. (2000) The outer mebrane of *Pseudomonas aeruginosa* NCTC 6749 contributes to its tolerance to the essential oil of *Melaleuca alternifolia* (tea tree oil). *Letters in Applied Microbiology* **30**, 294–297.

MANOU, L., BOUILLARD, L., DEVLEESCHOUWER, M.J. and BAREL, A.O. (1998) Evaluation of the preservative properties of *Thymus vulgaris* essential oil in topically applied formulations under a challenge test. *Journal of Applied Microbiology* **84**, 368–376.

MCCLURE, P.J., COLE, M.B., DAVIES, K.W. and ANDERSON, W.A. (1993) The use of automated turbidimetric data for the construction of kinetic models. *Journal of Industrial Microbiology* **12**, 277–285, 4, 3.

MEJLHOLM, O. and DALGAARD, P. (2002) Antimicrobial effect of essential oils on the seafood spoilage micro-organism *Photobacterium phosphoreum* in liquid media and fish products. *Letters in Applied Microbiology* **34**, 27–31.

MENON, V.K. and GARG, S.R. (2001) Inhibitory effect of clove oil on *Listeria monocytogenes* in meat and cheese. *Food Microbiology* **18**, 647–650.

MOKEN, M.C., MCMURRY, L.M. and LEVY, S.B. (1997) Selection of multiple-antibiotic resistant (mar) mutants of *Escherichia coli* by using the disifectant pine oil: roles of the mar and acrABloci. *Antimicrobial Agents and Chemotherapy* **41**, 2770–2772.

MOUREY, A. and CANILLAC, N. (2002) Anti-Listeria monocytogenes activity of essential oils components of conifers. *Food Control* **13**, 289–292.

NAGARAJ, R. (2001) Glycation and oxidative stress. *7th Int. Symposium on the Mailard Reaction*, Kumamoto, Japan.

NAGY, J.G. and TENGERDY, R.P. (1967) Antibacterial action of essential oils of Artemis as an ecological factor. *Applied Microbiology* **15**, 819–821.

NELSON, R.R.S. (2000) Selection of resistance to the essential oil of *Melaleuca alternifolia* in *Staphylococcus aureus*. *Journal of Antimicrobial Chemotherapy* **45**, 549–550.

NYCHAS, G.J.E. (1995) Natural antimicrobials from plants. In *New Methods of Food Preservation*, Gould, G.W (ed.), pp. 58–89. Blackie Academic Professional, London.

NYCHAS, G.J.E. and TASSOU, C.C. (2000) Preservatives: traditional preservatives – oils and spices. In *Encyclopedia of Food Microbiology*, Robinson, R., Batt, C. and Patel, P. (eds), pp. 1717–1722. Academic Press, London.

OYEDEJI, A.O., EKUNDAYO, O., OLAWORE, O.N., ADENIYI, B.A. and KOENIG, W.A. (1999) Antimicrobial activity of the essential oils of five *Eucalyptus* species growing in Nigeria. *Filoterapia* **70**, 526–528.

OZCAN, M., CHALCHAT, J.C. and AKGUL, A. (2001) Essential oil composition of Turkish mountain tea (*Sideritis* spp.). *Food Chemistry* **75**, 459–463.

PASTER, N., JUVEN, B.J., SHAAYA, E., MENASHEROV, M., NITZAN, R., WEISSLOWICZ, H. and RAVID, U. (1990) Inhibitory effect of oregano and thyme essential oils on moulds and foodborne bacteria. *Letters in Applied Microbiology* **11**, 33–37.

PASTER, N., MENASHEROV, M., RAVID, U. and JUVEN, B. (1995) Antifungal activity of oregano and thyme essential oils applied as fumigants against fungi attacking stored grain. *Journal of Food Protection* **58**, 81–85.

PATTE, J. (1996) Biosynthesis of threonine and lysine. In *Escherichia coli* and *Salmonella*, Frederci M. and Neidhardt F. (eds), 2nd Edition, pp. 528–541. ASM Press, Washington DC.

PATTNAIK, S., RATH, C. and SUBRAMANYAM, V.R. (1995a) Characterization of resistance to essential oils in a strain of *Pseudomonas aeruginosa* (VR-6). *Microbios* **81**, (326) 29–31.

PATTNAIK, S. SUBRANANYAM, V.R. and RATH, C.C. (1995b) Effect of essential oils on the viability and morphology of *Escherichia coli* (SP-11) *Microbios* **84**, 195, 199.

PATTNAIK, S., SUBRAMANYAM, V.R., KOLE, C.R. and SAHOO, S. (1995c) Antibacteial activity of essential oils from Cymbopogon: inter- and intra-specific differences. *Microbios* **84**(341), 239–245.

PEREZ, C., AGNESE, A.M. and CABRERA, J.L. (1999) The essential oil of *Senecio graveolens (Compositae)*: Chemical composition and antimicrobial activity tests. *Journal of Ethnopharmacology* **66**, 91–96.

PRINDLE, R.F. and WRIGHT, E.S. (1977) Phenolic compounds. In *Disinfection, Sterilisation & Preservation*, Block, S.S. (ed.), pp. 219–251. Lea & Febiger, Philadelphia.

RAMADAN, F.M., EL-ZANFALY, R.T., EL-WAKEIL, F.A. and ALLIAN, A.M. (1972) On the antibacterial effects of some essential oils I. Use of agar diffusion method. *Chem. Mikrobiol. Technol. Lebensm.* **2**, 51–55.

ROBINS, M.M, BROCKLEHURST, T.F. and WILSON, P.D.G (1994) Food structure and the growth of pathogenic bacteria. *Food Technology International Europe*, 31–36.

SALMERON, J. JORDANO, R. and POZO, R. (1990) Antimycotic and antiaflatoxigenic activity of oregano (*Origanum vulgare*, L.) and Thyme (*Thymus vulgaris*, L.). *Journal of Food Protection* **53**, 697–700.

SANKARAN, R. (1976) Comparative antimicrobial action of certain antioxidants and preservatives. *Journal of Food Science and Technology* **13**, 203–204.

SHELEF, L.A (1983) Antimicrobial effects of spices. *Journal of Food Safety* **6**, 29–44.

SHELEF, L.A., NAGLIK, O.A. and BOGEN, D.W. (1980) Sensitivity of some common food-borne bacteria to the spices sage, rosemary and allspice. *Journal of Food Science* **45**, 1042–1044.

SIKKEMA, J., DE BONT, J.A.M. and POOLMAN, B. (1995) Mechanisms of membrane toxicity of hydrocarbons. *Microbiology reviews* **59**, 201–222.

SKANDAMIS P.N. (2001) Effect of oregano essential oil on spoilage and pathogenic microorganisms in foods Ph.D Thesis Agricultural University of Athens.

SKANDAMIS, P.N. and NYCHAS, G-J.E. (2000) Development and evaluation of a model predicting the survival of *Escherichia coli* O157:H7 NCTC 12900 in homemade eggplant salad at various temperatures, pHs, and oregano essential oil concentrations. *Applied and Environmental Microbiology* **66**, 1646–1653.

SKANDAMIS, P. and NYCHAS, G-J,E. (2001) Effect of oregano essential oil on microbiological and physicochemical attributes of mince meat stored in air and modified atmospheres *Journal of Applied Microbiology* **91**, 1011–1022.

SKANDAMIS, P. and NYCHAS G-J.E. (2002a) Essential oils; can be considered 'smart' enough for their potential to be used in active packaging system? In *Joint Meeting of the SFAM & DSM 'Frontiers in Microbial fermentation and Preservation'*, Wageningen, The Netherlands 9–11 January 2002.

SKANDAMIS, P.N. and NYCHAS, G-J.E. (2002b) Preservation of fresh meat with active and modified atmosphere packaging conditions *International Journal of Food Microbiology* **79**, 35–43.

SKANDAMIS, P.N., MICHAILIDOU, E. and NYCHAS G.-J.E. (1999a) Modelling the effect of oregano (*Origanum vulgare*) on the growth/survival of *Escherichia coli* O157:H7 in broth and traditional Mediteranean foods *International Congress on 'Improved Traditional Foods for the next Century'*, Valencia 28–29/1999, Spain, pp. 270–273.

SKANDAMIS, P., TASSOU, C.C. and NYCHAS, G.-J.E. (1999b) Potential use of essential oils as food preservatives, In *17th International Symposium of the International Committee on Food Microbiology and Hygiene (ICFMH)*, Tuijtelaars, A.C.J., Samson, R.A., Rombouts, F.M. and Notermans, S. (eds), Veldhoven, The Netherlands, 13–17 September 1999, pp. 300–303.

SKANDAMIS, P., TSIGARIDA, E. and NYCHAS, G.-J.E. (2000) Ecophysiological attributes of *Salmonella typhimurium* in liquid culture and within gelatin gel with or without the addition of oregano essential oil. *World Journal of Microbiology and Biotechnology* **16**, 31–35.

SKANDAMIS, P., ELIOPOULOS, V. and NYCHAS, G.-J.E. (2001a) Effect of essential oils on survival *of Escherichia coli* O157:H7 NCTC 12900 and *Listeria monocytogenes* in traditional Mediterranean salads. Annual Conference of Society for Applied Microbiology, Swansea, UK 1–17 July 2001.

SKANDAMIS, P., KOUTSOUMANIS, K., FASSEAS, K. and NYCHAS, G.-J.E. (2001b) Evaluation of the inhibitory effect of oregano essential oil on *Escherichia coli* O157:H7, in broth culture with or without EDTA, using viable counts, turbidity and impedance, *Italian Journal of Food Science & Technology* **13**, 65–75.

SKANDAMIS, P., TSIGARIDA, E. and NYCHAS, G.-J.E (2002a) The effect of oregano essential oil on survival/death of *Salmonella typhimurium* in meat stored at 5 °C under aerobic, vp/map conditions. *Food Microbiology* **19**, 97–103.

SKANDAMIS, P., TSIGARIDA, E. and NYCHAS, G.-J.E. (2002b) Effect of conventional and natural preservatives on the death/survival of *Escherichia coli* O157:H7 NCTC 12900 in traditional Mediterranean salads; In *Joint Meeting of the SFAM & DSM 'Frontiers in Microbial fermentation and Preservation'*, Wageningen, The Netherlands, 9–11 January 2002.

SKANDAMIS, P.N., DAVIES, K.W., MCCLURE, P.J., KOUTSOUMANIS, K. and TASSOU, C. (2002c) A vitalistic approach or non-thermal inactivation of pathogens in traditional Greek salads. *Food Microbiology* **19**, 405–421.

SMITH-PALMER, A., STEWART, J. and FYFE, L. (1998) Antimicrobial properties of plant essential oils and essences against important food-.borne pathogens. *Letters in Applied Microbiology* **26**, 118–122.

SMITH-PALMER, A., STEWART, J. and FYFE, L. (2001) The potential application of plant essential oils as natural food preservatives in soft cheese. *Food Microbiology* **18**, 463–470.

STECCHINI, M.L., SARAIS, I. and GIAVEDONI, P. (1993) Effect of essential oils on *Aeromonas hydrophila* in a culture medium and in cooked pork. *Journal of Food Protection* **56**, 406–409.

STECCHINI, M.L., DEL TORRE, M., SARAIS, I., SARO, O., MESSINA, M. and MALTINI, E. (1998) Influence of structural properties and kinetic constraints on *Bacillus cereus* growth. *Applied and Environmental Microbiology* **64**, 1075–1078.

TAHARA, T., OSHIMURA,M., UMEZAWA, C. and KANATANI, K. (1996) Isolation, partial characterization and mode of action of acidocin J1132, a two-component bacteriocin produced by *Lactobacillus acidophilus* JCM 1132. *Applied and Environmental Microbiology* **62**, 892–897.

TAKAHASHI, Y., INABA, N., KUWAHARA, S. KUKI, W., YAMANE, K. and MURAKAMI, A. (2002) Rapid and convenient method for preparing aurapten-enriched product from hassaku peel oil. Implication of cancer-preventive food additives. *Journal of Agricultural Food Chemistry* **50**, 3193–3196.

TASSOU, C.C. (1993) Microbiology of olives with emphasis on the antimicrobial activity of phenolic compounds. Ph.D Thesis, University of Bath, Bath, UK.

TASSOU, C.C. and NYCHAS G.-J.E. (1995a) Inhibition of *Salmonella enteritidis* by oleuropein in broth and in a model food system. *Letters in Applied Microbiology* **20**, 120–124.

TASSOU, C.C. and NYCHAS, G.-J.E (1995b) The inhibitory effect of the essential oils from basil and sage in broth and in food model system. In *Developments in Food Science 37; Food Flavors: Generation, Analysis and Process Influence*, Charalambous, G. (ed.), pp. 1925–1936. Elsevier, New York.

TASSOU, C.C. and NYCHAS, G.-J.E. (1995c) Antimicrobial activity of the essential oil of mastic gum (*Pistachia lentiscus* var.*chia)* on Gram-positive and Gram-negative bacteria in broth and in model food system. *International Biodeterioration & Biodegradation* **36**, 411–420.

TASSOU, C.C., DROSINOS, E.H. and NYCHAS, G.-J.E. (1995) Effects of essential oil from mint *(Mentha piperita*) on *Salmonella enteritidis* and *Listeria monocytogenes* in model food systems at 4 and 10 °C. *Journal of Applied Bacteriology* **78**, 593–600.

TASSOU, C.C., DROSINOS, E.H. and NYCHAS, G.-J.E. (1996) Inhibition of the resident microbial flora and pathogen inocula on cold fresh fillets in olive oil, oregano and lemon juice under modified atmosphere or air. *Journal of Food Protection* **59**, 31–34.

TASSOU, C.C., DROSINOS, E.H. and NYCHAS, G.-J.E. (1997) Antimicrobial effect of carob (*Ceratonia siliqua*) extract against food related bacteria in culture media and model food systems. *World Journal of Microbiology and Biotechnology* **13**, 479–481.

TASSOU, C.C., KOUTSOUMANIS, K., SKANDAMIS, P. and NYCHAS, G.-J.E. (1999) Novel combinations of natural antimicrobial systems for the improvement of quality of agro-industrial products. Pp. 51–52 *International Congress on 'Improved Traditional Foods for the next Century*', Valencia, Spain, 28–29 November 1999.

TASSOU, C.C., KOUTSOUMANIS, K. and NYCHAS, G.-J.E. (2000) Inhibition of *Salmonella enteritidis* and *Staphylococccus aureus* in nutrient broth by mint essential oil. *Food Research International* **33**, 273–280.

THANGADURAL, D., ANITHA, S., PULLAIAH, T., REDDY, P.N. and RAMACHAMDRAIAH, O.S., (2002) Essential oil Constituents and *in vitro* antimicrobial activity of *Decalepis hamiltonii* roots against foodborne pathogens. *Journal of Agricultural and Food Chemistry* **50,** 3147–3149.

THOMAS, L.V, WIMPENNY, W.T. and BARKER, G.C. (1997) Spatial interactions between subsurface bacteria colonies in a model system: a territory model describing the inhibition of *Listeria monocytogenes* by a nisin-producing lactic acid bacterium. *Microbiology* **143**, 2575–2582.

TOKUTAKE, N., MIYOSHI, H. and IWAMURA, H. (1992) Effects of phenolic respiration inhibitors on cytochrome bc1 complex of rat-liver mitochondria. *Bioscience Biotechnology and Biochemistry* **56**, 919–923.

TRANTER, H.S., TASSOU C.C. and NYCHAS, G.-J.E. (1993) The effect of the olive phenolic compound, oleuropein, on growth and enterotoxin B production by *Staphylococcus aureus. Journal of Applied Bacteriology* **74**, 253–260.

TSIGARIDA, E., SKANDAMIS, P. and NYCHAS, G.-J.E. (2000) Behaviour of *Listeria monocytogenes* and autochthonous flora on meat stored under aerobic, vacuum and modified atmosphere packaging conditions with or without the presence of oregano essential oil at 5 °C. *Journal of Applied Microbiology* **89**, 901–909.

ULTEE, A., KETS, E.P.W. and SMID, E.J. (1999) Mechanisms of action of carvacrol on the food-borne pathogen *Bacillus cereus. Applied and Environmental Microbiology* **65**, 4606–4610.

ULTEE, A., SLUMP, R. A., STEGING, G. and SMID, E.J. (2000) Antimicrobial activity of carvacrol on rice. *Journal of Food Protection* **63**, 620–624.

ULTEE, A., BENNIK, M.H.J. and MOEZELAAR, R. (2002) The phenolic hydroxyl group of carvacrol is essential for action against the food-borne pathogen *Bacillus cereus*. *Applied Environmental Microbiology* **68**, 1561–1568.

WANDA, P., CUPP, J., SNIPES, W., KEITH, A., RUCINSKY, T., POLISH, L. and SANDS, J. (1976) Inactivation of the enveloped bacteriophage o6 by butylated hydroxytoluene and butylated hydroxyanisole. *Antimicrobial Agents and Chemotherapy* **10**, 96.

WILKINS, K.M. and BOARD R.G. (1989) Natural antimicrobial systems. In *Mechanisms of Action of Food Preservation Procedures*, Gould, G.W. (ed.), Chapter 11, pp. 285–362, Elsevier, London.

WILSON, P.D.G., BROCKLEHURST, T.F., ARINO, D., THUAULT, M., JAKOBSEN, M., LANGE, J.W.T., FARKAS, J., WIMPENNY, J.W.T. and VAN IMPE, J.F. (2002) Modelling microbial growth in structured foods: towards a unified approach. *International Journal of Food Microbiology* **75**, 273–289.

10

Natural antimicrobials in postharvest storage of fresh fruits and vegetables

A. Ippolito and F. Nigro, University of Bari, Italy

10.1 Introduction

Fresh fruits and vegetables are susceptible to microbial attack after harvest due to loss of their natural resistance and their high water and nutrient content (Eckert, 1991). Postharvest losses may reach up to 50% and more of the field depending on the crop, harvest methods, length of storage, marketing conditions, etc. (FAO, 1981). Fungal diseases are the major cause of losses in the postharvest industry and to consumers (Sommer *et al.*, 1992). Infections occurring either before harvest, between flowering and fruit maturity or during harvest, handling and storage are currently controlled with chemical fungicides (Eckert *et al.*, 1994). However, the use of synthetic fungicides has many limitations and disadvantages. These include progressively restrictive legislation, social rejection due to toxicological problems affecting humans and the environment, and development of resistance. Owing to their toxicity and oncogenic risk, their use in food products to prevent microbial spoilage is in general no longer recommended. Furthermore, the market for postharvest fungicides is relatively small and it has become difficult to sustain the costs of new registration or to support previous registrations. In addition, the rapid rise in demand for organically produced fruits and vegetables is increasing the demand for natural pesticides (Suslow, 2000). These issues have elicited widespread interest in the development of alternatives to synthetic fungicides for controlling postharvest diseases of fresh fruits and vegetables. Currently, no natural antimicrobial compound is used in industrial practice. However, natural compounds active against plant pathogens have served as models for the synthesis of fungicides with low toxicity, e.g. the phenyl pyrroles derived from pyrrolnitrin, a metabolite produced *Pseudomonas* spp. (Nevill *et al.*, 1988) and the strobilurins derived from a substance produced by the fungus *Strobilurus tenacellus* (Ammermann *et al.*, 1992; Godwin *et al.*, 1992).

The natural antimicrobial compounds reviewed in this chapter have been classified into those of plant, microbial and animal origin. The review focuses on those natural compounds that seem most promising for future development in the control of postharvest diseases. Other reviews of interest have focused on biological antagonists (Wilson and Wisniewski, 1989; Wilson *et al.*, 1994; El Ghaouth and Wilson, 1995; Mari and Guizzardi, 1997) or on specific groups of natural compounds (Arras, 1999).

10.2 Compounds of plant origin

The plant kingdom is recognized as an enormous source of antimicrobial compounds with applications in the food, cosmetic and drug industries. However, relatively little effort has been devoted to the development of plant-derived compounds as substitutes for synthetic fungicides.

10.2.1 Plant extracts

There is a large body of literature on the *in vitro* effects of plant extracts on foliar and root pathogens; Table 10.1 shows some examples. In contrast, relatively few reports of *in vivo* trials against postharvest diseases are available (Table 10.2).

The antimicrobial activity of allicin in garlic (*Allium sativum* L.) and other *Allium* species (onion (*A. cepa* L.) and leek (*A. porrum* L.)) has been described in Chapter 2, Section 2.9.1. Another compound in garlic, ajoene, has also been reported as antifungal (Yoshida *et al.*, 1987). Other compounds such as phenolics (Cao *et al.*, 1996; Yin and Tsao, 1999) and antifungal proteins such as allivin (Van Damme *et al.*, 1993; Wang and Ng, 2001) are also thought to be responsible for the inhibition of fungi. Extracts of *Allium* species inhibit the *in vitro* growth of *Aspergillus parasiticus*, *A. niger, A. flavus* and *A. fumigatus* (Sharma *et al.*, 1981; Yin and Tsao, 1999), and many other spoilage fungi of grains, legumes and processed food. Generally, the activity of *Allium* extracts diminishes during storage and is lost by heating (Sharma *et al.*, 1981; Yin and Tsao, 1999); crude juices or aqueous extracts were more active than ethyl acetate, ether, chloroform or ethanol extracts (Sharma *et al.*, 1979; Abdou *et al.*, 1992; Lima, Nigro and Ippolito, unpublished data). To the best of our knowledge, the only application of *Allium* extracts on fresh fruits against postharvest diseases has been that of Ark and Thompson (1959); they obtained complete control of brown rot of peaches caused by *Monilinia fructicola* by treatment with 5, 10 and 20% of a deodorized aqueous extract of commercial powdered garlic.

Of the 19 aqueous extracts of leaves and stems from plants indigenous to Mexico, 8 were active *in vitro* against sporulation of *Rhizopus stolonifer* (Bautista-Baños *et al.*, 2000); three of these extracts (from *Annona cherimola*

Table 10.1 Compounds of plant origin active *in vitro* against postharvest pathogenic fungi*

Plant source	Extract type or compound	Target organism	References
Adenocalyma, Allium, Tulbaghia, Capsicum	Aqueous extract	*B. cinerea*	Wilson *et al.* (1997)
Allium	Aqueous extract	*A. parasiticus, A. niger, A. flavus, A. fumigatus*	Sharma *et al.* (1981); Yin and Tsao (1999)
Aloe	Aqueous extract	*P. digitatum, P. expansum, B. cinerea, A. alternata*	Saks and Barkai-Golan (1995)
Brassica	Glucosinolates	*M. laxa, B. cinerea, R. stolonifer, M. piriformis, P. expansum*	Mari *et al.* (1993; 1996; 2002)
Citrus	Essential oils	*P. expansum, P. italicum, P. digitatum*	Caccioni *et al.* (1995); Venturini *et al.* (2002)
Citrus	Coumarins	*A. niger, P. glaucum, R. nigricans, P. digitatum, P. italicum, B. cinerea, Fusarium* spp.	Jurd *et al.* (1971a,b); Recio *et al.* (1989); Kwon *et al.* (1997) Rodov *et al.* (1995); Angioni *et al.* (1998); Ojala *et al.* (2000)
Helicrysum	Galangin	*P. digitatum, P. italicum*	Afolayan and Meyer (1997)
Malus	Chlorogenic acid	*P. vagabunda*	Lattanzio *et al.* (2001)
Origanum, Inula, Mentha	Petroleum extracts	*B. cinerea, A. solani, Penicillium* sp., *Cladosporium* sp., *F. oxysporum*	Abou Jawdah *et al.* (2002)
Origanum	Essential oils	*A. niger, P. citrophthora*	Elgayyar *et al.* (2001); Arras (1988)
Several genera	Propolis	*B. cinerea, P. expansum*	Lima *et al.* (1998); La Torre *et al.* (1990); Meir *et al.* (1998)
	Jasmonates: jasmonic acid, methyl jasmonate	*B. cinerea*	
	Phenols: cinnamic acid derivatives	*S. sclerotiorum, F. oxysporum*	Lattanzio *et al.* (1994a)
	2,5-dimetho-xybenzoic acid	*B. cinerea, R. stolonifer*	Lattanzio *et al.* (1994b)
	Flavonoids	*Penicillium* spp.	Lattanzio *et al.* (1994b)

Table 10.1 Compounds of plant origin active *in vitro* against postharvest pathogenic fungi*

Plant source	Extract type or compound	Target organism	References
Thymus	Essential oils	*P. italicum, P. digitatum, B. cinerea, A. citri, A. alternata, F. oxysporum*	Arras and Grella, (1992); Arras *et al.* (1995); Reddy *et al.* (1998); Arras and Usai, 2001)
Fruit tissues of several genera	Acetaldehyde	*B. cinerea, R. stolonifer*	Avissar *et al.* (1990)
	Benzaldehyde, methyl salicylate, Ethyl benzoate	*M. frutticola, M. laxa, B. cinerea*	Wilson *et al.* (1987; Tonini and Caccioni (1990)
	(*E*)-2-hexenal	*B. cinerea*	Fallik *et al.* (1998)
	Low molecular weight aldehydes	*Alternaria* spp., *B. cinerea, C. herbarum, P. expansum, Rhizopus* spp.	Mattheis and Roberts (1993)
Vitis and other genera	Acetic acid	*B. cinerea, P. expansum*	Stadelbacher and Prasad, 1974; Sholberg and Gaunce (1995)

* The *in vitro* antimicrobial properties of plant-derived compounds against microorganisms other than postharvest pathogens are shown in Table 9.3 (Chapter 9).

Table 10.2 Examples of successful control of postharvest diseases by natural compounds of plant origin

Commodity	Disease	Active principle	References
Apples	Gleosporium rot (lenticel spot)	Phloridzin and chlorogenic acid	Lattanzio *et al.* (2001)
Cherries, sweet	Botrytis rot	Thyme essential oil	Chu *et al.* (1999)
	Penicillium rot	Low molecular weight aldehydes	Mattheis and Roberts (1993)
Ciruela	Rhizopus rot	*Annona cherimola, Bromelia hemisphaerica,* and *Carica papaya* extracts	Bautista-Baños *et al.* (2000)
Grapefruit	Penicillium rot	*Aloe vera* gel	Saks and Barkai-Golan (1995)
Grapefruit	Penicillium rot	Jasmonate	Droby *et al.* (1999)
Peaches	Monilinia rot	*Allium* extract	Ark and Thompson (1959)
Peaches, nectarines, plums	Monilinia and Rhizopus rots	Benzaldehyde, hexanal	Caccioni *et al.* (1995b)
Peaches, apples, grapes, stone fruits, strawberries	Botrytis, Penicillium, Rhizopus and Monilinia rots	Acetic acid and vinegar	Sholberg and Gaunce (1995, 1996; Sholberg *et al.* (1996, 2000)
Pears	Botrytis, Monilinia, Mucor and Penicillium rots	Isothiocyanates	Mari *et al.* (1996, 2002)
Strawberries	Botrytis rot	2,5-Dimethoxybenzoic acid	Lattanzio *et al.* (1994b)
Strawberries, raspberries, apples, grapes	Botrytis, Rhizopus, Aspergillus, Alternaria and Penicillium rots	Acetaldehyde	Prasad and Stadelbacher (1973, 1974); Stadelbacher and Prasad (1974); Pesis and Avissar (1990); Avissar and Pesis (1991)
Strawberries, blackberries, grapes, stone fruits	Botrytis, Rhizopus and Monilinia rots	Hexanal, 1-hexanol, (*E*)-2-hexenal, (*Z*)-6-nonenal, (*E*)-3-nonen-2-one, methyl salicylate, methyl benzoate	Archbold *et al.* (1997, 1999); Hamilton-Kempt *et al.* (1992); Nandi and Fries (1976); Caccioni *et al.* (1997)
Sponge-gourd	Soft rot caused by *F. scirpi* and *H. spiciferum*	Aqueous leaf extracts of *Azadirachta indica, Datura fistulosa, Muraya exotica, Lantana camara, Ocimum sanctum* and *Catharantes roseus*	Ahmad and Prasad (1995)

M., *Bromelia hemisphaerica* L. and *Carica papaya* L.) were also active *in vivo* on 'ciruela' fruits (*Spondias purpurea* L.) against *R. stolonifer*. Interestingly, the leaf extract of *Casimiroa edulis* Llav. et Lex. was not active *in vitro* but completely inhibited disease development on fruits. Extracts from the leaf pulp of *Aloe vera*, commonly known as *Aloe vera gel* (Saks and Barkai-Golan, 1995) have been tested *in vitro* against four common postharvest pathogens *P. digitatum, P. expansum, B. cinerea*, and *Alternaria alternata*, the first and the last being most sensitive. Dipping of *P. digitatum*-inoculated grapefruits in solutions containing 1 mg/l of gel delayed lesion development and significantly reduced the incidence of infection. Aqueous leaf extracts of *Azadirachta indica* Adv. Juss., *Datura fistulosa* L. *Muraya exotica* L. *Lantana camara* L. *Ocimum sanctum* L. and *Catharantes roseus* L. almost completely inhibited the spread of soft rot diseases caused by *Fusarium scirpi* and *Helminthosporium spiciferum* on *Luffa cylindrica* L. (sponge-gourd) when applied after infection. However, activity was less evident when treatment with leaf extracts was done before infection with the pathogens (Ahmad and Prasad, 1995). In general, the *in vitro* activity of leaf extracts on mycelial growth, mycelial dry weight, and spore germination agreed with *in vivo* activity.

10.2.2 Propolis

Propolis (bee glue) is a resinous or sometimes wax-like compound collected by *Apis mellifera* bees from plant parts and used for the construction and maintenance of their hives. The chemical composition, as well as its colour and aroma, change according to the geographical zones (Bankova *et al.*, 1998). Hydroquinones, caffeic acid and its esters, and quercetin are the most common constituents of propolis (Greenaway *et al.*, 1991; Burdock, 1998). Many of these are also present in foods, food additives and/or are Generally Recognized As Safe (GRAS). The pharmacological and antimicrobial activities of propolis are well documented (La Torre *et al.*, 1990; Marcucci, 1995; Burdock, 1998; Ota *et al.*, 2001). Antimicrobial action is linked to the polar compounds, mainly flavonoids, phenolic acids and their esters (Ghisalberti, 1979). In fact, the flavonoids galangin, pinocembrin and pinostrobin, and the ferulic and caffeic acids are the most effective antibacterial compounds occurring in propolis (Marcucci, 1995). Despite this, there are few reports of propolis applications for controlling plant pathogens. In trials aimed at assessing the compatibility of postharvest antagonistic yeasts with additives and agrochemicals, Lima *et al.* (1998) found that propolis inhibited both antagonists (*Rhodotorula glutinis, Cryptococcus laurentii* and *Aureobasidium pullulans*) and plant pathogens (*B. cinerea* and *P. expansum*), suggesting that it could be used as a natural fungicide.

10.2.3 Jasmonates

Jasmonic acid (JA) and its volatile esterified derivative, methyl jasmonate (MeJA), are naturally occurring lipids of the plant cell membranes, synthesized via the lipoxygenase pathway. Application of JA or MeJA to plants induces the expression of genes involved in defence mechanisms (Gundlach *et al.*, 1992; Doares *et al.*, 1995; Sharan *et al.*, 1998). When tested *in vitro*, neither JA nor MeJA had any direct antifungal effect on *P. digitatum* spore germination or germ tube elongation (Droby *et al.*, 1999). However, jasmonates significantly reduced green mould of grapefruit caused by *P. digitatum* after either natural or artificial inoculation. The most effective concentration for reducing decay in cold-stored fruit or after artificial inoculation of wounded fruit was 10 μmol/l. These results suggest that jasmonates act as resistance inducers in the fruit. The involvement of phytoalexins cannot be excluded considering that MeJA induces the syntheses of scopoletin and scopolin in tobacco cell cultures (Sharan *et al.*, 1998). MeJA was also tested for postharvest control of grey mould disease in cut rose cultivars. Pulsed treatment of the flowers with 200 μM MeJA induced systemic protection against *B. cinerea*, following either natural or artificial infection. However, for local protection on flower petals following simultaneous application of *B. cinerea* conidia and MeJA, at least 300 μM MeJA was required (Meir *et al.*, 1998).

10.2.4 Essential oils

The antimicrobial activity of essential oils (EOs) against important plant and human pathogens, as well as food spoilage organisms, has been studied extensively (Chapter 9). Recently, there has been a renewed interest in the application of these substances to control plant pathogens and postharvest diseases in particular (Arras and Usai, 2001; Aligiannis *et al.*, 2001; Thangadurai *et al.*, 2002). The role played by these substances in the plant has not been fully elucidated; however, it is likely that most of them are involved in chemical defence mechanisms against phytopathogenic micro-organisms (Mihaliak *et al.*, 1991).

Among the many EOs tested *in vitro* and *in vivo* against postharvest pathogens, those from plants of the genus *Thymus* have been particularly active. Thyme EOs have been tested on *P. italicum, P. digitatum, B. cinerea, Alternaria citri, A. alternata, Fusarium oxysporum*, and *R. stolonifer* (Arras and Grella, 1992; Arras *et al.*, 1995; Reddy *et al.*, 1998; Arras and Usai, 2001); concentrations of 200–250 ppm of *T. capitatus* EO completely inhibited growth of the fungi (Arras and Usai, 2001). With *T. vulgaris* extracts, *B. cinerea* and *R. stolonifer* were inhibited by more than 50% (Reddy *et al.*, 1998). On strawberries, *T. vulgaris* EO reduced decay due to *B. cinerea* and *R. stolonifer* by up to 76% (Reddy *et al.*, 1998). Vapours of thyme EO reduced grey mould development on *Botrytis*-inoculated sweet cherries (Chu *et al.*, 1999) and brown rot in apricots and plums (Liu *et al.*, 2002).

Generally, the fungicidal activity of EOs observed *in vitro* was not reproduced *in vivo* or *in situ* because of the volatile nature of the constituents. Arras and Usai (2001) have overcome this problem by combining thyme EO with vacuum at 0.5 atm (see also Section 10.6). Carvacrol has been identified as the substance responsible for the antimicrobial activity in the EO of *T. capitatus* (Arras and Grella, 1992). Thymol, carvacrol and linalool were the active agents in *T. vulgaris* (Reddy *et al.*, 1998). However, other minor components can also contribute synergistically to the antimicrobial activity of an EO (Lattaoui and Tantoui-Elaraki, 1994; Cosentino *et al.*, 1999; Karaman *et al.*, 2001). The EO of oregano (*Origanum* spp.) containing thymol and carvacrol was reported as very active *in vitro* against several mycotoxigenic fungi (Elgayyar *et al.*, 2001; Lambert *et al.*, 2001) and against some citrus disease agents (Arras, 1988).

Among the complex constituents of citrus EOs, the terpene citral (3,7-dimethyl-2,6-octadiennal) is known to have strong antifungal properties (Rodov *et al.*, 1995; see also Chapter 11). Recent *in vitro* studies have demonstrated that citral inhibited *P. expansum*, *P. italicum* and *P. digitatum*, responsible for severe storage rot of apples and citrus (Caccioni *et al.*, 1995a, 1998; Venturini *et al.*, 2002). However, because of its phytotoxicity, citral may be difficult to use on fresh fruits and vegetables (Rodov *et al.*, 1995). The strong aroma of EOs limits their application in foods, as already discussed in Chapter 9.

10.2.5 Plant phenolic compounds

Phenolic compounds confer flavour and colour to plants, fruits and vegetables. In addition, they defend the plant against attack by microorganisms, insects and herbivores. The role of phenolics in plant physiology has been comprehensively reviewed in recent years, e.g. by Lattanzio *et al.* (1994a), Rhodes (1994), Dixon *et al.* (1995), Boudet *et al.* (1995) and Hammerschmidt (1999). The genomics of the phenylpropanoid pathway and the plant defence mechanisms have also been reviewed (Dixon *et al.*, 2002). In this section, the focus is on the relationship between phenolic compounds and their role in preventing postharvest diseases of fresh fruits and vegetables.

All fruits and vegetables contain biologically active phenolic compounds (Spanos and Wrolstad, 1992; Shahidi and Naczk, 1995), most of which have not been fully exploited as natural antifungal substances and alternative means to control postharvest diseases (Wilson and Wisniewski, 1989). Accumulation of phenolic compounds depends on the species and/or variety, the physiological state, and the geographical location of the plant (Lattanzio, 1988; Mueller Harvey and Dhanoa, 1991). Concentrations of phenolic compounds are generally higher in young fruits and tissues (Macheix *et al.*, 1990; Lattanzio *et al.*, 1994a). In fruits, the total phenol content falls during growth, but two distinct phenomena can be observed: the level continues to fall steadily, as in the case of white-coloured species and varieties (e.g. white grape cultivars), or it rises at the end of maturation as in the case of red fruits in which anthocyanins or flavonoids accumulate (Macheix *et al.*, 1990).

Phenolic compounds that inhibit the growth of fungi may be present in healthy, unchallenged fruits and vegetables (preformed antimicrobial compounds), or may be found only in fruit tissues which have either been infected by pathogens or weakly stressed (phytoalexins) (Paxton, 1981; Kuć, 1995). The first group includes simple phenols, phenolic acids, flavonols, some isoflavones and dihydrocalchones (phloridzin). The second group includes phenolic phytoalexins, isoflavonoids, flavans, stilbenes, phenanthrenes, pterocarpans and furocoumarins (Lattanzio *et al.*, 2001). All these compounds originate through different branches of the so-called 'general phenylpropanoid pathway', leading to the formation of a series of hydroxycinnamic acids and derivatives with antimicrobial activity. *In vitro* trials have shown that the hydroxycinnamaldehydes possess higher antifungal and antibacterial activity than hydroxycinnamic acids and hydroxycinnamyl alcohols (Barber *et al.*, 2000).

Caffeic and coumaric acid are cinnamics widely distributed in apples, pears and grapes. They occur naturally in combination with other compounds, usually in the form of esters. The ester of caffeic with quinic acid, chlorogenic acid, is a classic example. Among a group of cinnamic acid derivatives (Table 10.3) tested *in vitro* for their activity against several postharvest pathogens, chlorogenic and ferulic acid were strong inhibitors of *F. oxysporum* and *S. sclerotiorum*, respectively (Lattanzio *et al.*, 1994a). The presence of a hydroxyl group in the aromatic ring of cinnamic acid (i.e. *p*-coumaric acid, 3-coumaric acid and 2-coumaric acid) increased the antifungal activity against *F. oxysporum* and *Alternaria* spp. An additional hydroxyl group in the benzene ring did not increase antifungal activity. In contrast, the presence of a methoxy group increased activity compared with coumaric acids: ferulic acids were the better inhibitors among the cinnamic derivatives. Benzoic derivatives were the best inhibitors of *B. cinerea, P. digitatum, S. sclerotiorum, F. oxysporum* and *Alternaria* spp. The presence of an additional hydroxyl group in the ring of p-hydroxybenzoic acid improved the antifungal activity of the mono-phenol (Table 10.3). When a methoxy group was introduced in the aromatic ring of the mono-phenol, vanillic and isovanillic acids, antifungal activity decreased. The presence of two methoxy groups in the molecule (i.e. 2,3-, 2,4- and 2,5-dimethoxybenzoic acid) increased the antifungal activity up to 75% or more.

The antifungal activity of 2,5-dimethoxybenzoic acid (DMBA) in controlling postharvest pathogens has been tested both *in vitro* and *in vivo*. *In vitro* studies demonstrated that DMBA inhibited both spore germination and mycelial growth of *B. cinerea* and *R. stolonifer*. Spraying or dipping of strawberries in 0.01 M DMBA solutions supplemented with 0.05% Tween for 1 min reduced storage decay at both 20 and 3 °C (Lattanzio *et al.*, 1994b). DMBA levels decreased rapidly during the first three days of storage and this was more pronounced in fruits stored at 20 °C. At the end of the storage period, less than 15% of the applied phenolic remained on the strawberries.

The antifungal activity of phenolics is associated with their lipophilicity and/or the presence of a hydroxyl group. Lipophilic properties allow penetration of biological membranes while hydroxyl groups help to uncouple oxidative phosphorylation (Lattanzio *et al.*, 1994a).

Among a group of phenolics from apples (Golden Delicious), only chlorogenic acid inhibited *Phlyctaena vagabunda* spore germination and mycelial growth *in vitro*, while (+)-catechin, (−)-epicatechin, phloridzin and

Table 10.3 Mycelial growth inhibition or stimulation (%) of four storage fungi by phenolic compounds determined *in vitro*

Compound	% change in growth of fungal species			
	S. sclerotiorum	*Alternaria* sp	*B. cinerea*	*P. digitatum*
Flavonoids				
(+)-Catechin	3	−11	0	−17
Apigenin-7-glucoside	−8	−2	−5	−13
Kaempferol-3-rutinoside	−11	−2	−20	0
Luteolin-7-glucoside	−2	−7	−2	−21
Quercetin-3-rhamnoside	0	−11	−7	−25
Quercetin-3-rutinoside	12	−16	12	−17
Myricetin	−12	−2	16	−14
Cinnamic derivatives				
Cinnamic acid	−44	−9	−9	−18
4-Coumaric acid	−33	−32	−18	−27
3-Coumaric acid	−12	−27	−4	−14
2-Coumaric acid	−31	−47	−15	7
Caffeic acid	−18	−27	0	−36
Dihydrocaffeic acid	−24	−27	−34	−41
Ferulic acid	−69	−34	−32	−31
3,5-Dimetoxycinnamic acid	−15	−8	−7	0
Sinapic acid	−7	−9	−7	−29
5-Caffeoylquinic acid	−7	−45	0	−8
1,3-Dicaffeoylquinic acid	−12	−7	3	−14
Benzoic derivatives				
4-Hydroxybenzoic acid	−26	−37	11	−15
2,3-Dihydroxybenzoic acid	−10	−74	−2	−11
3,4-Dihydroxybenzoic acid	−33	−22	−3	−14
2,5-Dihydroxybenzoic acid	−80	−60	−38	−18
Gallic acid	−29	−38	10	−17
Vanillic acid	−8	−3	−2	0
2,3-Dimethoxybenzoic acid	−58	−76	−20	−24
2,4-Dimethoxybenzoic acid	−61	−81	−48	−37
2,5-Dimethoxybenzoic acid	−100	−79	−87	−60
Syringic acid	−16	−43	10	−6

Source: adapted from Lattanzio *et al.* (1994b).

quercetin glycosides showed no activity (Lattanzio *et al.*, 2001). Similarly, *in vitro* testing of catabolic phloridzin derivatives (phloretin, phloroglucinol, phloretic acid and *p*-hydroxybenzoic acid) indicated no inhibitory effects on mycelial growth of *P. vagabunda*. However, both phloridzin and chlorogenic acid in combination with polyphenol oxidase inhibited *P. vagabunda* in quiescent infections associated with immature and ripening apple fruit during cold storage (Lattanzio *et al.*, 2001).

Flavonoids are a large group of secondary plant metabolites that are widely distributed throughout the plant kingdom. To date, about 8000 varieties of flavonoids have been reported (Pietta, 2000). They are characterized by a common benzo-γ-pyrone structure and play important roles in plant growth and development and in defence against microorganisms and pests (Harborne and Williams, 2000). Flavonoids have been reported to exhibit antiallergenic, antiviral, antifungal, anti-inflammatory and vasodilating activities (Colerige Smith *et al.*, 1980; Bors *et al.*, 1990). Since they show low toxicity in mammals, some flavonoids are used in human medicine (Cesarone *et al.*, 1992; Hertog *et al.*, 1993; Pietta, 2000).

Galangin (3,5,7-trihydroxyflavone) extract from *Helicrysum aureonitens*, used topically by indigenous populations of South Africa, is active against *P. digitatum* and *P. italicum* (Afolayan and Meyer, 1997). Lattanzio *et al.* (1994a) have shown that flavanoids have low antifungal activity at the concentration range 0.1 to 0.01 mM. The highest mycelial inhibition was observed with apigenin-7-glucoside and kaempferol-3-rutinoside, which differ in their polarity (Table 10.3). All the flavonoids tested, except for kaempferol-3-rutinoside, showed appreciable inhibitory activity against *Penicillium* spp. (Table 10.3). These results suggest that the combination of flavonoid/fungus/host is of fundamental importance in inhibiting growth. The major role of flavonoids has been linked with the resistance mechanisms in the host, acting as phytoanticipins or phytoalexins.

Among the constitutive secondary metabolites, those occurring in citrus fruits have been widely investigated. Tangeretin and naringin act as first and second defence barriers against fungal attack since polymethoxylated flavones (tangeretin) are mainly localized in the outermost tissue of the fruit (the flavedo) while flavanones (naringin) are located in the albedo (Kanes *et al.*, 1992). Other secondary compounds induced after infection, such as coumarins, have also been identified in citrus fruits (Angioni *et al.*, 1998; Arcas *et al.*, 2000). Coumarins are phenolic substances containing fused benzene and alfa-pyrone rings. Some citrus species accumulate coumarins such as xanthyletin, seselin and scoparone when infected by pathogenic fungi (Afek and Sztejnberg, 1993; Stange *et al.*, 1993). The nature of the coumarin biosynthesized in this process varies within a species according to the pathogen. For example, *Citrus limon* L. accumulated scoparone after inoculation with *P. digitatum* (Kim *et al.*, 1991); however, there was no significant accumulation of any antifungal compounds in the tissues of lemons inoculated with *Geotrichum candidum* (Baudoin and Eckert, 1985). Although coumarin inhibits the germination of spores of *A. niger, P. glaucum* and *R. nigricans*, other 4-hydroxycoumarins are generally ineffective against fungi. Antibacterial and antifungal effects have been found for umbelliferone, and scopoletin (Jurd *et al.*, 1971a,b; Recio *et al.*, 1989; Kwon *et al.*, 1997). *In vitro* tests indicated that pure coumarins have a very modest activity against *A. niger*. In contrast, scopoletin promoted the growth of *B. cinerea* while *Fusarium* proved to be the most sensitive among the pathogenic fungi (Ojala *et al.*, 2000).

Limettin (5,7-dimethoxycoumarin) and 5-geranoxy-7-methoxycoumarin and isopimpinellin (5,8-dimethoxypsoralen) were shown to be effective antifungal compounds (Rodov *et al.*, 1995). 7-Geranoxycoumarin showed inhibitory activity against *P. italicum in vitro* which was comparable to other naturally occurring compounds like scoparone and scopoletin. When applied to grapefruit at 500 mg/l, the antifungal activity of 7-geranoxycoumarin against *P. italicum* was comparable to that of scoparone and scopoletin. However, its activity was much higher against *P. digitatum*, the major postharvest pathogen of citrus fruit (Angioni *et al.*, 1998). Scoparone and scopoletin are phytotoxic compounds, causing browning and necrosis of the rind at very low concentrations (50 mg/l). In contrast, 7-geranoxycoumarin had no adverse effects on the citrus peel when applied at higher concentrations (Rodov *et al.*, 1995; Angioni *et al.*, 1998). All these compounds have potential for development as alternatives to control the postharvest decay of citrus fruits since it has been demonstrated that they can be induced by a number of physical and biological treatments (Ben-Yehoshua *et al.*, 1988, 1992; Kim *et al.*, 1991; Wilson *et al.*, 1994).

10.2.5 Glucosinolates

Glucosinolates (GLs) are sulfur-containing plant secondary metabolites occurring in every organ of cruciferous crops (*Brassicaceae*) and a restricted number of other plant families, among which *Capparaceae* and *Caricaceae* are the most important (Fahey *et al.*, 2001). The glucosinolate molecule consists of a β-thioglucose moiety, a sulphonated oxime group and a variable side chain, derived from an amino acid. The most numerous GLs are those containing either straight or branched carbon chains. GLs are always found associated with thioglucosidases ('myrosinases') which gives rise to products such as nitriles, thiocyanates and isothiocyanates (Mithen, 2001), depending on both the substrate and reaction conditions, especially pH. In controlled pH conditions (near neutral) the GL breakdown products are predominantly isothiocyanates (Gil and MacLeod, 1980). These products have a wide range of biological activity including both negative and positive nutritional attributes and effects on insects and herbivores. Recent reports about the potential anticarcinogenic activity of GL degradation products (Verhoeven *et al.*, 1997) have stimulated interest in their possible use as food additives. Many GL breakdown products are toxic to microorganisms and it has been suggested that these compounds may play a role in plant disease resistance (Mithen *et al.*, 1986; Doughty *et al.*, 1996; Manici *et al.*, 1997).

The activity of six GLs (glucoraphenin, gluconapin, sinigrin, glucotropaeolin, sinalbin and rapeseed glucosinolates) against *B. cinerea, R. stolonifer, Monilia laxa, Mucor piriformis* and *P. expansum* has been tested *in vitro* and *in vivo* by Mari *et al.* (1993, 1996). They found that the six native GLs were ineffective in inhibiting germination of the pathogens whereas all the derived isothiocyanates inhibited germination with variable intensity. The isothiocyanates from glucoraphenin, sinigrin and sinalbin totally inhibited the germination of five pathogens. None of the compounds inhibited the mycelial growth of *M. piriformis* and *R. stolonifer*, whereas isothiocyanates from glucoraphenin were the most effective against *P. expansum, B. cinerea* and *M. laxa*. The volatile compounds obtained from the enzymatic hydrolysis of sinigrin and gluconapin (2-propenyl and 3-butenyl isothiocyanates, respectively) strongly inhibited germination and/or mycelial growth of *M. laxa, B. cinerea* and *P. expansum* (Mari *et al.*, 1993). The same GLs and isothiocyanates were tested *in vivo* using two pear varieties (Conference and Kaiser). Isothiocyanates from glucoraphenin were the most effective against *M. laxa, B. cinerea* and *M. piriformis* after six days at 20 °C, showing a significant reduction in lesion diameter in artificially inoculated pears. The concentration of the glucoraphenin-derived isothiocyanates strongly affected the antifungal activity: the highest value tested (3.6 mg/ml) completely inhibited lesion development by *M. laxa* even when a spore suspension of 10^6 cells/ml was used. Moreover, isothiocianate stopped *M. laxa* infections already underway, showing a curative effect up to 40 h after inoculation (Mari *et al.*, 1996).

The activity of allyl-isothiocyanates (AITC) vapour from pure sinigrin or from *Brassica juncea* defatted meal against the decay caused by *P. expansum* on pears has also been tested (Mari *et al.*, 2002). The best decay control was obtained by exposing fruits for 24 h at 20 °C in an atmosphere enriched with 5 mg/l of AITCs. The extent of control depended on the inoculum density of the pathogen. The results suggested that effective use of this compound would depend on the extent of existing contamination of fruit surfaces, packhouse working lines, floating water, etc. Moreover, AITC treatments were effective up to 24 h for Conference and 48 h for Kaiser pears after inoculation with thiabendazole-resistant *P. expansum* and reduced the incidence of blue mould in both cultivars by 90% (Mari *et al.*, 2002). The availability of compounds possessing curative effects could overcome one of the major limitations of alternative methods in controlling postharvest decay, namely their inefficacy against active infections. Residue concentrations of AITC in the skin and pulp of the treated pears was extremely low, suggesting the absence of any implications for human health (Mari *et al.*, 2002). However, further evidence is necessary to validate the effectiveness of this compound in pome and stone fruits and in large-scale treatments.

10.3 Volatile compounds

Volatile compounds from plants can be either inhibitory or stimulatory to fungal growth and/or spore formation and germination (Fries, 1973; French and Wilson, 1981). Resistance of strawberry fruit to rot in high-CO_2 storage has been ascribed to the production of high levels of acetaldehyde and ethyl acetate by the fruit in response to these conditions. Fumigation with acetaldehyde at 0.1–1% has resulted in inhibition of spore germination and reduced mould development on strawberries (Prasad and Stadelbacher, 1974; Pesis and Avissar, 1990) raspberries (Prasad and Stadelbacher, 1973), apples (Stadelbacher and Prasad, 1974), grapes (Avissar and Pesis, 1991) and sweet cherries (Mattheis and

Roberts, 1993). However, in some cases acetaldehyde-induced phytotoxicity symptoms (Stadelbacher and Prasad, 1974; Stewart *et al.*, 1980; Avissar and Pesis, 1991; Mattheis and Roberts, 1993) and altered fruit sensory traits (Pesis and Avissar, 1990; Avissar and Pesis, 1991) depending on concentration and exposure duration. The mode of action of acetaldehyde has not been fully elucidated; however, there is evidence that it causes membrane disruption followed by leakage of electrolytes, reducing sugars and amino acids from the cells (Avissar *et al.*, 1990).

In vitro trials using 16 volatile compounds from peach and plum have demonstrated the high efficacy of ethyl benzoate, methyl salicylate and benzaldehyde in inhibiting the growth of *B. cinerea* and *M. fructicola*; benzaldehyde totally inhibited spore germination of *B. cinerea* at 25 μl/l and germination of *M. fructicola* at 125 μl/l (Wilson *et al.*, 1987). Tonini and Caccioni (1990) reported similar results for stone fruit. Caccioni *et al.* (1995b) reported that, among eight volatile compounds forming the characteristic aroma of ripe stone fruits, benzaldehyde was, at 5000 ppm, one of the most active in reducing decay by *M. laxa* and *R. stolonifer* in inoculated peaches, nectarines, and plums. Also among nine low molecular weight aliphatic aldehydes produced by sweet cherries, acetaldehyde, together with propanal and butanal, significantly reduced decay of *P. expansum*-inoculated sweet cherries (Mattheis and Roberts, 1993), acetaldehyde being the most active.

Archbold *et al.* (1997) have reported that hexanal, 1-hexanol, (*E*)-2-hexenal, (*Z*)-6-nonenal and (*E*)-3-nonen-2-one, and the aromatic compounds methyl salicylate and methyl benzoate had potential as postharvest fumigants for the control of *B. cinerea* on strawberries, blackberries, and grapes at concentrations as low as 2–12 μl/250 ml. Later, the same authors (Archbold *et al.*, 1999) showed that one of the compounds, (*E*)-2-hexenal, was effective against grey mould on seedless table grapes but complete mould suppression was not achieved as the level of hexenal declined during the course of the trial. Headspace analyses have shown that (*E*)-2-hexenal concentrations below 0.5 μmol/l stimulated *B. cinerea* mycelial development *in vitro*, while concentrations above it inhibited growth of the mould (Fallik *et al.*, 1998). The antifungal activity of hexanal has been studied on several host-pathogen combinations (Hamilton-Kemp *et al.*, 1992; Nandi and Fries, 1976; Caccioni *et al.*, 1997) and it proved to be active on stone fruits against *R. stolonifer* and *M. laxa*. At 2500 ppm, hexanal produced the same fungistatic effect as 5000–10 000 ppm of benzaldehyde, but at higher concentrations it was phytotoxic (Caccioni *et al.*, 1995b). Methyl-salicylate demonstrated some fungistatic activity, but it also gave an unpleasant odour to the fruit (Caccioni *et al.*, 1995b).

The use of natural, volatile aroma compounds as antimicrobial fumigants is an interesting field of investigation still not fully explored. In general, these substances have limited toxicity to mammals and a degree of volatility that allows their application in fumigation of cold storage rooms or in ‘active packaging’ (Toray Research Center, 1991).

Acetic acid, the principal organic compound of vinegar (reviewed in Chapter 6), could also be used in the vapour phase to control postharvest pathogens. Early studies demonstrated its efficacy against conidia of *M. fructicola* on peaches (Roberts and Dunegan, 1932). As little as 1.4 mg/l of acetic acid vapour prevented decay of peaches inoculated with conidia of *M. fructicola* or *R. stolonifer*; fumigation with 2.0 or 4.0 mg/l acetic acid before wounding prevented decay in apples contaminated with *B. cinerea* or *P. expansum*, respectively (Sholberg and Gaunce, 1995, 1996). Acetic acid was lethal at 0.1 and 0.15% to *B. cinerea* and *P. expansum*, respectively, while 0.7 and 2% acetaldehyde was required to achieve the same effect (Avissar *et al.*, 1990; Stadelbacher and Prasad, 1974). At concentrations of 0.18–0.27% (vol/vol), acetic acid controlled *Botrytis* and *Penicillium* decay on two Canadian table grape varieties, to the same extent as SO_2, with no adverse effects on fruit composition (°Brix, titratable acidity and pH) (Sholberg *et al.*, 1996). However, acetic acid, like other short-chain organic acids, can be extremely phytotoxic in the vapour form, depending on concentration and exposure (Sholberg, 1998). The phytotoxic effects were eliminated by using heat-vapourized vinegar on stone fruits, strawberries, and apples but the volume needed to reduce decay in these fruits was high at 36.6 μl/l of air (Sholberg *et al.*, 2000). Since vinegar and its active component, acetic acid, do not penetrate into the fruits, they do not control latent or quiescent infection (Sholberg and Gaunce, 1996). Although informal tasting of the treated fruits has not identified any off-odours, rigorous sensory evaluation of acetic acid-treated fruits has not been undertaken to date (Sholberg *et al.*, 1996, 2000).

Vapours of ethanol (EtOH) to control postharvest decay have not been investigated extensively but aqueous ethanol has been used as a dip treatment to control brown rot and *Rhizopus* rot of peaches (Ogawa and Lyda, 1960; Feliciano *et al.*, 1992) and table grapes (Lichter *et al.*, 2002) or in combination with hot water to control postharvest decay of table grapes, lemons and stone fruits (Smilanick *et al.*, 1995; Margosan *et al.*, 1997) with varying degrees of success. Ethanol vapours inhibited decay of oranges by *P. italicum* and *P. digitatum* after five days of continuous exposure (Yuen *et al.*, 1995). No other use of ethanol vapours has been reported, probably because of concern about its inflammability and explosive potential under high pressures.

10.4 Compounds of microbial and animal origin

Examples of decay control achieved using compounds of microbial or animal origin are shown in Table 10.4. Iturin, an antibiotic produced by several strains of *Bacillus subtilis*, has been effective in controlling brown rot of peaches (Gueldner *et al.*, 1988). Similarly, pyrrolnitrin, purified from a strain of *Pseudomonas cepacia*, has provided effective control of grey mould on raspberry (Goulart *et al.*, 1992), blue mould and grey mould on apples and pears (Janisiewicz *et al.*, 1991), and has delayed rot of strawberries (Takeda *et al.*,

Table 10.4 Examples of successful control of postharvest diseases by natural compounds of microbial or animal origin

Commodity	Disease	Compound	Source	References
Kiwifruits	*Botrytis* rot	Six-pentil-2-pyrone	*Fusarium semitectum*	Poole and Whitmore (1997)
Peaches	*Monilinia* rot	Iturin	*B. subtilis*	Gueldner *et al.* (1988)
Raspberry, apples, Pears, cut rose flowers	*Botrytis, Penicillium*, and *Rhizopus* rots	Pyrrolnitrin	*P. cepacia*	Goulart *et al.* (1992); Janisiewicz *et al.* (1991); Takeda *et al.* (1990); Hammer *et al.* (1993)
Table grapes	*Sour* rot	Xanthan gum	*X. campestris*	Ippolito *et al.* (1998)
Tomatoes, bell peppers, cucumbers, strawberries, table grapes, mangoes, sweet cherries, oranges	*Botrytis, Penicillium, Rhizopus* and *Colletotrichum* rots	Chitosan Glycol-chitosan	Arthropod exoskeleton, fungal cell wall	El Ghaouth (1994); Bégin *et al.* (2001); Reddy *et al.* (2000); Romanazzi *et al.* (1999a,b, 2000a,b, 2001, 2002, 2003)

Table 10.5 Compounds of microbial and animal origin active *in vitro* against postharvest pathogens.

Compound	Source	Pathogen	References
Fusapyrone, deoxyfusapyrone	*F. semitectum*	*Penicillium* spp., *A. alternata, B. cinerea, Fusarium* spp	Altomare *et al.* (2000)
Chitosan	Arthropod exoskeleton, fungal cell walls	*B. cinerea, C. gloeosporioides, R. stolonifer, M. laxa, F. oxysporum* and *A. alternata*	Allan and Hadwiger (1979); El Ghaouth *et al.* (1997, 1992a,b,c); Benhamou (1992); Romanazzi *et al.* (2001)

1990) and prevented postharvest rot on cut rose flowers (Hammer *et al.*, 1993). However, the potential for development of antibiotic resistance precludes a more widespread use of these compounds.

Six-pentyl-2-pyrone, a secondary metabolite of *Trichoderma* fungi, applied topically at 0.4–0.8 mg/fruit controlled *B. cinerea* rot on kiwifruit (Poole and Whitmore, 1997). This compound occurs naturally in ripe peaches and nectarines (Horvat *et al.*, 1990) and it is an approved food flavouring additive (Oser *et al.*, 1984). Fusapyrone and deoxyfusapirone, two a-pyrones originally isolated from cultures of *Fusarium semitectum*, but also produced by *Alternaria, Aspergillus, Penicillium*, and *Trichoderma* (Table 10.5) inhibited the growth of postharvest pathogens as well as mycotoxigenic and human filamentous fungi *in vitro* (Altomare *et al.*, 2000).

Xanthan gum, an industrial polysaccharide produced by fermentation of *Xanthomonas campestris* (Kennedy and Bradshaw, 1984), reduced the sour rot of grapes when applied prior to harvest at a concentration of 0.05% (w/v) (Ippolito *et al.*, 1998). Sour rot is a disease even more severe than grey mould in southern Italy. Xanthan gum was more effective than the antagonist *Aureobasidium pullulans* (10^7 cells/ml), $CaCl_2$ (1% w/v) or the fungicide procymidone at 100 g/ml. No synthetic fungicide is currently available to combat sour rot. Therefore, xanthan gum could be an interesting compound to be validated in large-scale trials. The mode of action of xanthan gum has not been elucidated but it is conceivable that the nature of the polysaccharide and its coating properties are involved. Xanthan gum is Generally Recognized As Safe (GRAS) and is used in food, non-foods and cosmetics as an emulsifier, thickener and stabilizing agent (Fennema, 1996).

The antimicrobial properties of chitosan in foods have been reviewed in detail elsewhere in this book (Chapter 8). When applied as a coating on fruits and vegetables, chitosan delayed ripening, reduced decay, interfered with fungal growth, and activated defence mechanisms in the plant tissues (Table 10.4). Because of its film-forming ability, chitosan delayed ripening by acting as a barrier to gas diffusion (El Ghaouth *et al.*, 1991, 1992a). Tomatoes, bell peppers, cucumbers and strawberries coated with chitosan had reduced weight loss and respiration rates, improved appearance and extended shelf-life (El Ghaouth *et al.*, 1992a). The compound has never been reported to induce symptoms of phytotoxicity on treated fruits and vegetables. The polymer was effective in reducing decay of other fruits including table grapes, mangoes, sweet cherries and oranges (El Ghaouth, 1994; Bégin *et al.*, 2001; Romanazzi *et al.*, 2001). The range of chitosan application has varied from 0.1% to 2%; the higher the concentration, the greater the control of various diseases.

In the above examples, chitosan has been applied as a postharvest treatment; however, the few examples of its application in preharvest treatment of strawberries, sweet cherries and table grapes to control postharvest decay have also been very promising (Romanazzi *et al.*, 1999b, 2000a,b, 2001, 2002; Reddy *et al.*, 2000). Preharvest spray of chitosan to reduce storage decay of table grapes seems the best way to apply this compound since exposure to postharvest liquid-based treatments is not advisable, as it could damage the bloom (Ippolito and Nigro, 2000). The only effect on grape berry appearance was a slight shining using the highest concentration of chitosan (1%), while no effects were visible at lower doses (0.5–0.1%), which were as effective as the higher one (Romanazzi *et al.*, 2002). Preharvest application of chitosan is also advisable against *Botrytis* rot on strawberries. Field application of the polymer during flowering can avoid infection via senescent floral parts that later develop into active rots on ripe fruits (Powelson, 1960; Romanazzi *et al.*, 2000a). On strawberries, preharvest application of glycolchitosan, a water-soluble form of chitosan, gave similar results to chitosan (Romanazzi *et al.*, 1999a).

The mode of antifungal action of chitosan is still not well understood. In *B. cinerea, R. stolonifer* and *F. oxysporum*, chitosan caused cellular leakage and morphological alterations consisting of excessive branching and cell wall damage (Benhamou, 1992; El Ghaouth *et al.*, 1992b), presumably via its action on chitin deacetylase (El Ghaouth *et al.*, 1992c). In tissues of chitosan-treated bell peppers, *B. cinerea* hyphae displayed various levels of cellular disorganization, ranging from wall loosening to cytoplasm disintegration (El Ghaouth *et al.*, 1997).

The eliciting property of chitosan has been demonstrated in several postharvest commodities. Induction of antifungal hydrolases such as beta-1,3-glucanase, chitinase and chitosanase have been observed in strawberries, tomatoes and bell peppers (Wilson *et al.*, 1994); in tomato and bell pepper, the activity of these enzymes remained high up to 14 days after treatment. On table grapes, chitosan enhanced the activity of phenylalanine ammonia-lyase (PAL) (Romanazzi *et al.*, 2002). Moreover, it elicited phytoalexin formation in carrot roots, thus restricting *B. cinerea* infection (Reddy *et al.*, 1999). The induction of lytic enzymes, PAL and phytoalexins in harvested tissues by prestorage treatment with chitosan could supply the tissue with weapons capable of restricting fungal colonization; this could be important in retarding the resumption of quiescent and latent infections which typically become active when tissue resistance declines. Chitosan treatment also stimulates various structural defence barriers in host tissues such as thickening of the cell wall, formation of papillae, and deposition of electron opaque materials in the intercellular spaces, presumably being antifungal phenolic-like compounds (Wilson *et al.*, 1994).

As the endogenous microflora on fruit surfaces may play an important role in antagonism to pathogens (Wilson and Wisniewski, 1994), treatments of fruits and vegetables should avoid any negative effect on the naturally occurring microflora (Nigro *et al.*, 1998, 2000). Chitosan applied as pre- and postharvest treatment on table grapes did not impair the naturally occurring yeasts and yeast-like fungi, among which antagonistic microrganisms are common (Ippolito *et al.*, 1997). On the contrary, chitosan treatment reduced the propagules of filamentous fungi naturally occurring on berries (Romanazzi *et al.*, 2002). As naturally occurring filamentous fungi on table grapes include decay-causing species such as *B. cinerea, Penicillium* spp., *Aspergillus* spp. and *Cladosporium* spp., it has been hypothesized that their reduction can cooperate in lowering rot incidence during storage (Ippolito *et al.*, 1998).

10.5 Resistance

In developing new methods for controlling postharvest diseases of fruits and vegetables, the potential for development of resistance to natural antimicrobials needs to be considered. It is recognized that resistance has less of a chance to develop towards substances that have more than one mode of action, as is the case for most of the compounds reviewed above.

There are several examples of pathogens producing enzymes able to degrade phytoalexins of its host plant to various degrees. Stilbene-type phytoalexins are degraded by culture filtrates of *B. cinerea* in the grapevine–*Botrytis* interaction but the secretion by the fungus of the phytoalexin-detoxifying enzyme is delayed compared with the accumulation of phytoalexins in the grapevine. The ability to degrade phytoalexins is a contributory, but not a primary, determinant of successful parasitism in the grapevine–*Botrytis* interaction (Jeandet *et al.*, 1993). The net accumulation of phytoalexins within the plant tissues infected by various pathogens is probably controlled by a balance which may result, on the one hand, from the ability of the host to produce phytoalexins and, on the other hand, the ability of the pathogen to detoxify such compounds (Vanetten *et al.*, 1989).

10.6 Additive and synergistic combinations

The successful commercial control of postharvest diseases of fruits and vegetables must be extremely efficient, in the range of 95–98%, unlike the control of tree, field crop or soil-borne diseases (Droby, 2001). None of the natural antimicrobial systems investigated to date offer consistent postharvest disease control comparable to that obtained with synthetic fungicides. Attempts to surmount the variable performance and enhance the efficacy of natural compounds have led to the development of combined approaches based on additive and synergistic effects.

While many synergistic combinations of antimicrobials have been identified *in vitro* and in prepared foods (e.g. nisin and garlic, or nisin and phenolic compounds discussed in Section 2.9.1; potassium sorbate and vanillin or citral discussed in Section 11.2), relatively few investigations of multifactorial systems have focused on organisms or conditions relevant to postharvest storage of fruits and vegetables. Synergistic effects have been reported for combinations of *Allium sativum* and *A. cepa* bulb extracts. A mixture containing 0.25% garlic and 0.75% onion bulb extract (total conc. 1%) was most inhibitory to *in vitro* growth of *Alternaria* spp. (Bokhary, 1985). Similarly, mixtures of *Allium* plant extracts and acetic acid were more active against *A. flavus, A. niger* and *A. fumigatus* than single treatments, especially when treatment was further combined with high temperature (Yin and Tsao, 1999).

An interesting combination of different approaches has been the use of sub-atmospheric pressure with essential oils. The fungitoxic activity of *T. capitatus* EOs (75, 150 and 250 ppm) on oranges inoculated with *P. digitatum* and placed in 10-litre desiccators was weak at atmospheric pressure (3 to 10% inhibition), while under vacuum (0.5 bar), conidia mortality on the exocarp reached 90–97% (Arras and Usai, 2001). Under subatmospheric conditions (0.2–0.8 atm) two phytoalexins (scoparone and scopoletin) were elicited on orange and mandarin fruits; the biosynthesis of these compounds was also stimulated in fruits treated with thyme oil vapours at a concentration of 50–100 ppm. The simultaneous use of thyme oil and hypobaric pressure had a synergistic effect eliciting five times as much scoparone (Arras, 1999). This was attributed to increased contact between the EO, the pathogen's conidia and the host tissue.

A synergistic effect was observed in sweet cherries under sub-atmospheric pressure in combination with chitosan (Table 10.6). The extent of decay inhibition was, on average, 20% with sub-atmospheric pressure alone, 65% with chitosan treatment alone, and 83–89% when both treatments were applied (Romanazzi *et al.*, 2003). In this work, Limpel's formula, as described by Richer (1987), was used to determine synergistic interactions between chitosan and hypobaric treatment. The machinery for rapid vacuum cooling of fruit and vegetables is already in use in some packinghouses; therefore, it might be feasible to add some of the natural antifungal compounds reported above to improve the effectiveness of the treatment.

'Bioactive coating' is a preparation consisting of a water-soluble form of chitosan (glycolchitosan) and an antagonistic yeast. This bioactive coating was superior to *Candida saitoana* and glycolchitosan in controlling decay of several varieties of sweet orange, lemons, and apples, and the control level was comparable to that achieved with imazalil or thiabendazole (El Ghaouth *et al.*, 2000). Unlike C. *saitoana* alone, which showed a poor performance on late season fruit, the bioactive treatments offered consistent control of decay on Washington navel and Eureka lemons in early and late season.

Table 10.6 Effect of combinations of chitosan and hypobaric treatments on the percentage of rotted sweet cherries

Disease	Pressure (atm)	Chitosan concentration (%)	
		1.0	0.0
Brown rot	0.25	11.3†	44.0
	0.50	6.0†	35.3
	1.00 (control)	15.3	55.3
Grey mould	0.25	6.0	7.8
	0.50	4.0	7.5
	1.00 (control)	8.7	28.0
Total rots*	0.25	26.7	49.3
	0.50	13.3†	42.0
	1.00 (control)	30.7	78.7

* Total rots include grey mould, brown rot, *Rhizopus* rot, *Alternaria* rot, blue mould and green rot. The interaction of pressure with chitosan was significant at the 5% level for brown rot, grey mould, and total rots.
† Synergistic effect, according to Limpel's formula.

Source: adapted from Romanazzi *et al.* (2003).

Xanthan gum has also been evaluated in combination with *A. pullulans* to control postharvest table grape and strawberry rots; on both commodities, the activity of the antagonist was significantly improved when applied in combination with the polysaccharide at 0.5% (w/v) (Ippolito *et al.*, 1997, 1998). The higher activity of *A. pullulans* combined with xanthan gum on table grapes and strawberries has been related to its greater survival on the fruit surface, probably because of its coating properties.

Ethanol applied at concentrations between 8% and 20% was not effective in reducing grey mould incidence and severity on apples and pears. Likewise, several strains of *Saccharomyces cerevisiae* (10^8 CFU/ml) were not effective. However, a combination of the two reduced disease incidence by over 90% (Mari and Carati, 1997). Similar results were obtained against green mould on lemon, combining a heated solution of ethanol (10%, 45 °C) with curing, a physical treatment consisting in keeping fruits at relatively high temperature and humidity (e.g. 32 °C, 95–98% RH), and combining ethanol with the biocontrol yeast *Candida oleophila*. Infections were reduced from 82% (control), 17% (ethanol alone) and 40% (yeast alone) to 3.5% (ethanol-curing) and 3.3% (EtOH-yeast), with no appreciable differences compared with the fungicide imazalil (Lanza *et al.*, 1997).

The above examples clearly show the advantages in using combination strategies to control postharvest diseases of fresh fruit and vegetables. Many other possible combinations could be explored, such as the use of antagonistic microorganisms, low doses of fungicides, physical means (UV radiation, modified atmospheres, etc.), organic and inorganic salts, nutrients, mixtures of natural substances with different modes of action, etc. The complexity of the mode of action that combined alternatives can display should also make the development of pathogen resistance more difficult.

10.7 Extent of take-up by industry

In spite of widespread public concern regarding the use of synthetic fungicides, natural fungicides from plant, animal and microbial sources have, at present, made little impact in the marketplace. The success of large-scale studies on microbial antagonists has generated interest by several agro-chemical companies and currently four antagonistic microorganisms to control postharvest diseases of fresh fruit and vegetables are commercially available: *C. oleophila* (Aspire^{TM}), *Cryptococcus albidus* (YieldPlus^{TM}) and two strains of *Pseudomonas syringae* (Biosave-110TM) (http://www.oardc.ohio-state.edu/apsbcc/productlist.htm, 2002).

Although the commercial formulation of natural compounds to fight postharvest diseases is still in an early phase of development, it seems they have potential for market expansion. Indeed, for certain substances the take-up by industries should be relatively easy, as for those already tested in other systems, especially human medicine and the food industry. For example, chitosan, owing to its lipid-binding capacity and hypocholesterolaemic action, is used in human medicine for slimming diets; a wide variety of plant extracts and essential oils are available without prescription through health food stores, herbalists, vitamin retailers, etc., and are commonly used as culinary herbs and spices; xanthan gum is a compound commonly used in food, cosmetic and pharmaceutical industries. Essential oils have a long history of global use by the food and fragrance industries and, recently, in the field of aromatherapy; they are readily available at low to moderate cost. Moreover, in some countries, some of these compounds are exempt from the usual data requirements for registration. American companies taking advantage of this situation have been able to bring pesticides based on essential oils to market in a far shorter time period than would normally be required for a conventional pesticide. This is the case of Cinnamite^{TM} and Valero^{TM}; both of these are miticide/fungicides for glasshouse and horticultural crops based on cinnamon oil with cinnamaldehyde as the active ingredients (Isman, 2000).

Possible barriers to the commercial development of natural compounds as antimicrobials for controlling postharvest diseases of fruit and vegetables are:

- expensive procedures for extracting the active compounds;
- the need for chemical standardization, stability and quality control;
- extended studies on toxicological aspects for specific compounds;
- production at competitive costs with existing pesticides;
- the lack of studies on development of resistance;
- difficulties in registration as pesticides;
- efficacy sometimes not consistent and acceptable unless in complex integrated approaches;

- restricted market confined to postharvest environment;
- scarce interest by companies in testing for botanicals since it is still unclear whether proprietary claims can be made.

Nevertheless, these issues should not deter the search for novel combinations of antimicrobials from animal, plant and microbial sources. Possible withdrawal of most of the conventional pesticides currently registered for postharvest disease control may change the situation radically, providing new perspectives for the development and commercialisation of pesticides based on natural substances.

10.8 Concluding remarks

What is needed is a change in philosophy of companies and growers, still rooted to the concept that a 'stand-alone' treatment has to control a disease completely. In addition, consumers need to consider the benefits of products without residues and with improved quality. A more sustainable approach against postharvest diseases needs to be based on the use of multifaceted control strategies, including the use of natural compounds, microbial antagonists, physical methods, induced and genetic resistance, low doses of fungicide, etc.

The ability to control previously established infections in the postharvest environment is of crucial importance. Under commercial conditions, the application of postharvest treatment may be delayed for hours or even days after harvest, allowing the pathogen to penetrate the flesh where control becomes very difficult. Therefore, compounds able to control incipient, latent and quiescent infections should be preferred. Another interesting feature for some of these compounds is the possibility of acting in the vapour phase; besides their efficacy, their use is promising because of their capability to easily penetrate the mass of stacked commodities in a cold room without further manipulation. This characteristic, exploitable only in a confined environment, seems the most appropriate treatment for products such as strawberries and table grapes, where postharvest handling reduces their market appeal.

Emerging new biocontrol methods are still evaluated by comparison with synthetic pesticides. This is an approach that needs debunking. Natural antimicrobials may require different application timing, depending on the intrinsic characteristics and the mode of action of the compound. Since the majority of natural antimicrobials have no toxic effects on mammals and have low phytotoxicity, it should be possible to apply such compounds in a wider range of concentrations and application schedules.

Many natural antimicrobial compounds have been tested and some are very promising; however, there is an inestimable number of other substances yet to be discovered in nature's store. Many substances effective in apparently unrelated systems could be tested against postharvest diseases. A rational way to get results at the lowest cost/benefit ratio should be via a collaborative effort between plant pathologists, physicians, chemists and companies to develop safe commercial products. Their potential benefits for human health, e.g. phenolics as antioxidants, should further increase their approval and speed-up demand.

10.9 References

ABDOU I A, ABDOU-ZEID A A, EL-SHERBEENY M R, and ABOU-EL-GHEAT Z H (1992), 'Antimicrobial activity of *Allium sativum, A. cepa, Barbanus sativus, Capsicum frutescens, Eruca sativa*, and *Allium kurrat* on bacteria', *Qual Plant Mater Veget*, **22**, 29–35.

ABOU JAWDAH Y, SOH A, and SALAMEH A (2002), 'Antimycotic activity of selected plant flora, in Lebanon, against phytopathogenic fungi', *J Agric Food Chem*, **50**, 3208–3213.

AFEK U and SZTEJNBERG A (1993), 'Temperature and gamma irradiation effects on scoparone, a citrus phytoalexin conferring resistance to *Phytophthora citrophthora*', *Phytopathology*, **83**, 753–758.

AFOLAYAN A J and MEYER J J M (1997), 'The antimicrobial activity of 3,5,7-trihydroxyflavone isolated from the shoots of *Helicrysum aurenitens', J Ethnopharmacol,* **57**, 177–181.

AHMAD S K and PRASAD J S (1995), 'Efficacy of foliar extracts against pre- and post-harvest diseases of sponge-gourd fruits', *Letters Appl Microbiol*, **21**, 375–385.

ALIGIANNIS N, KALPOUTZAKIS E, CHINOU I B, MITAKOU S S, GIKAS E, and TSARBOPOULOS A (2001), 'Composition and antimicrobial activity of the essential oils of five taxa of *Sideritis* from Greece', *J Agric Food Chem*, **49**, 811–815.

ALLAN C R and HADWIGER L A (1979), 'The fungicidal effect of chitosan on fungi of varying cell wall composition', *Exp Mycol*, **3**, 285–287.

ALTOMARE C, PERRONE G, ZONNO M C, EVIDENTE A, PENGUE R, FANTI F, and PONELLI L (2000), 'Biological characterization of fusapyrone and deoxyfusapyrone, two active metabolite of *Fusarium semitectum'*, *J Nat Prod*, **63**, 1131–1135.

AMMERMANN E, LORENZ G, and SCHLEBERGER K (1992), 'A broad-spectrum fungicide with a new mode of action', *Brit Crop Prot Conf Pests Dis*, **1**, 403–410.

ANGIONI A, CABRAS P, D'HALLEWIN G, PIRISI F M, RENEIRO F, and SCHIRRA M (1998), 'Synthesis and inhibitory activity of 7-geranoxycoumarins against *Penicillium* species in citrus fruit', *Phytochemistry*, **47**, 1521–1525.

ARCAS M C, BOTÍA J M, ORTUÑO A M, and DEL RÍO J A (2000), 'UV irradiation alters the levels of flavonoids involved in the defence mechanism of *Citrus aurantium* fruits against *Penicillium digitatum'*, *Eur J Plant Pathol*, **106**, 617–622.

ARCHBOLD D D, HAMILTON KEMP T R, BARTH M M, and LANGLOIS B E (1997), 'Identifying natural volatile compounds that control grey mould (*Botrytis cinerea*) during postharvest storage of strawberry, blackberry, and grape', *J Agr Food Chem*, **45**, 4032–4037.

ARCHBOLD D D, HAMILTON-KEMP T R, CLEMENTS A M, and COLLINS R W (1999), ' Fumigating 'Crimson seedless' table grapes with (*E*)-2-hexenal reduces mould during long-term postharvest storage', *HortSci*, **34**, 705–707.

ARK P A and THOMPSON J P (1959), 'Control of certain diseases of plants with antibiotics from garlic (*Allium sativum* L.)', *Plant Dis Rep*, **43**, 276–282.

ARRAS G (1988), 'Antimicrobial activity of various essential oil against citrus fruit disease agents', in Goren R, Mendel K, *Proceedings of the Sixth International Citrus*

Congress, Balaban Publishers, Philadelphia, USA, 787–793.

ARRAS G (1999), 'Postharvest response of citrus fruit diseases to natural compounds', in Schirra M, *Advances in Postharvest Diseases and Disorders Control of Citrus Fruit*, Pandalai Publisher, Research Singnpost, Trivandrum, India, 123–131.

ARRAS G and GRELLA G E (1992), 'Wild Thyme, *Thymus capitatus*, essential oil seasonal changes and antimycotic activity', *J Hort Science*, **67**, 197–202.

ARRAS G and USAI M (2001), 'Fungitoxic activity of twelve essential oils against four postharvest citrus pathogens: chemical analysis of *Thymus capitatus* (L.) Hofmgg oil and its effect in subatmospheric pressure conditions', *J Food Protec*, **64**, 1025–1029.

ARRAS G, AGABBIO M, PIGA A, and D'HALLEWIN G (1995), 'Fungicide effect of volatile compounds of *Thymus capitatus* essential oil', *Acta Hortic*, **379**, 593–600.

AVISSAR I and PESIS E (1991), 'The control of postharvest decay in table grape using acetaldehyde vapors', *Ann Appl Biol*, **118**, 229–237.

AVISSAR I, DROBY S, and PESIS E (1990), 'Characterization of acetaldehyde effects on *Rhizopus stolonifer* and *Botrytis cinerea', Ann Appl Biol*, **116**, 213–220.

BANKOVA V, BOUDOUROVA-KRASTEVA G, POPOV S, SFORCIN J M, and CUNHA FUNARI S R (1998), 'Seasonal variations of the chemical composition of Brazilian propolis', *Apidologie*, **29**, 361–367.

BARBER M S, MCCONNELL V S, and DECAUX B S (2000), 'Antimicrobial intermediates of the general phenylpropanoid and lignin specific pathways', *Phytochemistry*, **54**, 53–56.

BAUDOIN A B A M and ECKERT J W (1985), 'Development of resistance against *Geotrichum candidum* in lemon peel injuries', *Phytopathology*, **75**, 174–179.

BAUTISTA-BAÑOS S, HERNÀNDEZ-LÒPEZ M, DìAZ-PÉREZ J C and CANO-OCHA C F (2000), 'Evaluation of the fungicidal properties of plant extracts to reduce *Rhizopus stolonifer* of 'ciruela' fruit (*Spodias purpurea* L) during storage', *Postharvest Biol* Tec, **20**, 99–106.

BÉGIN A, DUPUIS I, DUFAUX M and LEROUX G (2001), 'Use of chitosan to control growth of *Colletotrichum gloeosporioides in vitro* and on stored mangoes', in Muzzarelli R A A, *Chitin Enzymology* 2001, Atec, Grottammare, Italy, 163–170.

BENHAMOU N (1992), 'Ultrastructural and citochemical aspects of chitosan on *Fusarium oxysporum* f sp *radici-lycopersici*, agent of tomato crown rot', *Phytopathology*, **82**, 1185–1190.

BEN-YEHOSHUA S, SHAPIRO B, KIM J J, SHARONI J, CARMELI S, and KASHMAN Y (1988), 'Resistance of citrus fruit to pathogens and its enhancement by curing', in Goren R, Mendel K, *Proceedings of the Sixth International Citrus Congress*, Balaban Publishers, Philadelphia, USA, 1371–1379.

BEN-YEHOSHUA S, RODOV V, KIM J J, and CARMELI S (1992), 'Preformed and induced antifungal materials of citrus fruits in relation to the enhancement of decay resistance by heat and ultraviolet treatments', *J Agric Food Chem*, **40**, 1217–1221.

BOKHARY, H A (1985), 'Effects of mixtures of *Allium sativum* L. and *Allium cepa* L. bulb extracts on growth of *Alternaria* spp. and bacteria', *J College Sci*, **16**, 87–97.

BORS W, HELLER W, MICHEL C, and SARAN M (1990), 'Flavonoids as antioxidants: determination of radical scavenging efficiencies', *Methods Enzymol*, **186**, 343–355.

BOUDET A M, LAPIERRE C, and GRIMA-PETTENATI J (1995), 'Tansley Review No. 80: biochemistry and molecular biology of lignification', *New Phytol*, **129**, 203–236.

BURDOCK G A (1998), 'Review of the biological properties and toxicity of bee propolis', *Food Chem Toxic*, **36**, 347–363.

CACCIONI D R L, DEANS S G, and RUBERTO G (1995a), 'Inhibitory effect of citrus fruit essential oil components on *Penicillium italicum* and *P. digitatum'*, *Petria*, **5**, 177–182.

CACCIONI D R L, TONINI G, and GUIZZARDI M (1995b), 'Antifungal activity of stone fruit aroma compounds against *Monilinia laxa* and *Rhizopus stolonifer*: *in vivo* trials', *J Plant Dis Prot*, **102**, 518–525.

CACCIONI D R L, GARDINI F, LANCIOTTI R, and GUERZONI M E (1997), 'Antifungal activity of natural volatile compounds in relation to their vapor pressure', *Sci Alim*, **17**, 21–34.

CACCIONI D R L, GUIZZARDI M, BIONDI D M, RENDA A, and RUBERTO G (1998), ' Relationship between volatile components of citrus fruit essential oils and antimicrobial action on *Penicillium digitatum* and *P. expansum'*, *Int J Food Microbiol*, **43**, 73–79.

CAO G, SOFIC E, and PRIOR R L (1996), 'Antioxidant capacity of tea and common vegetables', *J Agric Food Chem* 44, 3426–3431.

CESARONE M R, LAURORA G, RICCI A, BELCACO G, and POMANTE P (1992), 'Acute effects of hydroxiethylrutosides on capillary filtration in normal volunteers, patients with various hypotension and in patients with diabetic micro angiopathy', *J Vas Dis*, **21**, 76–80.

CHU C L, LIU W T, ZHOU T, and TSAO R (1999), 'Control of postharvest grey mould rot of modified packaged sweet cherry by fumigation with thymol and acetic acid', *Can J Plant Sci*, **79**, 685–689.

COLERIGE SMITH P O, THOMAS P, SCURR J H, and DORMANDY J A (1980), 'Causes of various ulceration, a new hypothesis', *Br Med J*, **296**, 1726–1727.

COSENTINO S, TUBEROSO C I G, PISANO B, SATTA M, ARZEDI E, and PALMAS F (1999), '*In vitro* antimicrobial activity and chemical composition of sardinian thymus essential oils', *Lett Appl Microbiol*, **29**, 130–135.

DIXON R A, HARRISON M J, and PAIVA N L (1995), 'The isoflavonoid phytoalexin pathway: from enzymes to genes to transcription factors', *Physiol Plant*, **93**, 385–392.

DIXON R A, ACHNINE L, KOTA P, LIU C J, SRINIVASA REDDY M S, and WANG L (2002), 'The phenylpropanoid pathway and plant defence – a genomics perspective', *Mol Pl Pathol*, **3**, 371–390.

DOARES S H, SYROVETS T, WEILER E W, and RYAN C A (1995), 'The isoflavonoid phytoalexin pathway: from enzymes to genes to transcription factors', *Physiol. Plant*, **93**, 385–392.

DOUGHTY K J, BLIGHT M M, BOCK C H, FIELDSEND J K, and PICKETT J A (1996), 'Release of alkenyl isothiocyanates and other volatiles from *Brassica rapa* seedlings during infection by *Alternaria brassicae'*, *Phytochemistry*, **43**, 371–374.

DROBY S (2001), 'Enhancing biocontrol activity of microbial antagonists of postharvest diseases', in Vurro M, Gressel J, Butt T, Harman G, Pilgeram A, St Leger R, and Nuss D, *Enhancing Biocontrol Agents and Handling Risks*, Nato Science Series, vol 339, IOS Press, Amsterdam, 77–85.

DROBY S, PORAT R, COHEN L, WEISS B, SHAPIRO B, PHILOSOPH-HADAS S, and MEIR S (1999), 'Suppressing green mould decay in grapefruit with postharvest jasmonate application', *J Am Soc Hort Sci*, **124**, 184–188.

ECKERT J W (1991), 'Role of chemical fungicide and biological agents in postharvest disease control', in Wilson C L and Chalutz E, *Biological Control of Postharvest Diseases of Fruits and Vegetables: Workshop Proceedings*, USDA ARS-92, 14–30.

ECKERT J W, SIEVERT J R, and RATNAYAKE M (1994), 'Reduction of imazalil effectiveness against citrus green mould in California packinghouses by resistant biotypes of *Penicillium digitatum'*, *Plant Dis*, **78**, 791–794.

ELGAYYAR M, DRAUGHON F A, GOLDEN D A, and MOUNT J R (2001), 'Antimicrobial activity of essential oils from plants against selected pathogenic and saprophytic microrganisms', *J Food Prot*, **64**, 1019–1024.

EL GHAOUTH A (1994), 'Manipulation of defence system with elicitors to control postharvest diseases', in Wilson C L and Wisniewski ME, *Biological Control of Postharvest Diseases, Theory and Practices*, CRC Press, Boca Raton, Florida, 153–167.

EL GHAOUTH A and WILSON C L (1995), 'Biologically-based technologies for the control of postharvest diseases', *Postharvest New Inf*, **6**, 5–11.

EL GHAOUTH A, PONNAMPALAM R, CASTAIGNE F, and ARUL J (1991), 'Using chitosan coating to extend the storage life of tomatoes', *HortScience*, **27**, 1016–1122.

EL GHAOUTH A, ARUL J, and ASSELIN A (1992a), 'Potential use of chitosan in postharvest preservation of fruits and vegetables', in Brines J B, Sandford P A, and Zizachis J P, *Advances in Chitin and Chitosan*, Elsevier Applied Science, London, 45–52.

EL GHAOUTH A, ARUL J, ASSELIN A, and BENHAMOU N (1992b), 'Antifungal activity of chitosan on postharvest pathogen: induction of morphological and cytochemical alteration in *Rhizopus stolonifer*', *Mycol Res*, **96**, 769–779.

EL GHAOUTH A, ARUL J, GREINER J, and ASSELIN A (1992c), 'Effect of chitosan and other polyions on chitin deacetylase in *Rhizopus stolonifer*', *Exp Mycol*, **16**, 173–177.

EL GHAOUTH A, ARUL J, WILSON C, and BENHAMOU N (1997), 'Biochemical and cytochemical aspects of the interaction of chitosan and *Botrytis cinerea* in bell pepper fruit', *Postharvest Biol Tec*, **12**, 183–194.

EL GHAOUTH A, SMILANICK J L, BROWN G E, IPPOLITO A, and WILSON C L (2000), 'Application of *Candida saitoana* and glycolchitosan for the control of postharvest diseases of apple and citrus fruit under semi-commercial condition', *Plant Dis*, **84**, 243–248.

FAHEY J W, ZALCMANN A T, and TALALAY P (2001), 'The chemical diversity and distribution of glucosinolates and isothiocyanates among plants', *Phytochemistry,* **56**, 5–51.

FALLIK E ARCHBOLD D D, HAMILTON-KEMP T R, CLEMENS A M, COLLINS R W, and BARTH M E (1998), '(*E*)-2–hexenal can stimulate *Botrytis cinerea in vitro* and on strawberry fruit *in vivo* during storage', *J Am Soc Hort Sci*, **123**, 875–881.

FAO (1981), 'Food preservation in perishable crops', *FAO Agr Ser Bull*, **43.**

FELICIANO A, FELICIANO A J, VENDRUSCULO J, ADASKAVEG J E, and OGAWA J M (1992), 'Efficacy of ethanol in postharvest benomyl-DCNA treatment for control of brown rot of peach', *Plant Dis*, **76**, 226–229.

FENNEMA O R (1996) *Food Chemistry*, Marcel Dekker, New York.

FRENCH R C and WILSON C L (1981), 'The effect of nicotine and related compounds on germination and vacuolation in several species of rust uredospores', *Physiol Plant Pathol*, **19**, 201–207.

FRIES N F (1973), 'Effects of volatile organic compounds on the growth and development of fungi', *Trans Br Mycol Soc*, **60**, 1–21

GHISALBERTI E L (1979), 'Propolis: a review', *Bee Wld*, **60**, 59–84.

GIL V and MACLEOD A J (1980), 'The effects of pH on glucosinolate degradation by a thioglucoside preparation', *Phytochemistry,* **19**, 2547–2553.

GODWIN J R, ANTHONY V M, CLOUGH S M, and GODFREY C R A (1992), ' ICIA5504: a novel broad-spectrum systemic beta-methoxyacrylate fungicide', *Brit Crop Prot Conf Pests Dis*, **1**, 435–442.

GOULART BB L, HAMMER P E, EVENSEN K B, JANISIEWICZ W, and TAKEDA F (1992), 'Pyrrolnitrin, captan, benomyl, and high CO_2 enhanced raspberry shelf life', *J Am*

Soc Hortic Sci, **117**, 265–270.

GREENAWAY W, MAY J, SCAYSBROOK T, and WHATLEY F R (1991), 'Identification by gas chromatography-mass spectrometry of 150 compounds in propolis', *Z Naturforsch*, **46**, 111–121.

GUELDNER R C, REILLY CC, PUSEY P L, ARRENDALE R, HIMMELSBACH D S, and CUTLER H G (1988), 'Isolation and identification of iturin as antifungal peptides in biological control of peach brown rot with *Bacillus subtilis'*, *J Agric Food Chem*, **36**, 366–370.

GUNDLACH H, MULLER M J, KUTCHAN T M, and ZENK M H (1992), 'Jasmonic acid is a signal transducer in elicitor-induced plant cell cultures', *Proc Natl Acad Sci USA*, **89**, 2389–2393.

HAMILTON-KEMP T R, MCKRACKEN C T, LOUGRIN J H, ANDERSON R A, and HILDEBRAND D F (1992), 'Effects of some natural volatile compounds on the pathogenic fungi *Alternaria alternata* and *Botrytis cinerea'*, *J Chem Ecol*, **18**, 1083–1091.

HAMMER P E, EVENSEN K B, and JANISIEWICZ W (1993), 'Postharvest control of *Botrytis cinerea* on cut rose flowers with pyrrolnitrin', *Plant Dis*, **77**, 238–286.

HAMMERSCHMIDT R (1999), 'Phytoalexins: what have we learned after 60 years?', *Ann Rev Phytopathol*, **37**, 285–306.

HARBORNE J B and WILLIAMS C A (2000), 'Advances in flavonoid research since 1992', *Phytochemistry*, **55**, 481–504.

HATANAKA A (1993), 'The biogeneration of green odor by green leaves', *Phytochemistry*, **34**, 1201–1218.

HERTOG MG L, HOLLMAN P C H, KATAN M B, and KIOMHOUT D (1993), 'Intake of potentially anticarcinogenic flavonoids and their determinants in adults in the Netherlands', *Nutr Cancer*, **20**, 21–29.

HORVAT R J, CHAPMAN G W, ROBERTSON J A , MEREDITH F I, SCORZA R M, CALLAHAN A M, and MORGENS P (1990), 'Comparison of the volatile compounds from several commercial peach cultivars', *J Agric Food Chem*, **38**, 234–237.

IPPOLITO A and NIGRO F (2000), 'Impact of preharvest application of biological control agents on postharvest diseases of fresh fruits and vegetables', *Crop Prot,* **19**, 715–723.

IPPOLITO A, NIGRO F, LIMA G, ROMANAZZI G, and SALERNO M (1998), 'Xanthan gum as adjuvant in controlling table grape rots with *Aureobasidium pullulans', J Plant Pathol*, **80**, 258 (abstract).

IPPOLITO A, NIGRO F, ROMANAZZI G, and CAMPANELLA V (1997), 'Field application of *Aureobasidium pullulans* against *Botrytis* storage rot of strawberry' in Bertolini P, Sijmons P C, Guerzoni M E, Serra F, *COST 914–915*, *'Non conventional methods for* the *control of postharvest disease and microbiological spoilage*', Office for Official Publication of the European Communities, Luxemburg, 127–133.

ISMAN M B (2000),'Plant essential oils for pest and disease management', *Crop Prot*, **19**, 603–608.

JANISIEWICZ W, YOURMAN L, ROITMAN J, and MAHONEY N (1991), 'Postharvest control of blue mould and grey mould of apples and pears by dip treatment with pyrrolnitrin, a metabolite of *Pseudomonas cepacia', Plant Dis*, **75**, 490–494.

JEANDET P, SBAGHI M and BESSIS R (1993), 'The significance of stilbene-type phytoalexin degradation by culture filtrates of *Botrytis cinerea* in the vine-Botrytis interaction'. In Fritig B and Legrand M, *Mechanisms of Plant Defence Responses*, Kluwer Academic Publishers, Dordrecht, 84–98.

JURD L, CORSE J, KING A D, BAYNE H, and MIHARA K (1971a), 'Antimicrobial properties of

6,7-dihydroxy-, 7,8-dihydroxy-, 6-hydroxy- and 8-hydroxy-coumarins', *Phytochemistry*, **10**, 2971–2974.

JURD L, KING A D, and MIHARA K (1971b), 'Antimicrobial properties of umbelliferone derivatives', *Phytochemistry*, **10**, 2965–2970.

KANES K, TISSERAT B, BERHOW M, and VANDERCOOK C (1992), 'Phenolic composition of various tissues of Rutaceae species', *Phytochemistry*, **32**, 967–974.

KARAMAN S, DIGRAK M, RAVID U, and ILCIM A (2001), 'Antibacterial and antifungal activity of the essential oils of *Thymus revolutus* Celak from Turkey', *J Ethnopharm*, **76**, 183–186.

KENNEDY J F and BRADSHAW I J (1984), 'Production, properties and applications of xanthan'. In Bushell ME, *Progress in Industrial Microbiology: Modern Applications of Traditional Biotechnologies*, Vol 19, Elsevier, Amsterdam, 319–371.

KIM J J, BEN-YEHOSHUA S, SHAPIRO B, HENIS Y, and CARMELI S (1991), 'Accumulation of scoparone in heat-treated lemon fruit inoculated with *Penicillium digitatum* Sacc', *Plant Physiol*, **97**, 880–885.

KUĆ J (1995), 'Phytoalexins, stress metabolism and disease resistance in plants', *Ann Rev Phytopathol*, **33**, 275–297.

KWON Y S, KOBAYASHI A, KAJIYAMA S I, KAWAZU K, KANZAKI, H and KIM C M (1997), 'Antimicrobial constituents of *Angelica dahurica* roots', *Phytochemistry*, **44**, 887–889.

LAMBERT R J W, SKANDAMIS P N, COOTE P J, and NYCHAS G J E (2001), 'A study of the minimum inhibitory concentration and mode of action of oregano essential oil, thymol and carvacrol', *J Appl Microbiol*, **91**, 453–462.

LANZA G, DI MARTINO ALEPPO E, and STRANO M C (1997), 'Evaluation of integrated approach to control postharvest green mould of lemons', in Bertolini P, Sijmons PC, Guerzoni M E, Serra F, *COST 914–915, Non conventional methods for the control of postharvest disease and microbiological spoilage*, Luxemburg: Office for Official Publication of the European Communities, 111–114.

LA TORRE A, GUCCIONE M, and IMBROGLINI G (1990), 'Indagine preliminare sull'azione di preparati a base di propoli nei confronti di *Botrytis cinerea* della fragola', *Apicoltura*, **6**, 169–177.

LATTANZIO V (1988), 'Phenolics in fruit and vegetables: some nutritional and technological aspects', in Lintas C and Spadoni M A, *Food Safety and Health Protection*, CNR-IPRA, Rome, 25–36.

LATTANZIO V, CARDINALI A, and PALMIERI S (1994a), 'The role of phenolics in the postharvest physiology of fruits and vegetables: browning reactions and fungal diseases', *Ital J Food Sci*, **6**, 3–22.

LATTANZIO V, DE CICCO V, DI VENERE D, LIMA G, and SALERNO M (1994b), 'Antifungal activity of phenolics against different storage fungi', *Ital J Food Sci*, **6**, 23–30.

LATTANZIO V, DI VENERE D, LINSALATA V, BERTOLINI P, IPPOLITO A, and SALERNO M (2001), 'Low temperature metabolism of apple phenolics and quiescence of *Phlyctaena vagabunda'*, *J Agric Food Chem*, **49**, 5817–5821.

LATTAOUI N and TANTOUI-ELARAKI A (1994), 'Individual and combined antibacterial activity of essential oil components of three thyme essential oil', *Rivista italiana EPPOS*, **13**, 13–19.

LICHTER A, ZUTKHY Y, SONEGO L, DVIR O, KAPLUNOV T, SARIG P and BEN-ARIE R (2002), 'Ethanol controls postharvest decay of table grapes', *Postharvest Biol* Tec, **24**, 301–308.

LIMA G, DE CURTIS F, CASTORIA R, PACIFICO S, and DE CICCO V (1998), 'Additives and natural products against postharvest pathogens and compatibility with antagonistic yeasts', *J Plant Pathol*, **80**, 259 (abstract).

LIU W T, CHU C L, and ZHOU T (2002), 'Thymol and acetic acid vapors reduce postharvest brown rot of apricots and plums', *HortScience*, **37**, 151–156.

MACHEIX J J, FLEURIET A, and BILLOT J (1990), *Fruit Phenolics*, CRC Press, Boca Raton, Florida, USA.

MAEZAKI Y, TSUJI K, NAKAGAWA Y, KAWAI Y, AKIMOTO M, TSUGITA T, TAKEKAWA W, TERADA A, HARA H and MITSUOKA T (1993), 'Hypocholesterolemic effect of chitosan in adult males', *Biosci Biotech Biochem*, **57**, 1439–1444.

MANICI L M, LAZZERI L, and PALMIERI S (1997), '*In vitro* fungitoxic activity of some glucosinolates and their enzyme-derived products toward plant pathogenic fungi', *J Agric Food Chem*, **45**, 2768–2773.

MARCUCCI M C (1995), 'Propolis: chemical composition, biological properties and therapeutic activity', *Apidologie*, **26**, 83–99.

MARGOSAN D A, SMILANICK J L, SIMMONS G F, and HENSON D J (1997), 'Combination of hot water and ethanol to control postharvest decay of peach and nectarines', *Plant Dis*, **81**, 1405–1409.

MARI M and CARATI A (1997), 'Use of *Saccharomyces cerevisiae* with ethanol in the biological control of grey mould on pome fruits', in Bertolini P, Sijmons P C, Guerzoni M E, Serra F, COST 914–915, '*Non conventional methods for the control of postharvest disease and microbiological spoilage*', Office for Official Publication of the European Communities, Luxemburg, 85–91.

MARI M and GUIZZARDI M (1997), 'Protezione dei prodotti ortofrutticoli dopo la raccolta: tecniche di lotta biologica ai microrganismi fungini', *Inf Fitopatol*, **(9)**, 21–25.

MARI M, IORI R, LEONI O, and MARCHI A (1993), '*In vitro* activity of glucosinolate-derived isothiocyanates against postharvest fruit pathogens', *Ann Appl Biol*, **123**, 155–164.

MARI M, IORI R, LEONI O, and MARCHI A (1996), 'Bioassays of glucosinolate-derived isothiocyanates against postharvest pathogens', *Plant Pathol*, **45**, 753–760.

MARI M, LEONI O, IORI R, and CEMBALI T (2002), 'Antifungal vapour-phase activity of allyl-isothiocyanate against *Penicillium expansum* on pears', *Plant Path*, **51**, 231–236.

MATTHEIS J P and ROBERTS R G (1993), 'Fumigation of sweet cherry (*Prunus avium* 'Bing') fruit with low molecular weight aldehydes for postharvest decay control', *Plant Dis*, **77**, 810–814.

MEIR S, DROBY S, DAVIDSON H, ALSEVIA S, COHEN L, HOREV B, and PHILOSOPH-HADAS S (1998), 'Suppression of Botrytis rot in cut rose flowers by postharvest application of methyl jasmonate', *Postharvest Biol Tec*, **13**, 235–243.

MIHALIAK C A, GERSHENZO J, and CROTEAU R (1991), 'Lack of rapid monoterpene turnover in rooted plants, implication for theories of plant chemical defense', *Ecologia*, **87**, 373–376.

MITHEN R (2001), 'Glucosinolates – biochemistry, genetics and biological activity', *Pl Gr Reg*, **34**, 91–103.

MITHEN R F, LEWIS B G, and FENWICK G R (1986), '*In vitro* activity of glucosinolates and their products against *Leptosphaeria maculans*', *Trans Brit Mycol Soc*, **87**, 433–440

MUELLER HARVEY I, and DHANOA M S (1991), 'Varietal differences among sorghum crop residues in relation to their phenolic HPLC fingerprints and responses to different environments', *J Sci Food Agric*, **57**, 199–209.

NANDI B and FRIES N (1976), 'Volatile aldehydes, ketones, esters, and terpenoids as preservatives against storage fungi in wheat', *Z Pflk Pfls*, **83**, 284–294.

NEVILL D, NYFELER R and SOZZI D (1998), 'CGA 142705 a novel fungicide for seed treatment', *Brit Crop Prot Conf Pests Dis*, **1**, 65–72.

NIGRO F, IPPOLITO A and LIMA G (1998), 'Use of UV-C light to reduce Botrytis storage rot of table grapes', *Postharvest Biol Tec*, **13**, 171–181.

NIGRO F, IPPOLITO A, LATTANZIO V, DI VENERE D, and SALERNO M (2000), 'Effect of ultraviolet-C light on postharvest decay of strawberry', *J Plant Pathol*, **82**, 29–37.

OGAWA J M and LYDA S D (1960), 'Effect of alcohol on spores of *Sclerotinia fructicola* and other peach fruit rotting fungi in California', *Phytopathology*, **50**, 790–792.

OJALA T, REMES S, HAANSUU P, VUORELA H, HILTUNEN R, HAAHTELA K, and VUORELA P (2000), 'Antimicrobial activity of some coumarins containing herbal plants growing in Finland', *J Ethnopharm*, **73**, 299–305.

OSER B L, FORD R A, and BERNARD B K (1984), 'Recent progress in the consideration of flavoring ingredients under the Food Additive Amendments, 13, GRAS substances', *Food Technol*, **34**, 66–89.

OTA C, UNTERKIRCHER C, FANTINATO V, and SHIMIZU M T (2001), 'Antifungal activity of propolis on different species of *Candida*', *Mycoses*, **44**, 375–378.

PAXTON J D (1981), 'Phytoalexins – A working redefinition', *Phytopathol* Z, **101**, 106–109.

PESIS E and AVISSAR I (1990), 'Effect of postharvest application of acetaldehyde vapor on strawberry decay, taste, and certain volatiles', *J Sci Food Agric*, **52**, 377–385.

PIETTA P G (2000), 'Flavonoids as antioxidants', *J Nat Prod*, **63**, 1035–1042.

POOLE P R and WHITMORE K J (1997), 'Effect of topical postharvest application of 6-pentyl-2-pyrone on properties of stored kiwifruit', *Postharvest Biol Tec*, **12**, 229–237.

POWELSON, R.L. (1960), Initiation of strawberry fruit root caused by *Botrytis cinerea*, *Phytopathology*, **50**, 491–494.

PRASAD Y and STADELBACHER G J (1973), 'Control of postharvest decay of fresh raspberries by acetaldehyde vapor', *Plant Dis Rep*, **57**, 795–797.

PRASAD Y and STADELBACHER G J (1974), 'Effect of acetaldehyde vapor on postharvest decay and market quality of fresh strawberries', *Phytopathology*, **64**, 948–951.

RECIO M C, RÍOS J L, and VILLAR A (1989), 'A review of some antimicrobial compounds isolated from medicinal plants reported in the literature 1978–1988', *Phytother Res*, **3**, 117–125.

REDDY M V B, ANGERS P, GOSSELIN A, and ARUL J (1998), 'Characterization and use of essential oil from *Thymus vulgaris* against *Botrytis cinerea* and *Rhizopus stolonifer* in strawberry fruits', *Phytochemistry*, 47, 1515–1520.

REDDY M V B, BELKACEMI K, CORCUFF R, CASTAIGNE F, and ARUL J (2000), 'Effect of pre-harvest chitosan sprays on post-harvest infection by *Botrytis cinerea* and quality of strawberry fruit', *Postharvest Biol Tec*, **20**, 39–51.

REDDY M V B, CORCUFF R, KASAAI M R, CASTAIGNE F, and ARUL J (1999), 'Induction of resistance against grey mould rot in carrot roots by chitosan', *Phytopathology*, **89**, S6, (abstract).

RHODES M J C (1994), 'Physiological roles for secondary metabolites in plants: some progress, many outstanding problems', *Plant Mol Biol*, **24**, 1–20.

RICHER D L (1987), 'Synergism: a patent view', *Pesticide Sci*, **19**, 309–315.

ROBERTS J W and DUNEGAN J C (1932), 'Peach brown rot', *Technical Bulletin* No 328, US Dept of Agric, Washington DC.

RODOV V, BEN-YEHOSHUA S, FANG D Q, KIM J J, and ASHKENAZI R (1995), 'Preformed antifungal compounds of lemon fruit: citral and its relation to disease resistance', *J*

Agric Food Chem, **43**, 1057–1061.

ROMANAZZI G, IPPOLITO A, and NIGRO F (1999a), 'Activity of glicolchitosan on postharvest strawberry rot', *J Plant Pathol*, **81**, 237 (abstract).

ROMANAZZI G, SCHENA L, NIGRO F, and IPPOLITO A (1999b), 'Preharvest chitosan treatments for the control of postharvest decay of sweet cherries and table grapes', *J Plant Pathol*, **81**, 237 (abstract).

ROMANAZZI G, NIGRO F, and IPPOLITO A (2000a), 'Effectiveness of pre- and postharvest chitosan treatments on storage decay of strawberry', *Frutticoltura*, **62**(5), 71–75.

ROMANAZZI G, NIGRO F, LIGORIO A, and IPPOLITO A (2000b), 'Hypobaric and chitosan integrated treatments to control postharvest rots of sweet cherries', Proc *V EFPP* Congress, Taormina, Italy, 558–560.

ROMANAZZI G, NIGRO F, and IPPOLITO A (2001), 'Chitosan in the control of postharvest decay of some Mediterranean fruits', in Muzzarelli R A A, *Chitin Enzymology 2001*, Atec, Grottammare, Italy, 141–146.

ROMANAZZI G, NIGRO F, IPPOLITO A, DI VENERE D, and SALERNO M (2002), 'Effects of pre and postharvest chitosan treatments to control storage grey mould of table grapes', *J Food Science*, **67**, 1862–1867.

ROMANAZZI G, NIGRO F, and IPPOLITO A (2003), 'Short hypobaric treatments potentiate the effect of chitosan in reducing storage decay of sweet cherries', *Postharvest Biol Tec*, **20**(1) (in press).

SAKS Y and BARKAI-GOLAN R (1995), '*Aloa vera* gel activity against plant pathogenic fungi', *Postharvest Biol Tec*, **6**, 159–165.

SHAHIDI F and NACZK M (1995), *Food Phenolics: Sources, Chemistry, Effects, Applications*, Technomic Publishing Company, Inc., Lancaster, USA.

SHARAN M , TAGUCHI G, GONDA K, JOUKE T, SHIMOSAKA M, HAYASHIDA N, and OKAZAKI M (1998), 'Effects of methyl jasmonate and elicitor on the activation of phenylalanine ammonia-lyase and the accumulation of scopoletin and scopolin in tobacco cell cultures', *PlantSci*, **132**, 13–19.

SHARMA A, TEWARI G M, SHRIKHANDE A J, PADWAL-DESAI S R, and BANDYOPADHYAY A G (1979), 'Inhibition of aflatoxin-producing fungi by onion extracts', *J Food Sci*, **44**, 1545–1547.

SHARMA A, PADWAL-DESAI S R, TEWARI G M, and BANDYOPADHYAY G (1981), 'Factors affecting antifungal activity of onion extractives against aflatoxins-producing fungi', *J Food Sci*, **46**, 741–744.

SHOLBERG P L (1998), 'Fumigation of fruit with short-chain organic acids to reduce the potential of postharvest decay', *Plant Dis*, **82**, 689–693.

SHOLBERG P L and GAUNCE A P (1995), 'Fumigation of fruit with acetic acid to control postharvest decay', *HortScience*, **30**, 1271–1275.

SHOLBERG P L and GAUNCE A P (1996), 'Fumigation of stone fruit with acetic acid to control postharvest decay', *Crop Prot*, **15**, 681–686.

SHOLBERG P L, REYNOLDS A G, and GAUNCE A P (1996), 'Fumigation of table grapes with acetic acid to prevent postharvest decay', *Plant Dis*, **80**, 1425–1428.

SHOLBERG P L, HAAG P, HOCKING R, and BEDFORD K (2000), 'The use of vinegar vapor to reduce postharvest decay of harvested fruit', *HortScience*, **35**, 898–903.

SMILANICK J L, MARGOSAN D A, and HENSON D J (1995), 'Evaluation of heated solution of sulfur dioxide, ethanol, and hydrogen peroxide to control postharvest green mould of lemons', *Plant Dis*, **79**, 742–747.

SOMMER N F, FORTLAGE R J, and EDWARDS D C (1992), 'Postharvest diseases of selected commodities'. In Kader A A, *Postharvest Technology of Horticultural Crops*,

University of California, Div Agriculture and Natural Resources, Publication 3311, 117–160.

SPANOS G A and WROLSTAD R E (1992), 'Phenolics of apple, pear, and white grape juices and their changes with processing and storage. A review', *J Agric Food Chem*, **40**, 1478–1487.

STANGE JR R R, MIDLAND S L, ECKERT J W, and SIMS J J (1993), 'An antifungal compound produced by grapefruit and Valencia orange after wounding of the peel', *J Nat Prod*, **56**, 1627–1629.

STADELBACHER G J and PRASAD Y (1974), 'Postharvest decay control of apple by acetaldehyde vapor', *J Am Soc Hortic Sci*, **99**, 364–368.

STEWART J K, AHARONI Y, HARSTEN P L, and YOUNG D K (1980), 'Symptoms of acetaldehyde injury on head lettuce', *HortScience*, **15**, 148–149.

SUSLOW T (2000), *Postharvest Handling for Organic Crops*, University of California, Div Agriculture and Natural Resources, Publication 7254.

TAKEDA F, JANISIEWICZ W, ROITMAN J, MAHONEY N, and ABELES F B (1990), 'Pyrrolnitrin delays postharvest fruit rot in strawberry', *HortScience*, **25**, 320–322.

THANGADURAI D, ANITHA S, PULLAIAH T, REDDY P N, and RAMACHANDRAIAH O S (2002), 'Essential oil constituents *in vitro* antimicrobial activity of *Decalepis hamiltonii* roots against foodborne pathogens', *J Agric Food Chem*, **50**, 3147–3149.

TONINI G and CACCIONI D R L (1990), 'The effect of several natural volatiles on *Monilinia laxa*', Proc 'Qualità dei prodotti ortofrutticoli postraccolta', Fondazione Cesena Agri-cultura, 123–126.

TORAY RESEARCH CENTER INC. (1991) 'New development in functional packaging materials', in Eumura E, *Information on Frontier Technology and Future Trends*, Tokyo, Japan, 258–269.

TUNCEL, G and NERGIZ C (1993), 'Antimicrobial effect of some olive phenols in a laboratory medium', *Lett Appl Microbiol*, **17**, 300–302.

VAN DAMME E J M, WILLELM P, TORREKENS S, VAN LEUVEN F, and PEUMANS W J (1993) 'Garlic *Allium sativum* chitinases: characterization and molecular cloning', *Physiologica Planta*, **87**, 177–186.

VANETTEN H D, MATTHEWS D E, and MATTHEWS P S (1989), 'Phytoalexin detoxification: importance for pathogenicity and practical implications', *Ann Rev Phytopathol*, **27**, 143–164.

VENTURINI M E, BLANCO D, and ORIA R (2002), '*In vitro* antifiungal activity of several antimicrobial compounds against *Penicillium expansum*', *J Food Prot,* **65**, 834–839.

VERHOEVEN D T H, VERHAGEN H, GOLDBOHM R A, VAN DEN BRANDT PA, and VAN POPPEL G (1997), 'A review of mechanisms underlying anticarcinogenicity by Brassica vegetables', *Chem Biol Int*, **103**, 79–129.

WANG X G and NG T B (2001), 'Purification of allivin, a novel antifungal protein from bulbs of the round-cloved garlic', *Life Science*, **70**, 357–365.

WILSON C L and WISNIEWSKI M E (1989), 'Biological control of postharvest diseases of fruits and vegetables: an emerging technology', *Annu Rev Phytopathol*, **27**, 425–441.

WILSON C L and WISNIEWSKI M E (1994), *Biological Control of Postharvest Diseases – Theory and Practices*, CRC Press, Boca Raton, Florida.

WILSON C L FRANKLIN J D, and OTTO B (1987), 'Fruit volatiles inhibitory to *Monilinia fructicola* and *Botrytis cinerea*', *Plant Dis*, **71**, 316–319.

WILSON C L, EL GHAOUTH A, CHALUTZ E, DROBY S, STEVENS C, LU J Y, KHAN V, and ARUL J

(1994), 'Potential of induced resistance to control postharvest diseases of fruits and vegetables', *Plant Dis*, **78**, 837–844.

WILSON C L, SOLAR J M, EL GHAOUTH A, and WISNIEWSKI M E (1997), 'Rapid evaluation of plant extracts and essential oils for antifungal activity against *Botrytis cinerea*', *Plant Dis*, **81**, 204–210.

YIN M C and TSAO SM (1999), 'Inhibitory effect of seven Allium plants upon three *Aspergillus* species', *Int J Food Microbiol*, **49**, 49–56.

YOSHIDA S, KASUGA S, HAYASHI N, USHIROGUCHI T, MATSUURA H, and NAKAGAWA S, (1987), 'Antifungal activity of ajoene derived from garlic', *App Environ Microbiol*, **56**, 615–617.

YUEN C M C, PATON J E, HANAWATI R, and SHEN L O (1995), 'Effect of ethanol, acetaldehyde, and ethyl formate on the growth of *Penicillium italicum* and *P. digitatum* on oranges', *J Hortic Sci*, **70**, 81–84.

11

Plant antimicrobials combined with conventional preservatives for fruit products

S. M. Alzamora and S. Guerrero, University of Buenos Aires, Argentina and A. López-Malo and E. Palou, University of the Americas, Mexico

11.1 Introduction

The antimicrobial activity of some additives, combined or not, in water, buffers or broths is often a poor indication of performance in complex food systems. In the case of naturally occurring antimicrobials, extrapolation is even more difficult. In many cases, the antimicrobial activity is reduced due to interactions with food components (proteins, lipids, aldehydes and many macromolecules) and other preservation factors (Sofos *et al.*, 1998). Thus, concentrations required for inhibition or inactivation of microorganisms in real foods are considerably higher than in laboratory media and frequently above tolerable taste thresholds. Sensory changes in foods, however, can be minimized using multifactorial preservative systems including natural and conventional antimicrobials (Gould, 1996; López-Malo *et al.*, 2000; Roberts, 1989; Brul and Coote, 1999; Leistner, 2000).

Combinations of antimicrobials can be chosen because target microorganisms are resistant to inhibition or killing by conventional (legally approved and/or sensorially compatible) doses of single antimicrobials but against which the combination may exert antimicrobial activity. More frequently, antimicrobial combinations can be selected to provide broad spectrum coverage in the preservation of foods. However, little attention has been given to the study of the mechanisms of action or development of resistance in microorganisms of importance in fruit products, i.e. molds and yeasts. The lack of understanding of the relative contribution of factors in food preservation and safety is surprising because the need for an adequate database for developing safe combined preservation systems was pointed out many years ago (Roberts, 1989). The combined action of mixtures of conventional preservatives e.g. benzoates and sorbates, and natural antimicrobials forms the focus of this chapter.

The effect of garlic, onion and sodium benzoate on the fungi of ground pepper, cinnamon and rosemary was studied by Abdel-Hafez and El-Said (1997) in 30 Egyptian spice samples. A combination of sodium benzoate and garlic juice in the isolation medium inhibited osmophilic or osmotolerant and cellulose-decomposing fungi (strains of *Aspergillus flavus*, *A. niger* and species of *Cladosporium* and *Mycosphaerella*). Faid *et al.* (1995) reported that the minimum inhibitory concentration (MIC) of sorbic acid against yeasts isolated from spoiled almond paste was reduced in the presence of cinnamon. Sebti and Tantaoui-Elaraki (1994) tested sorbic acid, calcium propionate, benzoic acid, cinnamon powder and cinnamon water extract against 151 fungal isolates from pastilla papers (made from wheat flour, water, eggs, salt and vegetable oil). Cinnamon powder exerted good inhibitory action but imparted a dark color to the papers, while cinnamon extract did not inhibit fungal growth even at a concentration of 80 g/kg. Calcium propionate and benzoic acid failed to inhibit the fungi. However, all isolates were inhibited using a combination of sorbic acid (0.75 g/kg) and cinnamon water extract (20 g/kg cinnamon equivalent).

Azzous and Bullerman (1982) reported that clove was an efficient antimycotic agent against *A. flavus*, *A. parasiticus* and *A. ochraceus* and four strains of *Penicillium,* delaying mold growth by more than 21 days. Additive and synergic effects were also reported by these authors when 0.1% clove was combined with 0.1–0.3% potassium sorbate, delaying mold germination time.

This chapter is focused mainly on natural antimicrobials from plants and their application in the design of combined preservation technologies for shelf-stable and refrigerated fruit products. These technologies include different traditional and emerging preservation factors, such as reduction of water activity (a_w) and pH, mild heat treatment, ultrasound and addition of natural and synthetic preservatives. Studies in microbiological media and model fruit systems as well as some applications for preservation of real fruits are reviewed.

11.2 Combinations of natural and conventional antimicrobials for inhibiting microbial growth in laboratory media

The interaction of synthetic and natural preservatives against various fungi has been studied in model systems with environmental factors resembling those used in the preservation of high moisture fruit products (Alzamora *et al.*, 1995). The individual and combined effects of synthetic antimicrobials (sodium benzoate, sodium bisulfite and potassium sorbate) and naturally occurring antimicrobials (vanillin, eugenol, thymol, carvacrol and citral) on *A. flavus* in potato dextrose

Table 11.1 Minimum inhibitory concentrations (MICs)* of selected antimicrobials against *Aspergillus flavus* in PDA at a_w 0.99 or 0.95 and pH 3.5 or 4.5

Antimicrobial compound	MIC (ppm)			
	a_w 0.99		a_w 0.95	
	pH 3.5	pH 4.5	pH 3.5	pH 4.5
Vanillin	1300	1300	1300	1100
Eugenol	600	500	600	500
Carvacrol	300	400	200	300
Thymol	400	400	300	300
citral	1800	1800	1400	1600
sodium Benzoate	200	1500	300	1100
Sodium bisulfite	<100	900	300	700
Potassium sorbate	400	800	200	300

* MICs were defined as the minimum concentration required to inhibit mold growth for two months at 25 °C.

agar (PDA) at a_w 0.99 or 0.95 (adjusted with sucrose) and pH 4.5 or 3.5 (adjusted with hydrochloric acid) were evaluated by López-Malo (2000). The MICs (Table 11.1) varied from 200 ppm sodium benzoate at pH 3.5 to 1800 ppm citral at both pH 3.5 and 4.5. The effect of sodium benzoate, sodium bisulfite and potassium sorbate was strongly pH-dependent, being more noticeable at a_w 0.99. In contrast, natural antimicrobials showed less pH dependence. Lag time increased as antimicrobial concentrations in the agar increased. At a_w 0.99, *A. flavus* was more sensitive to thymol, eugenol, carvacrol, potassium sorbate and sodium benzoate (at pH 3.5) than to vanillin or citral.

Binary combinations of antimicrobials have been analyzed by transforming MIC data into fractional inhibitory concentrations (FICs), as defined by Davidson and Parish (1989):

$$FIC_A = \frac{\text{MIC of A in presence of B}}{\text{MIC of A}} \tag{11.1}$$

$$FIC_B = \frac{\text{MIC of B in presence of A}}{\text{MIC of B}} \tag{11.2}$$

The FIC Index was calculated as follows, using the FICs for individual antimicrobials:

$$FIC_{Index} = FIC_A + FIC_B \tag{11.3}$$

FIC isobolograms for combinations of vanillin and potassium sorbate (pH 3.5 or 4.5); citral and potassium sorbate (pH 3.5); and sodium benzoate and vanillin (pH 3.5) against *A. flavus* for different incubation times and a_w 0.99 are presented in Figs 11.1–11.3. Data on the straight line connecting unity (FIC = 1) on the *x* and *y*-axes indicate an additive effect; curves deviated to the left of the additive line indicate synergistic interaction and curves deviated to the right of

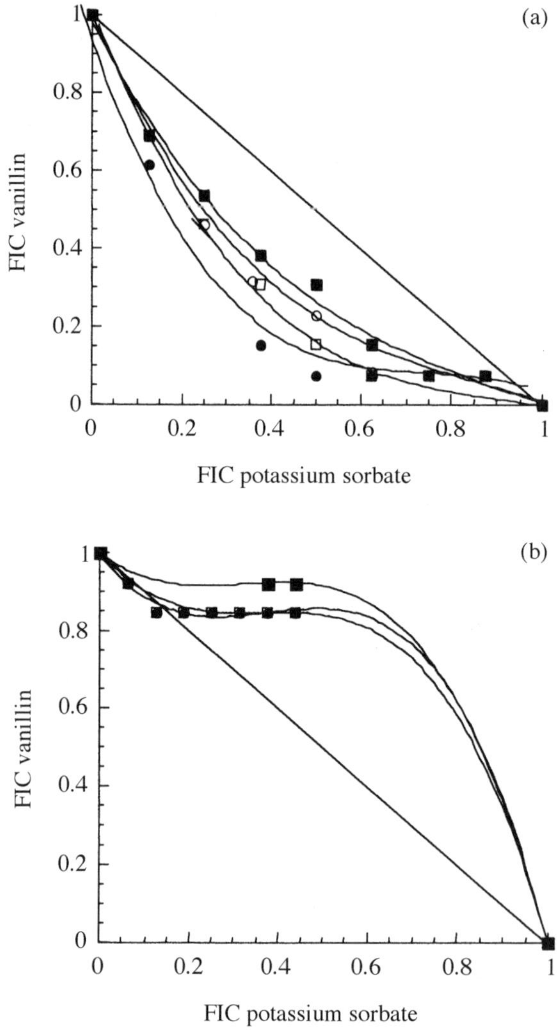

Fig. 11.1 FIC isobolograms for combinations of vanillin and potassium sorbate against *Aspergillus flavus* in potato dextrose agar (a_w 0.99) for ● 5 days, ❐ 10 days, ❍ 15 days and ■ 30 days at 25 °C. (a) pH 3.5 and (b) pH 4.5.

the additive line indicate antagonistic interaction. Similarly, a FIC index near 1 implicates additivity; < 1 implicates synergy; and >1 implicates antagonism (Davidson and Parish, 1989). At pH 3.5, several combinations of vanillin and potassium sorbate were synergistic, while at pH 4.5 most of the combinations were antagonistic (Fig. 11.1). In addition, Fig. 11.1(a) shows that, as incubation time increased, the synergistic interaction of vanillin with sorbate was lost and became additive. The same trend was observed for the initially synergistic sorbate + citral interaction (Fig. 11.2). The interaction vanillin + benzoate was antagonistic for most of the combinations studied (Fig. 11.3). Overall, the

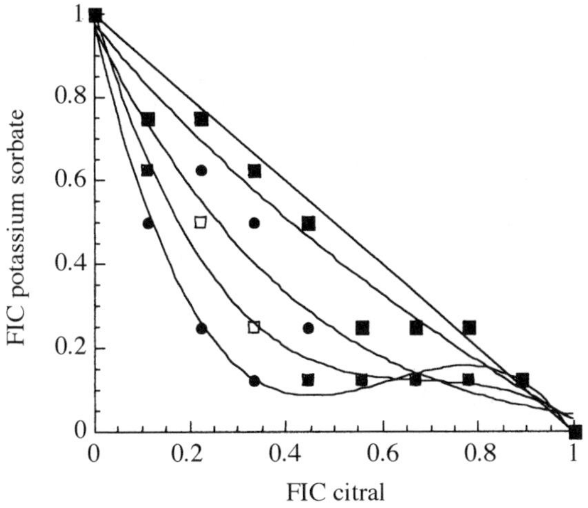

Fig. 11.2 FIC isobologram for combinations of citral and potassium sorbate against *Aspergillus flavus* in potato dextrose agar (a_w 0.99, pH 3.5) for ● 5 days, ❒ 10 days, ❍ 15 days and ■ 30 days at 25 °C.

following antimicrobial combinations were synergistic for at least 30 days at a_w 0.99 and pH 3.5: vanillin + sorbate, vanillin + eugenol, vanillin + citral, vanillin + thymol, vanillin + carvacrol; eugenol + carvacrol, eugenol + thymol and eugenol + citral. At pH 4.5, the combinations vanillin + benzoate, vanillin + citral, benzoate + thymol, benzoate + citral, sodium bisulfite + thymol, carvacrol + thymol, carvacrol + citral and thymol + citral were synergistic.

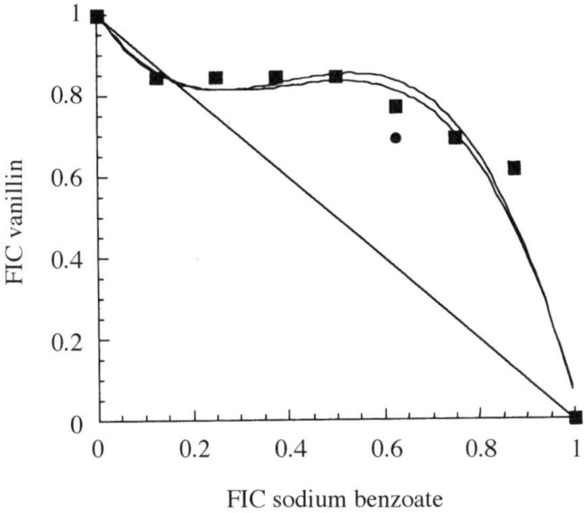

Fig. 11.3 FIC isobologram for combinations of sodium benzoate and vanillin against *Aspergillus flavus* in potato dextrose agar (a_w 0.99, pH 3.5) for ● 5 days and ■ 30 days at 25 °C.

Table 11.2 Minimum inhibitory concentrations (MICs)* of selected antimicrobials against *Zygosaccharomyces bailii* in PDA at a_w 0.99 or 0.95 and pH 3.5 or 4.5

Antimicrobial compound	MIC (ppm)			
	a_w 0.99		a_w 0.95	
	pH 3.5	pH 4.5	pH 3.5	pH 4.5
Vanillin	1000	800	800	650
Eugenol	250	50	100	100
Carvacrol	100	100	50	50
Thymol	100	100	50	100
Citral	1100	1100	>3000	1100
EDTA	600	600	2000	>3000
Sodium bisulfite	1100	800	100	100
Potassium sorbate	300	50	100	50

* MICs were defined as the minimum concentration required to inhibit yeast growth for 12 days at 25 °C.

In a similar study, the individual and combined activities of vanillin, eugenol, carvacrol, thymol, citral, potassium sorbate, sodium bisulfite and ethylenediaminetetraacetic acid (EDTA) were assessed against *Zygosaccharomyces bailii* in PDA at a_w 0.99 or 0.95 (reduced with sucrose) and pH 4.5 or 3.5 (adjusted with hydrochloric acid) (Rivera Carriles, 2002). The MICs for the different antimicrobials depended upon a_w and pH (Table 11.2). Citral showed the lowest activity against the yeast among the natural antimicrobials, followed by vanillin, while thymol and carvacrol were the the most effective in all conditions. Binary combinations of thymol, citral, eugenol or carvacrol with potassium sorbate were all synergistic except for the eugenol-containing agar at a_w 0.99 and pH 3.5, where an antagonistic interaction was observed. However, vanillin acting in the presence of EDTA, sorbate or sodium bisulfite exhibited antagonistic, synergistic or additive effects depending not only on a_w and pH but also on the relative amount of each antimicrobial in the binary mixture. Vanillin + citral, thymol + citral, carvacrol + citral, eugenol + citral and thymol + carvacrol combinations were synergistic at all conditions of pH and a_w and concentrations of antimicrobials in the mixture. At a_w 0.99 and pH 3.5, the combinations of vanillin + carvacrol, vanillin + thymol, vanillin + eugenol, thymol + eugenol and carvacrol + eugenol offered no advantage over the use of each antimicrobial alone or else gave rise to antagonistic effects.

Combinations of potassium sorbate and vanillin on the growth of *Penicillium digitatum*, *P. glabrum* and *P. italicum* in PDA at a_w 0.98 (adjusted with sucrose) and pH 3.5 (adjusted with citric acid) were synergistic for at least one month at 25 °C (Matamoros-León *et al.*, 1999). MICs for potassium sorbate were 150 ppm for *P. digitatum*; 200 ppm for *P. italicum* and 700 ppm for *P. glabrum* while for vanillin the MICs were 1100 ppm for *P. digitatum* and *P. italicum* and 1300 ppm for *P. glabrum*. When used in combination, the FICs were less than 1, indicating synergy.

These results show that it is not easy to anticipate synergistic effects in binary mixtures nor to explain the observed activity. Moreover, many combinations can be antagonistic. There are four mechanisms of antimicrobial interaction that produce synergism: (a) sequential inhibition of a common biochemical pathway, (b) inhibition of protective enzymes, (c) combinations of cell wall-active agents; and (d) use of cell wall-active agents to enhance the uptake of other antimicrobials (Eliopoulos and Moellering, 1991). Mechanisms of antimicrobial interaction that produce antagonism are less well-known and include: (a) combinations of bactericidal/fungicidal and bacteriostatic/fungistatic agents; (b) use of compounds that act on the same target of the microorganism; and (c) chemical (direct or indirect) interactions between the compounds (Larson, 1984). However, the specific modes of action of plant constituents on metabolic activities of microorganisms need further elucidation before results can be explained adequately.

11.3 Natural antimicrobials combined with ultrasonic treatment and conventional preservatives

The inactivation of microorganisms by ultrasonic waves of high intensity has been attributed to cavitation, which releases large amounts of energy, generating temperatures of 5000 K and shock waves with pressures of several atmospheres (Mason, 1990). Microbial inactivation by ultrasound depends on many factors including wave amplitude, temperature, volume of food processed, composition and physical properties of the food, microbial characteristics, etc.

Bacterial spores are much more resistant to ultrasound than vegetative cells and fungi are more resistant in general than vegetative bacteria (Alliger, 1975). Ultrasonic treatment at levels that cannot adversely modify nutritional and sensory properties of food is ineffective against microorganisms. Therefore it has been suggested that ultrasound could be more effective when used in combination with other stress factors. Accordingly, many studies reported in the literature combine ultrasound with heat (thermo-ultrasonication) or with heat under pressure (manothermosonication) for inactivation of some pathogenic and spoilage microorganisms (García *et al.*, 1989; Raso *et al.*, 1997, 1998; Guerrero *et al.*, 2001a).

In our laboratory, the sensitivity of *Saccharomyces cerevisiae* to sonication combined with moderate temperatures, addition of preservatives (synthetic and/or naturally occurring ones) and control of pH with citric acid was examined (Guerrero *et al.*, 2001 a,b,c). Non significant differences ($\alpha < 0.05$) between *D*-values at pH 3.0 or 5.6 in Sabouraud Broth were obtained when applying thermo-ultrasound alone (20 kHz, 95.2 μm wave amplitude, 45 °C, 15 min). Similarly, vanillin (800 ppm) did not significantly improve inactivation of *Sac. cerevisiae* by thermo-ultrasonic treatment at pH 5.6. However, when a lower amount of vanillin (500 ppm) was combined with citric acid to reduce pH to 3.0, there was synergism with thermo-ultrasonic treatment (Fig. 11.4a). This synergy

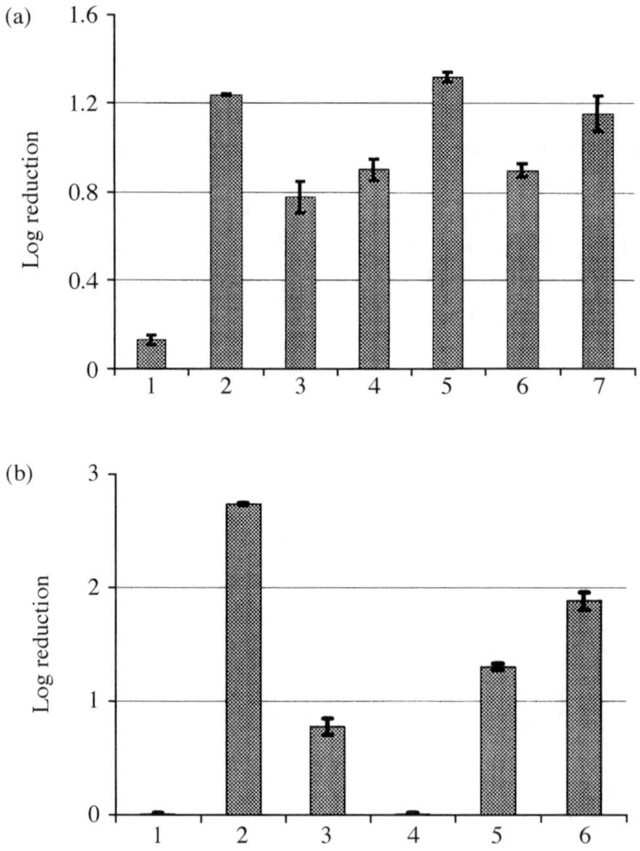

Fig. 11.4 (a) *Saccharomyces cerevisiae* log reductions obtained by application of ultrasound treatment (95 μm, 45 °C, 15 min) in laboratory media (pH 5.6) containing vanillin and/or EDTA and/or citric acid: (1) 1000 ppm EDTA; (2) 1000 ppm EDTA and ultrasound; (3) ultrasound; (4) citric acid (pH 3.0) and ultrasound; (5) 500 ppm EDTA, 500 ppm vanillin and ultrasound; (6) 800 ppm vanillin and ultrasound; (7) 500 ppm vanillin, citric acid (pH 3.0) and ultrasound. (Starting count $\approx 1 \times 10^6$ CFU/g). (b) *Saccharomyces cerevisiae* log reductions obtained by application of ultrasound treatment (95 μm, 45 °C, 20 min) in laboratory media (pH 5.6) containing cinnamon oil: (1) 50 ppm cinnamon type China; (2) 50 ppm cinnamon type China and ultrasound; (3) ultrasound; (4) 100 ppm cinnamon type Ceylon; (5) 100 ppm cinnamon type Ceylon and ultrasound; (6) 100 ppm cinnamon type Ceylon (32 h preincubation) and ultrasound. (Starting count $\approx 1 \times 10^6$ CFU/g).

was even greater when the medium was supplemented with EDTA (Guerrero *et al.*, 2001c). EDTA and vanillin both alter microbial cell permeability, leading to the loss of macromolecules from the interior and facilitating the penetration of the phenolic compound into the cell (Conner and Beuchat, 1984). The addition of the antimicrobials to the sonication medium also changed the inactivation kinetics of yeast from first-order to non-linear inactivation (combinations with cinnamon, vanillin or chitosan).

Greater inactivation was obtained when the yeast was exposed to the combined action of ultrasound and cinnamon at non growth-inhibitory concentrations (50 ppm). The effect was synergistic and depended on the type of cinnamon and on some process conditions (Fig. 11.4b). Cinnamon-type China was significantly more effective at 50 ppm than cinnamon-type Ceylon at 100 ppm, resulting in 2.5 log CFU/ml reductions in 20 min. The pre-incubation of the yeast in 100 ppm cinnamon type Ceylon for 35 h and subsequent ultrasonic treatment resulted in greater inactivation than exposure to cinnamon and ultrasound alone (Guerrero *et al.*, 2001c). This lipophilic compound could accumulate in the lipid bilayer of the cell, disturbing the function of the membrane or rupturing it (Brul and Coote, 1999).

11.4 Combination treatments for strawberry purée

Mild heat treatment, addition of ascorbic acid, reduction of a_w and control of pH have been combined with vanillin in minimally processed strawberry purée (Cerrutti *et al.*, 1997). Sucrose was added to freshly prepared strawberry purée to reduce initial a_w to 0.95. After addition of 500 ppm ascorbic acid, 1000 ppm calcium lactate and 3000 ppm vanillin, citric acid was used to adjust the pH to 3. The microbial stability of the purée was investigated by enumerating aerobic and anaerobic mesophilic bacteria, yeasts and molds, and by challenge testing with *Sac. cerevisiae*, *Zygosaccharomyces rouxii*, *Schizosaccharomyces pombe*, *Pichia membranaefaciens*, *Botrytis* sp., *Byssochlamys fulva*, *Bacillus coagulans* and *Lactobacillus delbrueckii.* The combination prevented the growth of native and inoculated flora for at least 60 days at room temperature. Control purées at low pH and a_w inhibited growth of *B. coagulans* and *Lac. delbrueckii* but vanillin addition exerted a bactericidal effect. Sensitivity to vanillin was dependent on the species. *Sac. cerevisiae* was the most sensitive and counts decreased abruptly in the presence of vanillin; *P. membranaefaciens* was the most resistant although growth was inhibited. The other yeasts gradually decreased in number during storage. Vanillin was effective against *Z. bailii* and *P. membranaefaciens*, both well-known preservative-resistant yeasts (Pitt and Richardson, 1973). In addition, vanillin controlled the growth of *Z. rouxii*, an osmotolerant spoilage yeast. *Byssochlamys fulva* and *Botrytis* sp. showed no growth in the preserved purée for 60 days. Although microbial behavior was similar at room and chill temperatures, color was better preserved at $\leq 10\,°C$. In conclusion, vanillin in combination with a slight reduction of a_w and regulation of pH with citric acid may be a promising technique for preserving strawberry products.

11.5 Combination treatments for banana purée

Castañón *et al.* (1999) have proposed a combined procedure to prepare minimally processed banana purée using vanillin. Sucrose and phosphoric acid (or citric acid) were added to banana purée containing ascorbic acid (approximately 0.2 ml/g) to reduce the initial a_w from 0.986 to 0.97 and to adjust the pH to 3.3, respectively. Vanillin (1000 or 3000 ppm) and/or 1000 ppm potassium sorbate were also added. The native flora (standard plate count, yeasts and molds) was enumerated periodically at different storage temperatures (15, 25 and 35 °C). The addition of 1000 ppm vanillin alone increased the lag phase up to 16 days at 15 °C, and the time to detect microbial spoilage was extended to around 21 days. In the presence of 3000 ppm vanillin or 1000 ppm potassium sorbate, no microbial growth was detected up to 60 days at any of the three temperatures studied. However, shelf-life of the purée was limited by browning. The sensorial aspects of multifactorial preservation are discussed in the following section.

11.6 Consumer acceptance and sensory evaluation of minimally processed fruits containing vanillin

Phenolic compounds are naturally present in fruit and vegetable products and sometimes in considerable quantities. Few of the natural phenols are toxic and only some cause allergic reactions in animals (Singleton, 1981). On the few occasions when phenolic compounds have been reported as harmful to humans, abnormal consumption of common phenols or consumption of rare phenols in the diet have been identified as possible causes. Fruits and vegetables and other sources of phenolic compounds in the diet, such as tea, may have beneficial effects in some disorders such as cancer, cirrhosis, emphysema and arteriosclerosis (Weisburger, 1992; Newmark, 1992; Nychas, 1995).

Vanillin (4-hydroxy-3methylbenzaldehyde) is a major constituent of vanilla beans. Like many other phenolic substances (see also Chapters 9 and 10), vanillin is a GRAS flavoring agent used in ice cream, soft drinks, confectionery, cookies, eggnogs, liquors, etc. (Dziezak, 1989). Its sensory characteristics are not unfamiliar to consumers. Vanillin not only imparts pleasant flavor notes but also acts as an antioxidant in complex foods containing polyunsaturated fatty acids (Burrí *et al.*, 1989).

Cerrutti and Alzamora (1996) have evaluated the sensory properties of apple, pear, plum and banana purées prepared without or with glucose to attain an a_w value equal to 0.95 and containing different concentrations of vanillin. A panel of untrained tasters scored the purées on a two category hedonic scale (agreeable and disagreeable), where agreeable purées had a pleasant vanillin flavor while maintaining the actual taste of the fruit. Identical scores were assigned to the different fruit purées, with or without the addition of glucose. The flavoring characteristics of vanillin were compatible with the fruits in concentrations up to 3000 ppm.

The color, flavor, odor and overall acceptability of banana purées at pH 3.4 and a_w 0.97 (adjusted with phosphoric acid and sucrose respectively) prepared with 3000 ppm vanillin or 1000 ppm potassium sorbate were evaluated by 50 untrained tasters using a 9 point hedonic scale (1: dislike extremely; 9: like extremely; Castañón *et al.*, 1999). The mean scores corresponded to products with a good overall acceptability with scores around 6 (like slightly). The purées were significantly different ($p < 0.05$) in odor (the one with vanillin was preferred) and flavor (the one with potassium sorbate was better) and there was no significant difference ($p < 0.05$) in color and overall acceptability. The shelf-life of these purées was determined on the basis of color change. Similar shelf-lives were obtained with vanillin and potassium sorbate at each storage temperature (50, 24 and 12 days at 15, 25 and 35 °C, respectively). In the presence of 3000 ppm vanillin or 1000 ppm potassium sorbate, the banana purée stored at 15, 25 and 35 °C presented several color changes that contributed to browning. After an induction period that depended on the storage temperature and the type of preservative, the browning index increased linearly until the upper limit of the method was reached. The rate of browning was greater in samples with vanillin than in those formulated with potassium sorbate, but the activation energy was 10% lower in the former system, suggesting a lesser temperature dependence for color changes in the purée with vanillin. Similarly, the net color difference (ΔE) remained constant during the first days of storage for a period of time that depended on the storage temperature, being shorter as temperature increased. After the induction period, ΔE increased rapidly during the first 15 and 27 days at 35 and 25 °C, respectively.

Leúnda *et al.* (1999) analyzed the color changes in minimally processed apple, pear and peach at 25 °C. The combined preservation process included the reduction of pH and a_w and addition of preservatives (vanillin and/or potassium sorbate) to ensure microbiological stability for at least three months at room temperature (Alzamora *et al.*, 1995; Cerrutti and Alzamora, 1996). Pulse vacuum osmotic dehydration (10 minutes, 20 mmHg) was used to impregnate the fruits with glucose or sucrose, ascorbic and citric acids, potassium sorbate and/or vanillin. Figure 11.5 shows the net color differences for a range of combinations. The magnitude and the pattern in the change of ΔE depended on the preservative and type of fruit. For apples, the greatest increase in ΔE was observed in samples containing both preservatives (vanillin and potassium sorbate) while apples containing only vanillin exhibited the lowest change for more than 90 days. For peach, the simultaneous inclusion of vanillin and potassium sorbate decreased the variation of net color differences, while for pear the opposite occurred. In peach and pear, ΔE values showed a pronounced increase after a short induction period and then remained approximately constant until the end of storage. In apple, there was an abrupt decrease in ΔE values at around 50 days of storage, which coincided with the limit of color acceptability. These results showed that combinations of vanillin and conventional preservation systems can be used to retain fresh-like characteristics, preserve sensory attributes and assure a reasonable shelf-life of fruit products.

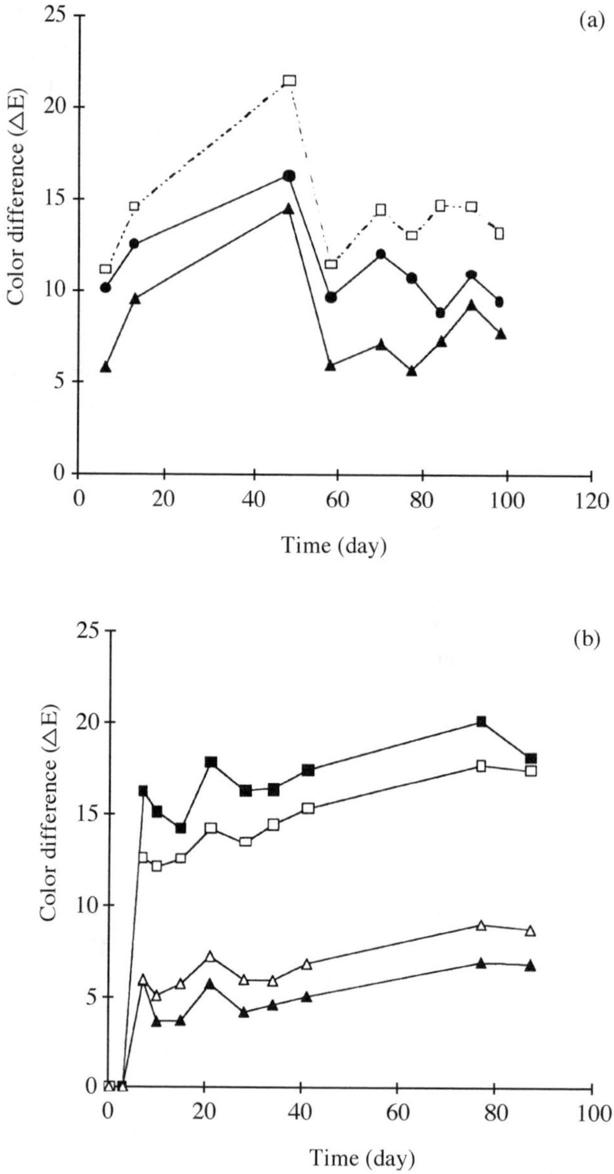

Fig. 11.5 Net color differences (ΔE) of minimally processed apple, peach and pear (a_w 0.98; pH 3.5; 250 ppm ascorbic acid) at 25 °C. (a) Apple (❑ 400 ppm potassium sorbate and 1300 ppm vanillin; ● 555 ppm potassium sorbate; ▲ 2540 ppm vanillin). (b) Peach (❑ 130 ppm potassium sorbate and 2000 ppm vanillin; ■ 130 ppm potassium sorbate). Pear (Δ 130 ppm potassium sorbate and 2000 ppm vanillin, ▲ 130 ppm potassium sorbate).

11.7 Future trends

Much further work on applications of different combinations of natural antimicrobials with conventional ones in different foods is still required. However, their mode of action in model systems and in food matrices is not well understood and represents a bottleneck for their application.

Most of the information available on the use of more than one antimicrobial has not been usable in formulating minimally processed combined techniques. Often, information includes only data from traditional challenge testing in specific foods, microbial presence or absence tests, or growth/no growth determinations. These isolated results do not allow us to compare quantitatively what happens in a food when the levels of the antimicrobials or the environmental stress factors are changed. The combination of conventional and natural antimicrobials can only be applied successfully if the interactions, which are dependent on many environmental factors (temperature, pH, water activity, non-thermal inactivation factors, food components, etc.) and exposure time, are well known.

Firstly, screening investigations and generation of databases with a quantitative basis should be initiated at model system level. Experience gained in developing predictive models can result in improved design, efficiency of data collection and precision of results (Alzamora and López-Malo, 2002). A systematic approach for planning, collection and appropriate processing of good quality data (even without integration of these data into mathematical models) would allow better assessment of the effect of antimicrobial mixtures in the presence of different extrinsic and intrinsic preservation factors. These data will enable a scientifically sound approach to the selection of antimicrobial 'equivalent mixtures' for the development of safe minimally processed fruit and vegetables. Secondly, the performance of selected mixtures must be investigated in specific foods. The collaboration of research institutions and industry will be key to the success of this work in the future.

A better understanding of microbial ecology and the physiological response of microorganisms to individual preservation factors as well as to combinations of natural and conventional antimicrobials in different food environments will offer new opportunities and provide greater precision for rational selection of antimicrobial combinations.

11.8 References

ABDEL-HAFEZ S I and EL-SAID AH M (1997), 'Effect of garlic, onion and sodium benzoate on the mycoflora of pepper, cinnamon and rosemary in Egypt', *Int Biodeterior Biodegrad*, **39**, 67–77.

ALLIGER H (1975), 'Ultrasonic disruption', *Am Lab*, **10**, 75–85.

ALZAMORA S.M and LÓPEZ-MALO A (2002), 'Microbial behavior modeling as a tool in the design and control of processed foods', in Welti Chanes J, Barbosa-Cánovas G and Aguilera J M, *Engineering and Food for the 21st Century*, Boca Raton, Florida.,

CRC Press, 631–650.

ALZAMORA S M, CERRUTTI P, GUERRERO S N and LÓPEZ-MALO A. (1995), 'Minimally processed fruits by combined methods', in Welti-Chanes J and Barbosa-Cánovas G, *Food Preservation by Moisture Control – Fundamentals and Applications.* International Symposium on the Properties of Water, Lancaster, Technomic Pub. Co, 463–492.

AZZOUS M A and BULLERMAN L B (1982), 'Comparative antimycotic effects of selected herbs, spices, plant components and commercial antifungal agents', *J Food Prot*, **45**, 1298–1301.

BRUL S and COOTE P (1999), 'Preservative agents in foods. Mode of action and microbial resistance mechanisms', *Int J Food Microbiol.*, **50** , 1–17.

BURRÍ J, GRAF M, LAMBELET P and LÖLIGER J (1989), 'Vanillin: more than a flavoring agent. A potent antioxidant', *J Sci Food Agric*, **48**, 49–56.

CASTAÑÓN X, ARGAIZ A and LÓPEZ-MALO A (1999), 'Effect of storage temperature on the microbial and color stability of banana puree with addition of vanillin or potassium sorbate', *Food Sci Technol Int*, **5**, 53–60.

CERRUTI P and ALZAMORA S M (1996), 'Inhibitory effects of vanillin on some food spoilage yeasts in laboratory media and fruit purées', *Int J Food Microbiol*, **29**, 379–386.

CERRUTI P, ALZAMORA S M and VIDALES S L (1997), 'Vanillin as an antimicrobial for producing shelf-stable strawberry purée', *J Food Sci*, **62**, 608–610.

CONNER D E and BEUCHAT L R (1984), 'Effects of essential oils from plants on growth of food spoilage yeasts', *J Food Sci*, **49**, 429–434

DAVIDSON P M and PARISH M E (1989), 'Methods for testing the efficacy of food antimicrobials', *Food Technol*, **43**(1), 148–155.

DZIEZAK J D (1989), 'Spices". *Food Technol.* **43**(1), 102–116.

ELIOPOULOS G M and MOELLERING R C (1991), 'Antimicrobial combinations', in Lorian V, *Antibiotics in Laboratory Medicine*, New York, Williams and Wilkins, 431–455.

FAID M, BAKHY K, ANCHAD M and TANTAOUI-ELARAKI A (1995), 'Almond paste: physicochemical and microbiological characterization and preservation with sorbic acid and cinnamon', *J Food Prot*, **58**, 547–550.

GARCÍA M L, BURGOS J, SANZ B and ORDÓÑEZ J (1989), 'Effect of heat and ultrasonic waves on the survival of two strains of *Bacillus subtilis*', *J Appl Bact*, **67**, 619–628.

GOULD G W (1996), 'Industry perspectives on the use of natural antimicrobials and inhibitors for food applications', *J Food Prot*, *Suppl*, 82–86.

GUERRERO S, LÓPEZ-MALO A and ALZAMORA S M (2001a), 'Effect of ultrasound on the survival of *Saccharomyces cerevisiae*: influence of temperature, pH and amplitude', *Innov Food Sci Emerg Technol*, **2**, 31–39.

GUERRERO S, TOGNON M and ALZAMORA, S M (2001b), 'Utilización de la Ecuación de Gompertz modificada para predecir el efecto combinado de ultrasonido, pH y algunos aditivos en la inactivación de *Saccharomyces cerevisiae*'. III Congreso Iberoamericano de Ingeniería de Alimentos .I Congreso Español de Ingeniería de Alimentos. Valencia, Spain.

GUERRERO S, TOGNON M and ALZAMORA, S M (2001c), 'Ultrasound and natural antimicrobials: inactivation of *Saccharomyces cerevisiae* by the combined treatment'. Institute of Food Technologists Annual Meeting, New Orleans, USA.

LARSON E (1984), *Clinical and Infection Control*, Boston, Blackwell Scientific Public.

LEISTNER L (2000), 'Hurdle technology in the design of minimally processed foods', in Alzamora, S M, Tapia M S and López-Malo A, *Minimally Processed Fruits and*

Vegetables. Fundamental Aspects and Applications, Maryland, Aspen Publishers Inc, 13–29.

LEÚNDA A., GUERRERO S N and ALZAMORA S M (1999), 'Efecto de la impregnación con vainillina en frutas mínimamente procesadas, in *Anales del VIII Congreso Argentino de Ciencia y Tecnología de Alimentos*, Santa Fé, Argentina, Asociación Argentina de Tecnólogos Alimentarios.

LÓPEZ-MALO A (2000), La PreservaciÓn Multiobjetivo de Alimentos: Efecto de Factores Tradicionales y Emergentes en la Respuesta de *Aspergillus flavus*. Universidad de Buenos Aires, Argentina, PhD Thesis.

LÓPEZ-MALO A, ALZAMORA S M and GUERRERO S N (2000), 'Natural antimicrobials from plants', in Alzamora, S M Tapia M S and López-Malo A, *Minimally Processed Fruits and Vegetables. Fundamental Aspects and Applications*, Maryland, Aspen Publishers Inc, 237–265.

MASON T (1990), *Sonochemistry: The Uses of Ultrasound in Chemistry*, Cambridge, Royal Society of Chemistry.

MATAMOROS-LEÓN B, ARGAIZ A and LÓPEZ-MALO A (1999), 'Individual and combined effect of vanillin and potassium sorbate on Penicillium digitarum, Penicillium glabrun, and Penicillium italicum growth', *J Food Prot*, **62**, 540–542.

NEWMARK H L (1992), 'Plant phenolic compounds as inhibitors of mutagenesis and carcinogenesis', in Huang M, Ho C, and Lee C Y, *Phenolic Compounds in Food and their Effect on Health II, Antioxidants and Cancer Prevention*, Washington, DC, ACS Symposium Series.

NYCHAS G J E (1995), 'Natural antimicrobials from plants', in Gould G W, *New Methods of Food Preservation*, Glasgow, Blackie Academic and Professional, 58–89.

PITT J I and RICHARDSON K C (1973), 'Spoilage by preservative-resistant yeasts', CSIRO *Food Res Q*, **33**, 80–85.

RASO J, CONDON S and SALA J F (1997), 'Microbial inactivation by ultrasonic waves under pressure and heat', *Appl Environm Microbiol*, **63**, 229–234.

RASO J, PAGAN R, CONDON S and SALA J F (1998), 'Influence of the temperature and pressure on the lethality of ultrasound', *Appl Environm Microbiol*, **64**, 465–471.

RIVERA CARRILES J (2002), Mezclas de antimicrobianos como agentes fungistáticos o funguicidas para la inhibición de *Zygosaccharomyces bailii*, Universidad de Puebla, México, MS Thesis

ROBERTS, T A (1989), 'Combinations of antimicrobials and processing methods', *Food Technol*, **42**, 156–163.

SEBTI F and TANTAOUI-ELARAKI A (1994), 'In vitro inhibition of fungi isolated from 'Pastilla' papers by organic acids and cinnamon', *Lebensm Wiss u-Technol*, **27**, 370–374.

SINGLETON V L (1981), 'Naturally occurring food toxicants: Phenolic substances of plant origin common in foods', in Chichester C O, Mrak E M and Stewart G F, *Advances in Food Research*, New York, Academic Press, vol. 27.

SOFOS J N, BEUCHAT L R, DAVIDSON P M and JOHNSON E A (1998), 'Naturally Occurring Antimicrobials in Food', Council for Agricultural Science and Technology. Task Force Report No. 132, USA.

WEISBURGER J H (1992), 'Mutagenic, carcinogenic and chemo preventive effects of phenols and catechols', in Huang M, Ho C, and Lee C Y, *Phenolic Compounds in Food and their Effect on Health II, Antioxidants and Cancer Prevention*, Washington, DC, ACS Symposium Series.

12

Edible coatings containing natural antimicrobials for processed foods

L. R. Franssen, General Mills Inc., USA and J. M. Krochta, University of California Davis, USA

12.1 Introduction

Edible coatings are used commercially to reduce moisture loss, prevent physical damage, enhance product appearance and carry food ingredients (Table 12.1). By incorporating antimicrobials, the functionality of edible coatings can be expanded to protect food products from microbial spoilage and extend shelf-life. Furthermore, edible antimicrobial coatings have the potential to enhance the safety of foods.

Selection of an appropriate coating material is influenced by intrinsic food properties (pH, water activity and composition) and extrinsic factors (temperature, relative humidity during processing, and storage). Coating materials that are currently used or being investigated include proteins, polysaccharides, lipids, waxes, and resins (Table 12.2) Additionally, materials may be combined with plasticizers, emulsifiers, and/or surfactants to enhance coating properties and the coating process (Baldwin, 1999).

In selecting an antimicrobial for addition to an edible coating, the effectiveness against the target microorganism is the primary consideration. Antimicrobial interactions with the coating material, food, and environment must also be investigated. Table 12.3 lists several categories and examples of antimicrobials that are potentially useful in edible films and coatings.

Table 12.1 Functions of edible coatings

- Barriers to moisture, oxygen, aromas, oils, etc.
- Enhancers of food integrity and appearance
- Carriers of antioxidants, nutrients, colors, flavors, and aromas
- Carriers of antimicrobials

Table 12.2 Materials used in formation of edible films and coatings

- ***Polysaccharides***: methyl cellulose (MC), carboxymethyl cellulose (CMC), hydroxypropyl cellulose (HPC), hydroxypropyl methyl cellulose (HPMC), starch, amylose, hydoxyproply amylose (HPA), alginate, carrageenan, pectinate, chitosan, etc.
- ***Proteins***: whey protein, casein, collagen, gelatin, egg white protein, keratin, fish myofibrillar protein, corn zein, wheat gluten, soy protein, pea protein, peanut protein, cottonseed protein, rice bran protein, etc.
- ***Waxes/lipids/resins:*** beeswax, candelilla wax, carnauba wax, cocoa butter, milkfat fractions, acetylated monoglycerides, fatty acids, shellac, etc.

Source: from Krochta *et al.* (1994), Debeaufort *et al.* (1998), Baldwin (1999) and Gennadios (2002).

Current antimicrobial edible coating research ranges from broad theoretical experiments to product-specific applications. Requirements for a practical and effective antimicrobial coating include: antimicrobial efficacy, controlled migration (diffusion) of the antimicrobial in the coating, good coating adhesion and interaction with the food product, and efficient coating application procedure. Evaluation of these requirements is key to success, and many different experimental methods are currently in use. Often in laboratory experiments, stand-alone coatings (films) containing a selected antimicrobial are prepared in order to measure the release and effectiveness of the antimicrobial independently of a specific food product. Using this approach, many experiments have been performed with the intent of improving our understanding of the film properties. Greater understanding of properties can enable subsequent design of coatings for food products. In this chapter, both laboratory studies of stand-alone films and combination studies involving coatings on model food systems and foods are reviewed.

Table 12.3 Some antimicrobials of potential use in edible coatings and films

Class	Examples
Organic acids	Acetic, benzoic, lactic, propionic, sorbic
Parabens	Methyl paraben
Fatty acid esters	Glyceryl monolaurate (monolaurin)
Polypeptides	Lysozyme, peroxidase, lactoferrin, nisin
Plant oils, spices, extracts	Cinnamon, sage, allicin
Nitrites	Potassium nitrite, sodium nitrite
Sulfites	Potassium sulfite, sodium sulfite
Other	Natamycin, EDTA

Source: from Brody *et al.* (2001), Davidson (1997) and Sofos *et al.* (1998).

12.2 Edible coatings and antimicrobials for food

Edible coatings have several unique advantages over synthetic films used to wrap food. A coating is 'applied to or made directly on foods whilst films are independent structures that can wrap food after their making' (Debeaufort *et al.*, 1998). Unlike a packaging film, an edible coating has an intimate and continuous association with the food until consumption. This is particularly advantageous with antimicrobial-containing coatings. Additionally, with extruded packaging films, the polymer extrusion process can destroy the antimicrobial. Dipping or spraying with antimicrobial solutions, while possible, limits the amount of preservative that can be used when compared to an antimicrobial-containing edible coating (Brody, 2001). Packaging can be removed by consumers whereas edible films cannot, thereby maximizing the product shelf-life. Additionally, coating materials may be combined with plasticizers and/or emulsifiers to enhance the coating properties and the coating process. Table 12.4 shows the functions of edible coatings in several product applicants.

Antimicrobial-containing coatings also have advantages over traditional applications by controlling the diffusion of preservatives. For example, preservatives are often dusted or sprayed on cheese surfaces to prevent mold growth during ripening (see, for example, Section 5.6.1 in Chapter 5). Over time, the antimicrobial may diffuse throughout the cheese, allowing mold to grow at the surface. An edible coating could maintain the necessary preservative concentration at the cheese surface for a longer period of time. Thus, lower levels of addition would be needed in an edible coating to achieve a targeted shelf-life, as compared to dusting or spraying antimicrobial on the cheese surface.

Many factors must be considered in developing an antimicrobial edible coating. The properties of the food, the coating, and the antimicrobial determine the ideal combination for an effective coating. Coating formulations can be applied by spraying, brushing, panning, dipping, or rolling. The type of food product determines the best coating method. For example, the most effective coatings for nuts are applied by panning, whereas cheese is often dipped or brushed in wax. The polarity of the coating material can also influence its behavior on a food product. Some films and coatings are ethanol-based, such as wheat protein, corn zein, and shellac coatings. These more hydrophobic coatings adhere well to foods with hydrophobic surfaces such as nuts. Water-based coatings may require a surfactant to adhere to hydrophobic surfaces. Additionally, the type of coating can affect the kosher status of a product or add an allergen to an otherwise allergen-free product.

While the target microorganism is the main determinant in selecting an antimicrobial, the food product pH, water activity, and composition can also influence efficacy. For example, organic acids are most effective at low pH and are, thus, most commonly used in lower pH foods (see also Chapter 6). The solubility and polarity of the preservative can also determine its compatibility

Table 12.4 Functions provided by edible coatings

	Confections and gum	Fresh produce	Snack foods	Fried foods	Encapsulated flavors	Nutritional supplements	Drug tablets
Moisture barrier	x	x	x	x	x	x	x
Oxygen barrier		x	x		x	x	x
Oil barrier	x		x	x	x	x	x
Aroma barrier	x	x	x		x	x	x
Modified release	x			x	x	x	x
Ingredient carrier, seasoning adhesive			x	x	x		
Color carrier	x			x		x	x
Mechanical integrity, Texture enhancer	x		x	x		x	x
Gloss	x	x		x		x	x

Source: adapted from Trezza (1999).

with the coating system. Stability during processing and storage also plays an important role.

12.3 Laboratory evaluation of antimicrobial-containing edible coatings and films

Numerous studies have been undertaken to determine the efficacy of antimicrobial-containing films and coatings on solid agar (by measuring zones of inhibition against selected microorganisms) and in liquid laboratory media (by determining the survival of target organisms in log CFU/ml). These methods are similar in principle to those used for evaluating antimicrobials in solution and their inherent advantages and disadvantages are discussed in Chapter 9 (Section 9.3). Some examples of growth inhibition achieved using antimicrobial-containing films in laboratory media are shown in Table 12.5. However, the results of these studies are difficult to compare directly because of the numerous differences in experimental conditions including film size and thickness, target microorganism, antimicrobial concentrations, incubation temperature and time, etc. Furthermore, although the results may suggest potential for application in foods, further trials in food models or real foods must be undertaken to confirm suitability.

In addition to antimicrobial activity *per se*, the release rate of a putative preservative from a film or coating may greatly influence its efficacy. Depending on the food application, it may be desirable to quickly release the antimicrobial to act throughout the food, or to slowly release it, maintaining a critical concentration at the surface of the product. Either way, it is useful to quantify the release rate of the coating. This can be done by determining the diffusion coefficient of an antimicrobial in a film or coating.

The diffusion coefficients of some antimicrobials, especially sorbic acid, in edible films are shown in Table 12.6. In these experiments, film composition was often altered by adding lipid or adjusting ingredient concentrations, and the effect on the antimicrobial diffusion coefficient was investigated. The addition of waxes or lipid components produced a decrease in sorbic acid diffusion in wheat gluten films (Redl *et al.*, 1996). Additionally, reduction in plasticizer content decreased the diffusion coefficient (diffusivity) of potassium sorbate in whey protein films (Franssen, 2002). The film material can also have an affect on antimicrobial diffusivity, as seen in the different sorbate diffusivities in wheat gluten and whey protein films (Redl *et al.*, 1996, Franssen, 2002). However, temperature may affect diffusion and therefore, results of experiments undertaken under different conditions may be difficult to compare.

12.4 Coatings on model food systems and foods

While testing the effectiveness of antimicrobial coatings on food products gives the most meaningful results, this approach can be the most difficult because of

Table 12.5 Examples of antimicrobial-containing films showing inhibition against microorganisms in laboratory media

Antimicrobial	Concentration	Polymer: plasticizer	*T* (°C)	Time (h)	Microorganism	Reference
Nisin	0–4.8% (by weight of film)	Corn zein: polyethylene	37	48	*Lac. plantarum*	Hoffman *et al.* (1997)
Nisin, lysozyme, EDTA	0–40 mg/g, 0–133 mg/g 0–30mM	Corn zein: glycerin, Soy protein: glycerin	37	48	*Lac. plantarum*	Padgett *et al.* (1998)
Nisin	0–5 mg/g,	Corn zein: glycerol	37	48	*Lac. plantarum*	Padgett *et al.* (2000)
p-Aminobenzoic acid. sorbic acid	0–1.5 %(w/v)	Whey protein: glycerol	35	24	*L. monocytogenes*, *E. coli*, *S. typhimurium*	Cagri *et al.* (2001)
Sodium benzoate, potassium sorbate	2–5%	Methylcellulose: Polyethylene glycol, chitosan	24	72	*P. notatum*, *R. rubra*	Chen *et al.* (1996)

Table 12.6 Diffusion coefficients (D) of antimicrobials in edible films

Antimicrobial	Polymer: plasticizer	Receiving solution	D $=10^{-12}$ m^2/s	T(°C)	Reference
Acetic acid	Chitosan	0.2 M sodium phosphate	2.59	24	Ouattara *et al.* (2000)
Propionic acid	Chitosan	0.2 M sodium phosphate	1.87	24	Ouattara *et al.* (2000)
Potassium sorbate	WPI: sorbitol (1.07:1)–(1.76:1)	water–glycerol $a_w = 0.8$	53.8–97.6	25	Ozdemir and Floros (2001)
Potassium sorbate	WPI: glycerol (1:1)–(10:1)	water–glycerol $a_w = 0.94$	11.6–92.4	24	Franssen (2002)
Sorbic acid	Wheat gluten: glycerol (5:1)	Distilled water	7.6	20	Redl *et al.* (1996)
Sorbic acid	Wheat gluten–beeswax: glycerol (5:1)	Distilled water	5.6	20	Redl *et al.* (1996)
Sorbic acid	Wheat gluten–AM: glycerol (5:1)	Distilled water	3.2	20	Redl *et al.* (1996)
Sorbic acid	Acetylated monoglyceride	Distilled water	0.27	20	Redl *et al.* (1996)
Sorbic acid	Beeswax	Distilled water	0.000 24	20	Redl *et al.* (1996)
Sorbic acid	Corn zein: glycerol (4:1)	50% water–glycerol	0.3–0.7	24	Torres *et al.* (1985)
Natamycin	WPI: glycerol (1:1)–(15:1)	Water–glycerol $a_w = 0.94$	0.063–0.378	24	Franssen (2002)

WPI = whey protein isolate.

the complexity of food systems. A coating method must be developed to coat the food product of interest evenly and effectively. The coating must withstand the product manufacturing and storage conditions. Some examples of studies done to assess the antimicrobial activity of coatings on specific food products are summarized in Table 12.7.

Coatings containing some antimicrobials can inhibit microbial growth and extend product shelf-life compared with uncoated controls. For example, in a study of citric, sorbic and acetic acid-containing hydroxypropyl methyl cellulose coatings on tomatoes, the coating alone resulted in a 2 log CFU/g reduction of *Salmonella montevideo* but addition of 0.4% sorbic acid to the coating led to a significant further reduction in counts. Insignificant effects were seen with other tested concentrations of sorbic acid or with the other acids. In a study of soy and whey protein-coated shrimp containing thyme oil and *trans*-cinnamaldehyde, a limit of acceptability for shrimp spoilage of *Pseudomonas putida* (10^7–10^8 bacteria/g) was used to determine shelf-life (Ouattara *et al.*, 2000). The coating resulted in a 12 day extension of shelf-life at 4 °C under aerobic conditions. The coating was also synergistic with irradiation (Chapter 13, Table 13.1). In another study, a nisin-containing corn zein film immersed in milk inoculated with *L. monocytogenes* was shown to reduce viable numbers up to 3.1 CFU/ml at 4 °C for up to 72 h (Orr *et al.*, 1998). Skim, low-fat and whole milk were tested, with skim milk exhibiting the largest reduction in bacterial counts. As with any test that uses a specific food, the results are only applicable to the specific food product.

12.5 Legislation and labeling

This section discusses legislation and labeling in the USA only. For information regarding labeling and legislation in Europe and worldwide, see Chapter 15. Edible coatings on foods are considered part of the food. Therefore, the coating ingredients must be declared on the label under the Federal Food, Drug, and Cosmetic Act (21 USC 343), which requires all ingredients to be labeled. However, there are two exemptions from the ingredient labeling requirement. First, section 403(i) of the act provides that flavorings, spices, and colors may be declared collectively without naming each one. Second, FDA regulations exempt incidental additives (e.g. processing aids) from ingredient declaration if they are present in a food at insignificant levels and do not have a technical or functional effect in the finished product (21 CFR 101.100; Anonymous, 2001). Since an antimicrobial edible coating would have ingredients with a functional effect, inclusion of the coating ingredients would be required on the label.

Many edible food coatings are made with proteins that can cause allergic responses in some consumers. Consumers with food allergies rely on accurate labels to avoid consuming foods that may contain allergens and therefore, accurate and clear labeling is critical. The FDA, in conjunction with several other groups, is currently working on guidelines to assist both food manufacturers and

Table 12.7 Studies of antimicrobials in edible coatings on foods

Antimicrobial	Concentration	Polymer: plasticizer	Food	Microorganism	Time of inoculation	T (°C)	Results	Reference
Benzoic acid	50–100 μg/g	Methylcellulose: glycerol	Fruit preserves	*Z. rouxii* *Z. mellis*	After coating	Ambient	Yeast inhibition	Chen *et al.* (1999)
Citric, acetic, sorbic acids	0.2–0.4%	Hydroxypropyl methylcellulose: polyethylene glycol	Tomatoes	*S. montevideo*	Before coating	30	~2 log reduction with coating	Zhuang *et al.* (1996)
Sodium benzoate, potassium sorbate	0.1–1.5%	Cellulose-based	Cut apples, cut potatoes	[Natural flora]	N/A	4	1 week shelf-life extension	Baldwin *et al.* (1996)
Sorbic acid	1 mg/g corn	Corn zein: glycerol	Cooked sweetcorn	*L. monocytogenes*	After coating	10	10-fold reduction	Carlin *et al.* (2001)
Sorbic acid	0.2%	Corn zein: glycerol	Cheese analog	*S. aureus*	After coating	30	>16 day stability	Torres and Karel (1985), Torres *et al.* (1989)
Nisin, EDTA, citric acid, lactic acid	0–500 μg/ml, 5.0 mM, 3.0%, 5.0%	Calcium alginate, agar	Poultry	*S. typhimurium*	Before coating	4	1.8–4.6 log reduction	Natrajan and Sheldon (2000)
Nisin, pediocin	10% (w/v), 7.75 $\mu g/cm^2$	Cellulose casings	Meats	*L. monocytogenes*	Before coating	4	Inhibition for 12 weeks	Ming *et al.* (1997)
Nisin	100 μg/ml	Calcium alginate	Ground beef	*B. thermosphacta*	Before coating	4	Limited, immediate protection	Cutter and Siragusa (1997)
Thyme oil, *trans*-cinnamaldehyde	0–1.5% (w/w) 0–0.3%	Whey protein:glycerol Soy protein: glycerol	Shrimp	*P. putida*	Before coating	4	12 day shelf-life extension	Ouattara *et al.* (2001)

allergic consumers. The most common eight allergens are peanuts, soybeans, milk, eggs, fish, crustacea (e.g. lobster, crab, shrimp), tree nuts (e.g. almonds, walnuts), and wheat (Anonymous, 2001). Several films and coatings have been successfully formed from milk protein (whey, casein), wheat protein (gluten), soy protein and peanut protein. The presence of a coating with a known allergen on a food must be clearly labeled. When coating a non-allergenic product with an allergenic material, particular consideration must be given to labeling to alert unsuspecting allergenic consumers. More information on allergen guidelines is available at http://www.cfsan.fda.gov/~dms/wh-alrgy.html.

There are specific regulations regarding the addition of preservatives to food. The Code of Federal Regulations (CFR) and the Compliance Policy Guide, designed to help understand the CFR, can be accessed at http://www.access.gpo.gov/cgi-bin/cfrassemble.cgi?title=200121/ and http://www.fda.gov/ora/compliance_ref/cpg/default.htm. In title 21 CFR 101.22, a chemical preservative is defined as 'any chemical that, when added to food, tends to prevent or retard deterioration thereof, but does not include common salt, sugars, vinegars, spices, or oils extracted from spices, substances added to food by direct exposure thereof to wood smoke, or chemicals applied for their insecticidal or herbicidal properties'. Deterioration of a food product includes undesirable microbial growth, flavor losses or changes, color losses or changes or other quality changes.

Regulations for specific preservative additives and food products can be found in 21 CFR 172. Often the regulations include purity, analytical methodology, labeling and maximum use or levels of preservatives. Some additives are considered Generally Recognized As Safe (GRAS), and they may be used with fewer restrictions in accordance with good manufacturing practice (GMP). A list of some of these additives can be found in 21 CFR 182 (Vetter, 2000). Several books are also available regarding the issue of food labeling and food regulations (Simmons, 1997; Vetter, 1996).

12.6 Consumer acceptance

There are presently no consumer acceptance tests or other sensory studies published on any foods with antimicrobial coatings. However, consumer studies have been conducted on edible coatings intended to perform other functions. In one study, attributes of whey-protein-coated and shellac-coated chocolate-covered almonds were evaluated for degree of liking. It was concluded that whey protein coatings could potentially be used as alternatives to shellac glaze (Lee *et al.*, 2002). In a sensory study, whey protein isolate and candelilla wax emulsion films were tested by a trained panel for milk odor, transparency/opaqueness, sweetness and adhesiveness (Kim and Ustunol, 2001). The films were found to have no milk odor and to be slightly sweet and adhesive. Candelilla wax rendered the films opaque while whey protein films with no wax were clear and transparent. Often the sensory attributes that are evaluated are specific to the coating or coated product. In another study of edible cellulose-based coatings on mini-peeled carrots, a trained panel tested samples for orange color intensity, white surface discoloration, slipperiness, and preference for appearance under fluorescent light. The carrots were also tested for fresh aroma, fresh flavor, sweetness, bitterness, harshness, succulence, and taste preference (Howard, 1996). Coated carrots resulted in less white surface discoloration and higher sensory scores for color intensity, fresh carrot aroma, fresh flavor, and overall acceptability than non-coated carrots.

12.7 Future prospects

While antimicrobial coatings may extend the shelf-life and protect food products, they are never substitutes for proper handling, storage and good manufacturing practices. Films or coatings made from natural antimicrobial materials would have the advantages of complete dispersion of the antimicrobial and little or no migration. Natural antimicrobials have both consumer and industry appeal and may have fewer restrictions than novel synthetic compounds for use in food applications.

The development of new coating processes and equipment in the food industry continues to present new applications for foods. The development of more accurate and more sensitive analytical methods for coatings and coated food products will aid understanding of systems at the molecular level. Advanced computer programs and sophisticated experimental data on antimicrobial diffusion can be used to construct mathematical models of complex coated food systems. Predictive diffusion models can then be used to determine product shelf-life. Such models will provide a quantitative understanding of food products and preservative behavior at the molecular level.

With growing concern for food safety and quality, increased awareness of environmental issues, and producer and consumer desire for extended product shelf-life, the need for natural antimicrobial coatings is increasing. As technology improves, the design and development of edible natural antimicrobial coatings will become more efficient and their application will become more widespread. The development of edible natural antimicrobial coatings has the potential to provide improved food safety, quality, and shelf-life.

12.8 References

ANONYMOUS (2001), Food safety and food labeling; presence and labeling of allergens in food, in *Food Safety and Food Labeling; Presence and Labeling of Allergens in Food,* Rockville, Food and Drug Administration, HHS.

BALDWIN, E. A. (1999), 'Surface treatments and edible coatings in food preservation', in Rahman, M. S., *Handbook of Food Preservation*, New York, Marcel Dekker, Inc.

BALDWIN, E. A., NISPEROS, M. O., CHEN, X. and HAGENMAIER, R. D. (1996), 'Improving storage life of cut apple and potato with edible coating', *Postharvest Biology and Technology,* **9**, 151–163.

BRODY, A. L. (2001), 'What's active in active packaging', *Food Technology,* **55**, 104–106.

BRODY, A. L., STRUPINSKY, E. R. and KLINE, L. R. (2001), *Active Packaging for Food Applications,* Lancaster, PA, Technomic Publishing Co., Inc.

CAGRI, A., USTUNOL, Z. and RYSER, E. T. (2001), 'Antimicrobial, mechanical, and moisture barrier properties of low pH whey protein-based edible films containing p-aminobenzoic or sorbic acids', *Journal of Food Science,* **66**, 865–870.

CARLIN, F., GONTARD, N., REICH, M. and NGUYEN-THE, C. (2001), 'Utilization of zein coating and sorbic acid to reduce *Listeria monocytogenes* growth on cooked sweet corn', *Journal of Food Science,* **66**, 1385–1389.

CHEN, M., YEH, G. H.-C. and CHIANG, B.-H. (1996), 'Antimicrobial and physiochemical properties of methylcellulose and chitosan films containing a preservative', *Journal of Food Processing and Preservation,* **20**, 379–390.

CHEN, M.-J., WENG, Y.-M. and CHEN, W. (1999), 'Edible coating as preservative carriers to inhibit yeast on Taiwanese-style fruit preserves', *Journal of Food Safety,* **19**, 89–96.

CRANK, J. (1975), *The Mathematics of Diffusion,* New York, Oxford University Press.

CUTTER, C. N. and SIRAGUSA, G. R. (1997), 'Growth of *Brochothrix thermosphacta* in ground beef following treatments with nisin in calcium alginate gels', *Food Microbiology,* **14**, 425–430.

DAVIDSON, P. M. (1997), 'Chemical preservatives and natural antimicrobial compounds', in Doyle, M. P., *Food Microbiology: Fundamentals and Frontiers*, Washington DC, ASM Press.

DEBEAUFORT, F., QUEZADA-GALLO, J.-A. and VOILLEY, A. (1998), 'Edible films and coatings: tomorrow's packagings: a review', *CRC Rev. Food Sci.,* **38**, 299–313.

FRANSSEN, L. R. (2002), in *Antimicrobial properties and preservative diffusion of preservative-containing whey protein coatings for cheddar cheese*, Ph.D. thesis, Food Science and Technology, University of California, Davis.

GENNADIOS, A. (ed.), (2002), *Protein-based films and coatings,* CRC Press, Boca Raton, FL.

HOFFMAN, K. L., DAWSON, P. L., ACTON, J. C., HAN, I. Y. and OGALE, A. A. (1997), 'Film formation effects on nisin activity in corn zein and polyethylene films', Research and Development Associates for Military Food and Packaging Systems, Inc., San Antonio, Texas, 238–287.

HOWARD, L. R. D. T. (1996), 'Minimal processing and edible coating effects on composition and sensory quality of mini-peeled carrots', *Journal of Food Science,* **61**, 643–651.

KIM, S. J. and USTUNOL, Z. (2001), 'Sensory attributes of whey protein isolate and candelilla wax emulsion edible films', *Journal of Food Science,* **66**, 909–911.

KROCHTA, J. M., BALDWIN, E. A. and NISPEROS-CARRIEDO, M. O. (eds.), (1994), *Edible Coatings and Films to Improve Food Quality,* Technomic Publishing Co., Inc., Lancaster, Pennsylvania.

LEE, S. Y., DANGARAN, K. L., GUINARD, J. X. and KROCHTA, J. M. (2002), 'Consumer acceptance of whey-protein-coated versus shellac coated chocolates', *Journal of Food Science*, **67**(7): 2764–2769.

MING, X., WEVER, G., AYRES, J. and SANDINE, W. (1997), 'Bacteriocins applied to food packaging materials to inhibit *Listeria monocytogenes* on meats', *Journal of Food Science,* **62**, 413–415.

NATRAJAN, N. and SHELDON, B. W. (2000), 'Inhibition of *Salmonella* on poultry skin using protein- and polysaccharide-based films containing a nisin formulation', *Journal of Food Protection,* **63**, 1268–1272.

ORR, R. V., HAN, I. Y., ACTON, J. C. and DAWSON, P. L. (1998), 'Effect of nisin in edible protein films on *Listeria monocytogenes* viability in milk', in *Effect of Nisin in Edible Protein Films on* Listeria monocytogenes *Viability in Milk,* San Antonio, TX, Research and Development Associates for Military Food and Packaging Systems, Inc., 300–305.

OUATTARA, B., SIMARD, R. E., PIETTE, G., BEGIN, A. and HOLLEY, R. A. (2000), 'Diffusion of acetic and propionic acids from chitosan-based antimicrobial packaging films', *Journal of Food Science,* **65**, 768–773.

OUATTARA, B., SABATO, S. F. and LACROIX, M. (2001), 'Combined effect of antimicrobial coating and gamma irradiation on shelf life extension of pre-cooked shrimp (*Penaeus* spp.)', *International Journal of Food Microbiology,* **68**, 1–9.

OZDEMIR, M. and FLOROS, J. D. (2001), 'Analysis and modeling of potassium sorbate diffusion through edible whey protein films', *Journal of Food Engineering,* **47**, 149–155.

PADGETT, T., HAN, I. Y. and DAWSON, P. L. (1998), 'Incorporation of food-grade antimicrobial compounds into biodegradable packaging films', *Journal of Food Protection,* **61**, 1330–1335.

PADGETT, T., HAN, I. Y. and DAWSON, P. L. (2000), 'Effect of lauric acid addition on the antimicrobial efficacy and water permeability of corn zein films containing nisin', *Journal of Food Processing and Preservation,* **24**, 423–432.

REDL, A., GONTARD, N. and GUILBERT, S. (1996), 'Determination of sorbic acid diffusivity in edible wheat gluten and lipid based films', *Journal of Food Science,* **61**, 116–120.

SIMMONS, R. A. (1997), 'Laws and regulations, United States', in Brody, A. L. and Marsh, K. S., *The Wiley Encyclopedia of Packaging Technology*, New York, John Wiley and Sons, Inc.

SOFOS, J. N., BEUCHAT, L. R., DAVIDSON, P. M. and JOHNSON, E. A. (1998), 'Naturally occurring antimicrobials in food', in *Naturally Occurring Antimicrobials in Food,* Ames, IA, Council for Agricultural Science and Technology (CAST).

TORRES, J. A. and KAREL, M. (1985), 'Microbial stabilization of intermediate moisture food surfaces III. Effects of surface preservative concentration and surface pH control on microbial stability of an intermediate moisture cheese analog', *Journal of Food Processing and Preservation,* **9**, 107–119.

TORRES, J. A., MOTOKI, M. and KAREL, M. (1985), 'Microbial stabilization of intermediate moisture food surfaces I. Control of surface preservative concentration', *Journal of Food Processing and Preservation,* **9**, 75–92.

TORRES, J. A., BOUZAS, J. O. and KAREL, M. (1989), 'Sorbic acid stability during processing and storage of an intermediate moisture cheese analog', *Journal of Food Processing and Preservation,* **13**, 409–415.

TREZZA, T. A. (1999), *Surface properties of edible, biopolymer coatings for foods: color, gloss, surface energy and adhesion.* Ph.D. thesis., University of California, Davis.

VETTER, J. (1996), *Food Laws and Regulation,* Manhattan, KS, American Institute of Baking.

VETTER, J. (2000), Personal communication to L. R. Franssen, Institute of Food Technologists, Adam's Mark Hotel, Dallas, Texas.

ZHUANG, R., BEUCHAT, L. R., CHINNAN, M. S., SHEWFELT, R. L. and HUANG, Y.-W. (1996), 'Inactivation of *Salmonella montevideo* on tomatoes by applying cellulose-based edible films', *Journal of Food Protection,* **59**, 808–812.

13

Natural antimicrobials in combination with gamma irradiation

B. Ouattara and A. A. Mafu, Agriculture and Agri-Food Canada

13.1 Introduction

Food irradiation is based on a physical phenomenon in which energy is generated and allowed to travel through space and matter. It refers to the application of this energy to sterilize or preserve food by destroying microorganisms, parasites, insects, or other pests. The type of radiation is called 'ionizing radiation' because of the production of electrically charged particles or ions. The most important sources of ionizing radiation are gamma rays from radionuclides such as ^{60}Co or ^{137}Cs, electron beams, and X rays. The benefits of food irradiation include reduction of food losses and expanding food trade, particularly in Asian countries, Latin America, the Middle East, and Africa (Loaharanu, 1994).

Although the effect of irradiation on the major food constituents (carbohydrates, proteins, and lipids) has been shown to be minor, the doses necessary to achieve complete microbial destruction may be higher than the threshold at which detectable 'irradiated flavor' occurs in food (Farkas, 1998). This has led to the demand by consumers that the lowest possible doses necessary to achieve desired effects should be used on a commercial basis. This chapter presents examples of combination treatments involving low-dose irradiation and natural antimicrobial compounds to improve the safety and quality of a variety of food products. These applications are not currently widely used but are promising for the future.

13.2 Gamma irradiation in food preservation

Although frequently termed 'a new technology', food irradiation has been used since the 1800s. From the reviews by Josephson (1983), Dielh (1990) and

Molins (2001a), the early history of food irradiation (1890s–1940s) was inseparably linked with the development of irradiation sources, followed by a period of intensive research and development (1940s–1970s) on the wholesomeness of irradiated foods. Since then, most investigations have been related to regulation and potential social and economic benefits. In 1980, the Joint Expert Committees sponsored by the United Nations Food and Agricultural Organization (FAO) and the International Atomic Energy Agency (IAEA) concluded that irradiated foods are safe and wholesome when treated at levels up to 10 kGy; hence, toxicological testing of foods so treated is no longer required (Anonymous, 1981). Recently, the increased demand for radiation-sterilized foods for military, athletic and hospital needs has led to more interest in recording data on high-dose irradiation (10–100 kGy) (Anonymous, 1999a; Molins, 2001b). A joint study group has concluded that doses greater than 10 kGy would greatly reduce potential microbial risk to the consumer without affecting the quality of food from a toxicological and nutritional point of view. Since 1960, several uses and applications have been permitted in many countries following Codex Alimentarius principles (Langunas-Solar, 1995). Currently, health and safety authorities in over 40 countries have approved irradiation of more than 60 different foods including spices, grains, chicken, beef, fruits and vegetables (Anonymous, 1999b). As of August 1999, over 30 countries are irradiating food for commercial purposes.

13.3 Combinations of low-dose irradiation with natural antimicrobial compounds

The doses at which detectable 'irradiated' flavors occur in beef and chicken are reported to be at or around 2.5 kGy (Farkas, 1998; Lacroix and Ouattara, 2000). However, according to Mahrour *et al.* (1998), doses over 2.5 kGy may be required to completely inactivate *Salmonella*. The biocidal efficacy of irradiation depends on the type of microorganism present in the food and the physico-chemical characteristics of the food. Furthermore, the free radicals generated by irradiation may, even at very low concentrations, react with myoglobin or hemoglobin to give undesirable colors (Giroux *et al.*, 2001; Kamarei *et al.*, 1979; Satterlee *et al.*, 1972). Combining low doses of irradiation with other preservation factors could be a solution to this problem (Patterson, 1990). Significant progress has already been made by using (i) irradiation and heat to destroy vegetative and spore-forming bacteria (Grant and Patterson, 1994; Gombas and Gomez, 1978; Thayer *et al.*, 1991) and (ii) irradiation and low temperatures to reduce the mobility of free radicals and their ability to interact with food constituents (Patterson, 2001; Stewart, 2001). The combination of irradiation with modified-atmosphere or vacuum packaging has also been investigated (Grant and Patterson, 1991; Thomas and Beime, 1999).

The effects of chemical preservatives in combination with irradiation are also well documented. These include nitrate and nitrites (Tritsch, 2000), chlorine and chlorine dioxide (Hagenmaier and Baker, 1998), and ozone (Yook *et al.*, 1998). Although in theory any antimicrobial agent could be used in combination with irradiation, natural antimicrobial compounds have recently received the most attention possibly because of their Generally Recognized As Safe (GRAS) status (Gould, 1996; Ouattara *et al.*, 1997). For further details on legislation, the reader is referred to Chapter 15.

13.3.1 Combination treatments with organic acids

Organic acids such as lactic and acetic acids are commonly used in food preservation (Chapter 6). However, evaluation of their radiosensitizing properties against foodborne microorganisms and their synergistic effect in combination with irradiation is a very recent development. Paul *et al.* (1998) showed that combining irradiation (2.5 kGy) with dipping in acetic acid solutions extended the shelf-life of buffalo meat by up to four weeks. Similarly, combinations of irradiation (2 kGy), pH reduction to 5.6 by ascorbic acid, and water activity reduction to 0.96 with sodium lactate extended the shelf-life of vacuum-packaged ready-to-cook minced meat stored at 2 °C by a factor of three (Farkas and Andrassy, 1996; Farkas *et al.*, 1996).

Giroux *et al.* (2001) evaluated the combined effect of ascorbic acid and gamma irradiation on microbial and sensory characteristics of beef patties during refrigerated storage and found enhanced inhibition of bacteria with a synergistic effect against both aerobic microorganisms and total coliforms. These results were consistent with those of Lee *et al.* (1999) who reported synergism between irradiation and some naturally occurring antioxidants such as ascorbyl palmitate, α-tocopherol and β-carotene on the microbial stability of ground beef. More recently, Bhide *et al.* (2001) demonstrated that treatment with propionic acid (1%), lactic acid (2%), and acetic acid (2%) followed by low-dose gamma irradiation (1 to 3 kGy) reduced numbers of the total microbial flora and *Bacillus cereus* more than irradiation alone. Reductions in bacterial counts ranged from 1.5 to 2.9 log CFU/g for total flora and 0.6 to 1.0 log CFU/g for *B. cereus*, depending on the irradiation dose and the acid treatment. Combinations of acid and irradiation also resulted in substantial extension of shelf-life (up to 33 days compared with 24 days for irradiation alone) at 5–7 °C. Singh *et al.* (1991) showed that addition of 0.1% sorbic acid to milk followed by irradiation at 2.5 kGy, preserved paneer for 30 days at 25–35 °C compared with only a few days for irradiation alone or acid treatment alone. These studies showed that gamma irradiation was more lethal to microorganisms at reduced pH.

13.3.2 Combination treatments with essential oils

The antimicrobial properties of essential oils from plants are reviewed in this book in Chapter 2 (in combination with nisin), Chapter 9 (against foodborne microorganisms *in vitro* and in processed foods), Chapter 10 (against fungi in postharvest storage of fruit and vegetables), Chapter 11 (against yeasts and moulds in lightly processed fruit products) and Chapter 14 (against spoilage moulds from bakery products). In general, the concentrations of active antimicrobial compounds in herbs and spices or their essential oils are too low to be used as preservatives without adverse effects on the sensory characteristics of the food (Naidu, 2000). However, they may contribute to a multifactorial preservation system. In particular, their combination with low-dose gamma irradiation has shown some potential. For example, Mahrour *et al.* (1998) investigated the combination of irradiation with marinating in a rosemary and thyme extract on the sensitivity of *Salmonella* on chicken at 4 °C. Their study demonstrated that the combination treatment extended the microbiological shelf-life of chicken by a factor of 7–8.

13.3.3 Combination treatments with edible antimicrobial-containing coatings

A review of antimicrobial-containing edible coatings is presented in Chapter 12. The combination of irradiation with edible antimicrobial coatings has also been investigated on various food products stored under refrigeration conditions (Ouattara *et al.*, 2002a,b,c). Some examples of results obtained using a combination of low-dose irradiation and various natural preservatives on microbial growth and shelf-life are shown in Table 13.1 and Fig. 13.1. Treatment with a protein-based coating containing 0.9% thyme oil and 1.8% *trans*-cinnamaldehyde and irradiated at 3 kGy inhibited both aerobic microorganisms and *Pseudomonas putida* on pre-cooked shrimp. Based on the onset of shrimp spoilage (established at 10^7 CFU/g), the shelf-life was 20–21 days for combined treatments (irrespective of thyme oil or *trans*-cinnamaldehyde concentration) whereas irradiation alone resulted in a maximum shelf-life of 12 days (Ouattara *et al.*, 2002a; 2002c). Similar results were obtained with ready-to-eat pizza

Table 13.1 Effect of low-dose irradiation combined with plant essential oils on shelf-life at 4 °C.

Food product	Treatment	Shelf-life (days)
Pre-cooked shrimp	Control	7
	Irradiation (1 kGy)	12
	Edible coating alone	8
	Edible coating + irradiation (1 kGy)	17
	Edible coating + essential oils + irradiation (1 kGy)	21
Ready-to-eat pizza	Control	3
	Irradiation (1 kGy)	7
	Antimicrobial coating*	10
	Antimicrobial coating* + Irradiation (1 kGy)	21

* Proprietary mixture from Bioenvelop Technologies.
Source: adapted from Ouattara *et al.* (2002a,b,c).

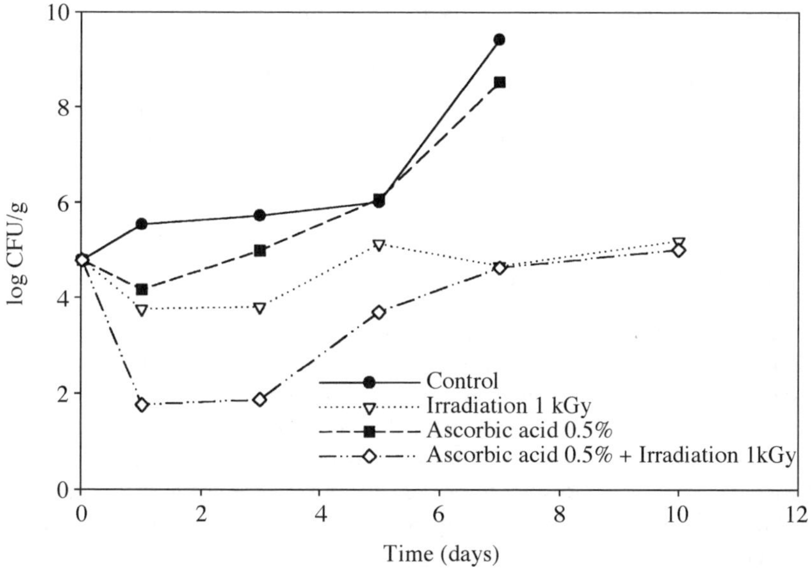

Fig. 13.1 Survival of microbial flora in ground beef treated with gamma-irradiation (1 kGy) and ascorbic acid (0.5%) at 4 °C. Adapted from Ouattara *et al.* (2002a,b,c).

coated with Longevita, a proprietary protein-based antimicrobial coating developed by Bioenvelop Technologies (Laval, Quebec). Combining this coating with low-dose gamma irradiation (1–2 kGy) resulted in a shelf-life of more than 21 days, compared with 12–14 days for irradiation alone at 1–2 kGy (Ouattara *et al.*, 2002a). Synergism was also reported in ground beef containing ascorbic acid and treated with a protein-based coating containing selected spice powders (Ouattara *et al.*, 2002b).

13.4 Natural antimicrobial compounds as antioxidants

Many natural antimicrobial compounds have other functional properties such as antioxidant activity, which may counteract the biocidal effects of irradiation. For example, antimicrobial phenolic compounds from rosemary, thyme, sage, cinnamanon, garlic and onions exhibit antioxidant activity (Baratta *et al.*, 1998; Gould, 1996; Lacroix *et al.*, 1997). The antioxidant activity of the essential oil from rosemary was greater than that of α-tocopherol and comparable to that of butylated hydroxytoluene (Baratta *et al.*, 1998). These compounds may reduce the efficacy of irradiation as a biocidal technique by scavenging reactive components (e.g. hydroxyl radicals and hydrogen peroxide), which are known to be directly responsible for the killing action of ionizing radiation (Dickson, 2001). This has previously been mentioned by Wills (1980), who examined several natural compounds (including vitamin E) in synthetic diet mixtures treated with gamma irradiation. Similarly, Stecchini *et al.* (1998) reported peroxy radical-scavenging activity in irradiated minced turkey meat supplemented with carnosine; in this system, the biocidal effect of irradiation on *Aeromonas hydrophila* was substantially reduced. These results suggest that natural food additives having both antimicrobial and antioxidant properties can protect microorgnisms in irradiated products by counteracting the toxic effect of free radicals.

13.5 Consumer acceptance

Recent attitude studies and market tests have shown that consumer concerns about pesticides, drug residues, growth hormones and chemical preservatives have increased during the last ten years. Chemical additives were considered as an extremely serious problem by over 20% of consumers in one survey in the USA, while less than 10% of them were concerned by the presence of natural toxicants in foods (Resurreccion *et al.*, 1995). However, the levels of antimicrobial compounds necessary to inhibit microorganisms in foods may also produce unacceptable sensory changes. Therefore, the future use of plant-derived flavoring substances as biopreservatives in foods will be in combination with other preservation methods, particularly gamma irradiation (Molins, 2001b; Patterson, 1990).

Giroux *et al.* (2001) have investigated the combined effect of ascorbic acid and gamma irradiation up to 4 kGy on the microbial and sensory characteristics of ground beef patties. A consistent reduction of total aerobic counts and coliforms was observed without any detrimental effect on the odor and taste evaluated on a hedonic scale. Furthermore, addition of ascorbic acid increased the red coloring of the meat, which is a major criterion used by consumers to assess the quality of meat and meat products. Ouattara *et al.* (2002a) showed that combined treatments with essential oils and gamma irradiation were effective in developing shelf-stable foods without affecting sensory parameters (odor, taste, or overall appearance).

The ultimate test is in the market place and irradiated food will not be a marketing success unless irradiation is accepted by consumers. According to the International Consultative Group on Food Irradiation (http://www.iaea.org./worldatom/Press/Booklets/foodirradiation.pdf), the commercial uptake of irradiation has been slow mainly because of consumer misunderstandings about the technology. A US Department of Agriculture (USDA) survey of community leaders' attitudes to and knowledge of food irradiation has shown that willingness to purchase irradiated products increased from 57% to 83% after exposure to a brief educational programme. Similar programmes conducted in South Africa have shown an increase from 15% to 54% in willingness to buy irradiated products. In 1996, irradiated products sold better in US supermarkets, capturing 73% of the market when priced 10% lower than the store's own brand, 58% when priced equally, and 31% when priced 10% higher (Anonymous, 1999b). This is consistent with other attitude surveys and market data obtained worldwide. In the last ten years, the acceptance of irradiated foods has increased in many countries including Argentina, Korea, China, Thailand and France. Recent studies reported by Molins (2001b) have confirmed that American consumers viewed food safety with increasing concern, and consequently, were willing to accept technologies such as food irradiation as long as they improved food safety.

13.6 Conclusion

It has been demonstrated that natural antimicrobial compounds can be combined advantageously with other preservation systems in order to reduce extreme use of a single treatment. From the information provided in this chapter, food irradiation appears to be suitable for this approach. It can be predicted that natural antimicrobial compounds combined with low-dose irradiation will have an increasing role to play in the future. Further scientific work is needed because bioactive compounds from natural sources may raise new safety issues, e.g. the formation of undesirable allergens, mutagens or toxins, and denaturation or reaction with other ingredients present in the food to form toxic by-products (Naidu, 2000). Furthermore, understanding the mechanisms of microbial inactivation in synergistic food preservation systems is very important since some bioactive molecules have been shown to protect microorganisms by counteracting the toxic effect of the free radicals produced by irradiation.

13.7 References

ANONYMOUS (1981). 'Wholesomeness of irradiated food', report of a joint FAO/IAEA/WHO Expert Committee on Food Irradiation. WHO Technical Report Series 659, World Health Organization, Geneva.

ANONYMOUS (1999a). High-dose irradiation: wholesomeness of food irradiated with doses above 10 kGy, report of a joint FAO/IAEA/WHO Study Group. WHO Technical Report Series 890, World Health Organization, Geneva.

ANONYMOUS (1999b). 'Facts about food irradiation: A series of facts sheets from the International Consultative Group on Food Irradiation'. International Atomic Energy Agency. Geneva. Internet web site: http://www.iaea.org/worldatom/Press/Booklets/foodirradiation.pdf

BARATTA, M.T, DORMAN, H.J.D., DEANS, S.G., FIGEIREDO, A.C., BARROSO, J.G., and RUBERTO, G. (1998). Antimicrobial and antioxidant properties of some commercial essential oils. *Flavour Fr. J.* **13**: 235–244.

BHIDE, M.R., PATURKAR, A.M., SHERIKAR, A.T., and WASKAR, V.S. (2001). Presensitization of microorganisms by acid treatments to low dose gamma irradiation with special reference to *Bacillus cereus*. *Meat Sci.* **58**: 253–258.

DICKSON, J.S. (2001). Radiation inactivation of microorganisms. In *Food Irradiation: Principles and Applications*. R.A. Molins Ed., John Wiley & Sons Inc., New York, pp. 23–35.

DIEHL, J.F. (1990). *Safety of Irradiated Foods*. Marcel Dekker, New York, pp. 1–7.

FARKAS, J. (1998). Irradiation as a method for decontaminating food: a review. *Int. J. Food Microbiol.* **44**: 189–204.

FARKAS, J. and ANDRASSY, E. (1996). Behavior of *Listeria monocytogenes* in an extended shelf life chilled meat product. *Acta Alimentaria*. 25: 185–187.

FARKAS, J., ANDRASSY, E., and HORTI, K. (1996). Extension of shelf life of a vacuum-packaged chilled meat product by combination of gamma irradiation, ascorbic acid and sodium lactate. *Acta Alimentaria* **25**: 181–184.

GIROUX, M., OUATTARA, B., YEFSAH, R., SMORAGIEWICZ, W., SAUCIER, L., and LACROIX, M. (2001). Combined effect of ascorbic acid and gamma irradiation on the microbial and sensorial characteristics of beef patties during refrigerated storage. *J. Agric. Food Chem.* **49**: 919–925.

GOMBAS, D.E. and GOMEZ, R.F. (1978). Sensitisation of Clostridium perfringens spores to heat by gamma irradiation. *Appl. Environ. Microbiol.* **36**: 403–407.

GOULD, G.W. (1996). Industry perspectives on the use of natural antimicrobials and inhibitors for food applications. *J. Food Prot. (Suppl.)*: 82–86.

GRANT, I.R. AND PATTERSON, M.F. (1991). Effect of irradiation and modified atmosphere on the microbiological safety of minced pork stored under temperature abuse conditions. *Int. J. Food Sci. Technol.* **26**: 521–533.

GRANT, I.R. and PATTERSON, M.F. (1994). Combined effect of gamma irradiation and heating on the destruction of *Listeria monocytogenes* and *Salmonella typhimurium* in cooked-chill roast beef and gravy. *Int. J. Food microbiol.* **27**: 117–128.

HAGENMAIER, R.D. and BAKER, R.A. (1998). Microbial population of shredded carrot in modified atmosphere packaging as related to irradiation treatment. *J. Food Sci.* **63**: 162–164.

JOSEPHSON, E.S. (1983). An historical review of food irradiation. *J. Food Safety* **5**: 161–165.

KAMAREI, A.R., KAREL, M., and WIERBICKI, E. (1979). Spectral studies on the role of ionizing radiation in color changes of rappertized beef. *J. Food Sci.,* **44**: 25–31.

LACROIX, M. and OUATTARA, B. (2000). Combined industrial processes with irradiation to assure innocuity and preservation of food products. A review. *Int. J. Food Microbiol.* **33**: 719–724.

LACROIX, M., SMORAGIEWICZ, W., PAZDERNIK, L., KONÉ, M.I., and KRZYSTRYNIAK, K. (1997). Prevention of lipid radiolysis by natural antioxidants from rosemary (*Rosmarinus officinalis* L.) and thyme (*Tymus vulgaris* L.). *Food Res. Int.* **30**: 457–462.

LAGUNAS-SOLAR, M.C. (1995). Radiation processing of foods: an overview of scientific principles and current status. *J. Food Prot.* **58**: 186–192.

LEE, J.-W., YOOK, H.-S., KIM, S.-A., LEE, K.-H., and BYUN, M.-W. (1999). Effects of antioxidants and gamma irradiation on the shelf life of beef patties. *J. Food Prot.* **62**: 619–624.

LOAHARANU, P. (1994). Cost–benefit aspect of food irradiation. *Food Technol.* **48**: 104–108.

MAHROUR, A., LACROIX, M., NKETSA-TABIRI, J., CALDERON, N., and LACROIX, M. (1998). Antimicrobial properties of natural substances in irradiated fresh poultry. *Radiation Physics and Chemistry*, **52**: 81–84.

MOLINS, R.A. (2001a). Historical notes on food irradiation. In *Food Irradiation: Principles and Application*, R.A Molins Ed., John Wiley & Sons Inc. New York, pp. 1–22.

MOLINS, R.A. (2001b). Irradiation of meats and poultry. In *Food Irradiation: Principles and Application*, R.A Molins Ed., John Wiley & Sons Inc. New York, pp. 131–191.

NAIDU, A.S. (2000). Overview. In *Natural Food Antimicrobial Systems*. A.S. Naidu Ed.,

CRC Press, New York. pp. 3–14.

OUATTARA, B., SIMARD, R.E., HOLLEY, R.A., PIETTE, G. J-P., and BÉGIN, A. (1997). Inhibitory effect of organic acids upon meat spoilage bacteria. *J. Food Prot.* **60**: 246–253.

OUATTARA, B. SABATO, S.F., and LACROIX, M. (2002a). Use of gamma-irradiation technology in combination with edible coating to produce shelf-stable foods. *Radiat. Phys. Chem.* **63**: 305–310.

OUATTARA, B. GIROUX, M., YEFSAH, R., SMORAGIEWICZ, W., SAUCIER, L., BORSA, J., and LACROIX, M. (2002b). Microbiological and biochemical characteristics of ground beef as affected by gamma irradiation, food additives and edible coating. *Radiat. Phys. Chem.* **63**: 299–304.

OUATTARA, B., SABATO, S.F., and LACROIX, M. (2002c). Combined effect of antimicrobial coating and gamma irradiation on shelf life of pre-cooked shrimp (*Penaeus* spp.). *Int. J. Food Microbiol.* **68**: 1–9.

PATTERSON, M. (2001). Combined treatments involving food irradiation. In *Food Irradiation: Principles and Applications*. R.A Molins Ed., John Wiley & Sons Inc., New York, pp. 313–327.

PATTERSON, M.F. (1990). A review: The potential for food irradiation. *Letts. Appl. Microbiol.* **11**: 55–61.

PAUL, P., CHAWLA, S.P., AND KANATT, S.R. (1998). Combination of irradiation with other treatments to improve the shelf life and quality of meat and meat products. In *Combination Processes for Food Irradiation*. International Atomic Energy Agency, Vienna, pp. 111–130.

RESURRECCION, A.V.A., CALVEZ, F.C.F, FLETCHER, S.M., and MISRA, S.K. (1995). Consumer attitudes towards irradiated foods: results of a new study. *J. Food Prot.* **58**: 193–196.

SATTERLEE, L.D., DUANE BROWN, W., and LYCOMETROS, C. (1972). Stability and characteristics of the pigment produced by gamma irradiation of metmyoglobin. *J Food Sci.* **37**: 213–217.

SINGH, L., MURALI, H.S., and SANKARAN, R. (1991). Extension of shelf life of paneer by sorbic acid and irradiation. *J. Food Sci. Technol. Ind.* **28**: 386–388.

STECCHINI, M.L., DEL-TORRES, M., SARAIS, I., FUOCHI, P.G., TUBARO, F., and URSINI, F. (1998). Carnosine increase the radiation resistance of *Aeromonas hydrophila* in minced turkey meat. *J. Food Sci.* **63**: 147–149.

STEWART, E. M. (2001). Detection methods for irradiated foods. In *Food Irradiation: Principles and Applications*. R.A Molins Ed., John Wiley & Sons Inc., New York pp. 347–385.

THAYER, D. W, SONGPRASERTEHAI, S. and BOYD. G. (1991). Effect of heat and ionising radiation on *Salmonella typhimurium* in mechanically deboned chicken meat. *J. Food Prot.* **54**: 717–718, 724.

THOMAS, C. and BEIME, D., (1999). Microbial safety in processed vegetables. *Int. Food Hyg.* **10**: 17–23.

TRITSCH, G.L. (2000). Food irradiation. *Nutrition* **16**: 698–701.

WILLS, E.D. (1980). Effects of antioxidants on lipid peroxide formation in irradiated synthetic diets. *Int. J. Radiat., Biol., Relat. Stud., Phys. Chem. Med.* **37**: 403–414.

YOOK, H.S., LIM, S., and BYUN, M.W. (1998). Changes in microbiological and physicochemical properties of bee pollen by application of gamma irradiation and ozone treatment. *J. Food Prot.* **61**: 217–220.

14

Natural antifungal agents for bakery products

N. Magan, M. Arroyo and D. Aldred, Cranfield University, UK

14.1 Introduction

Spoilage of bakery products by undesirable moulds is of major importance as they are intermediate moisture products where conducive conditions exist for growth, which can shorten the shelf-life of products (Earle and Putt, 1984). Moulds can cause significant economic losses, and some moulds can produce toxic secondary metabolites (mycotoxins), which can affect human health. Losses due to mould spoilage vary between 1 and 5% of products depending on season, type of product and method of processing and storage (Malkki and Rauha, 1978; Legan, 1993). Since fungal spores are ubiquitous in the atmosphere, they commonly contaminate bakery products during production from raw material to finished product. Some spoilage fungi have heat-tolerant spores, and many are xerotolerant or xerophilic and so are able to tolerate very dry conditions (Magan, 1997). Tolerance to a wide range of environmental conditions and their predominantly mycelial growth habit enable moulds to colonise food products rapidly, producing a battery of enzymes to utilise the food matrix. Thus, antifungal agents and/or environmental conditions that inhibit germination and growth are essential for maintaining shelf-life of many cereal-based bakery products.

While organic acids and their salts are the most commonly used preservatives in bakery products (see also Chapter 6) there is pressure from EU Directives to reduce the use of these and instead to use more natural preservatives, either alone or in combination with packaging systems. This chapter will consider the potential for using essential oils and antioxidants for control of spoilage moulds in bakery products, particularly bread, and will discuss the prospects for their use in industry.

14.2 Antimicrobial activity of essential oils and antioxidants against bakery moulds in laboratory media

In the last few years, great interest has emerged in the possible use of plant extracts and essential oils for food preservation. The antimicrobial properties of essential oils (EOs) against foodborne microorganisms in laboratory media and processed foods, against fungi associated with postharvest spoilage of fruits and vegetables, against yeasts and moulds in lightly-processed fruit products, and in combination with edible coatings and irradiation are reviewed in Chapters 9, 10, 11, 12 and 13, respectively. The focus in this chapter is on fungi of relevance to the spoilage of baked products.

Aspergillus flavus, one of the most toxigenic foodborne fungi in nature and a frequent contaminant of flour used for baking, has been reported as sensitive to some EOs (see Tables 9.3, 10.1 and 11.1). Dwivedy and Dubey (1993) studied the antifungal activity of several umbelliferous plant EOs against *Aspergillus* species and reported an important fungistatic effect of *Trachyspermum* seed essential oil at relatively low concentrations (< 500 ppm). However, in many studies carried out to date, the environmental factors appropriate to bakery products have not been simulated realistically. Furthermore, in some cases (Salmeron *et al.*, 1990), stimulation of growth of *Aspergillus* species has been demonstrated when extracts of thyme and oregano were incorporated into nutritive media. Some very useful studies have been published recently (Chapter 11) on the dose–response of *Asp. flavus* to vanillin, thymol, eugenol, carvacrol and citral and/or potassium sorbate/sodium benzoate (reviewed in Chapter 11 with examples in Table 11.1 and Figs 11.1–11.3).

Screening of 20 plant EOs for their activity against five fungal strains important as bakery contaminants, *Aspergillus ochraceus, Cladosporium herbarum, Penicillium corylophilum* and two strains of *Pen. verrucosum*, in a wheat-flour-based medium at 25 °C has shown that at least 500 ppm was required for control of growth. Indeed, only clove, thyme, bay and cinnamon EOs completely inhibited growth of all the species studied for 30 days (Table 14.1). The lag phase prior to growth was also extended in the presence of these EOs. This is an important indicator of the potential for extending shelf-life over a range of water activities and temperatures. Similarly, Patkar *et al.* (1993) found that 500 ppm of cinnamon EO completely inhibited growth of the aflatoxigenic strains of *Asp. flavus* in laboratory media for 7 days, whereas up to 1250 ppm of clove EO was necessary to exert the same inhibitory activity. Azzouz and Bullerman (1982) also reported a strong anti-mould activity of clove and cinnamon oils against several *Aspergillus* and *Pencillium* species.

Although many researchers have reported a strong inhibitory effect of thyme and cinnamon EOs on moulds, results are not always consistent. For instance, while 500 ppm of thyme oil completely inhibited growth of *Asp. ochraceus* on wheat flour agar for 30 days (Table 14.1, Arroyo and Magan, unpublished data), Paster *et al.* (1995) reported colonies on PDA of up to 35 mm in diameter in the presence of the same concentration of this EO. Özcan (1998) also reported

Table 14.1 Radial growth rate inhibition (%) of five fungal strains growing at 25 °C on 2% wheat flour agar in the presence of 500 ppm of 20 plant essential oils at a_w 0.95 and pH 6

	Growth rate inhibition (%)				
	Aspergillus ochraceus	*Cladosporium herbarum*	*Penicillium corylophium*	*Penicillium verrucosurm* M450	*Penicillium verrucosurm* PV3
Basil-lynalol type	−16.42	100.00	26.49	−11.48	−14.77
Basil-methyl chavicol type	−13.51	0.41	12.74	−11.48	−38.85
Bay	**100.00**	**100.00**	**100.00**	**100.00**	**100.00**
Cinnamon	**100.00**	**100.00**	**100.00**	**100.00**	**100.00**
Clove	**100.00**	**100.00**	**100.00**	**100.00**	**100.00**
Eucalyptus	−4.73	11.35	8.26	−25.47	−50.39
Ginger	−6.18	29.33	5.71	−3.15	−25.35
Grapefruit	−9.46	5.02	14.67	−12.79	−75.15
Lemongrass	−32.76	100.00	11.80	5.96	−10.36
Lime	−6.36	31.68	17.67	−11.32	−53.78
Majoram	−7.14	−4.11	6.16	−9.23	−51.73
Nutmeg	−10.21	69.85	7.91	−2.86	−50.24
Orange	−19.06	36.83	3.25	−14.84	−34.58
Peppermint	−19.06	34.11	5.62	−21.97	−18.54
Pine-syl	−8.28	12.68	0.96	−27.37	−49.41
Rosemary	−7.60	17.84	4.66	−17.46	−53.81
Sage	−19.56	33.00	9.24	−17.27	−71.97
Spearmint	−21.66	100.00	4.18	−28.24	−30.30
Sweetfennel	−21.04	61.56	20.99	−8.31	−41.56
Thyme	**100.00**	**100.00**	**100.00**	**100.00**	**100.00**

growth of *Asp. parasiticus* in laboratory media in the presence of 1% thyme (wild and black) oil. Although specific sensitivity to a certain plant EO may vary with the strain studied, these differences in sensitivity may be attributed principally to the variable composition of the EO, which can vary substantially with the plant origin and extraction method. For example, the results shown in Table 14.1 were obtained using a commercial brand of thyme oil from Spanish plants while Paster *et al.* (1995) extracted thyme oil themselves from the leaves of Israeli plants.

Although antioxidants are added to foods primarily to control chemical and physical deterioration associated with oxidation reactions, their antimicrobial activity has also recently been explored. For example, the inhibitory effects of antioxidants such as butylated hydroxyoluene (BHT), butylated hydroxyanisole (BHA), *n*-propyl gallate, propyl paraben and octyl galate have been investigated at a_w 0.95 and pH 6. Common bread contaminants such as *Cladosporium herbarum, Penicillium corylophilum, Pen. verrucosum* and *Aspergillus ochraceus* were completely inhibited by propyl paraben, BHA and octyl gallate at 500 ppm. Importantly, these agents also extended the lag phase, which is indicative of potential for extending shelf-life. Kudo and Lee (1998) compared the antifungal activity of propyl (C3), octyl (C8) and dodecyl (C12) gallates and reported that octyl gallate was the only active compound (minimum inhibitory concentration, MIC 25 ppm) against four fungal genera. Some recent studies have also suggested that mixtures of antioxidants may have more than just an additive effect against some spoilage moulds (Reynoso *et al.*, 2002). Mixtures of 0.5–1.0 mM of BHA and propyl paraben inhibited mycotoxigenic *Fusarium* species synergistically and also significantly reduced fumonisin formation.

It is notable that while some work has been done on the potential of using mixtures of antioxidants, practically no studies have examined combinations of antioxidants with EOs. However, the evidence from *in vitro* studies suggest that potential does indeed exist for exploiting both antioxidants and essential oils for use in bread and other bakery products.

14.3 Control of moulds in bakery products

Essential oils have been examined in two ways for control of spoilage moulds in bread. Firstly, the volatiles produced by EOs have been used in bread packaging, and, secondly, attempts have been made to incorporate low concentrations of EOs with the bread ingredients. Using the former approach, Nielsen and Rios (2001) examined the efficacy of volatiles in a modified atmosphere packaging system to control the spoilage fungi of rye bread. Mustard essential oil in the volatile phase at 1–10 μg/ml was most effective against spoilage fungi including *Pen. commune*, *Pen. roquefortii*, *Asp. flavus* and *Endomyces fibuliger*. Cinnamon, garlic and clove oils also controlled mould growth on slices of bread. Vanilla showed no inhibitory effects, and *Asp. flavus* was the most resistant of the species tested. Interestingly, the MIC depended on the active component allylthiocynanate and at least 3.5 μg/ml was required to achieve fungicidal effects. Introduction of small sachets or impregnated pads into packs (as, for example, in nisin applications shown in Table 3.4 and Fig. 3.1) could be an effective way of enabling the slow release of volatiles from EOs to control mould spoilage of bakery products, especially where this can be combined with modified atmospheres.

When incorporated directly into a bread formulation, the antimicrobial efficacy of EOs appears to be lower. For example, up to 1000 ppm of oregano EO was necessary to completely inhibit growth of *Asp. flavus* in bakery products for 21 days (Basilico and Basilico, 1999) while only 100 ppm was required to completely inhibit growth of *Asp. niger* and several *Penicillium* species over a period of 6 days *in vitro*. Chemically, EOs consist of a mixture of esters, aldehydes, ketones and terpenes. Although several studies have been carried out on the inhibitory effects of essential oil components (Lachowicz *et al.*, 1998; Saxena and Mathela, 1996; Mahmoud, 1994; Sinha and Gulati, 1990; Bilgrami *et al.*, 1992; see also Chapters 9–13), the exact role of these components in the overall antimicrobial activity of the oil is not clear. The components may act synergistically to bring about antimicrobial action or they may act antagonistically and lead to growth stimulation (French, 1985).

Over 20 EOs and 5 antioxidants have been investigated as novel means of controlling growth of the mycotoxigenic species *Fusarium culmorum* and *Penicillium verrucosum* on grain intended for flour and bakery prouction (Hope and Magan, 2003; Cairns and Magan, 2003). Only three EOs (bay, clove and cinnamon) and two antioxidants (propyl paraben and hydroxymethylanisole) were effective in controlling growth in the range 50–200 ppm at 15 ° and 25 °C at a_w 0.995–0.955. At 500 ppm, clove oil and BHA effectively controlled growth and both deoxynivalenol and nivalenol toxin production by *F. culmorum* at 15 and 25 °C. Similarly, effective control of *Pen. verrucosum* and of ochratoxin was achieved at 500 ppm. However, at intermediate concentrations (100 ppm), stimulation of mycotoxin formation was sometimes observed.

Direct incorporation of EOs (cinnamon, clove, thyme and mustard) into bread and bread analogues has shown reduced efficacy compared with that observed in laboratory media. The extent of growth control achieved was 50% for *Eurotium repens* by cinnamon and clove, and only about 25% for *Asp. ochraceus* by thyme EO. Practically no control of *Pen. verrucosum* and *Pen. corylophilum* with any of the EOs was observed. Figure 14.1 compares the efficacy of 1500 ppm of clove and mustard EOs with calcium propionate in bread analogues. This shows that at a high concentration (1500 ppm), mustard oil completely inhibited growth, while inhibition by clove oil was similar to that achieved by the traditional preservative against a range of target moulds. *Eurotium repens* was least affected by the treatments and *Pen. corylophilum* was most inhibited (Arroyo and Magan, unpublished data).

Reduced antimicrobial efficacy in foods is probably due to binding of EOs to proteins or fats (McNeil and Schmidt, 1993; Smid and Gorris, 1999). The requirement for higher doses means that the typical odour of EOs is also more

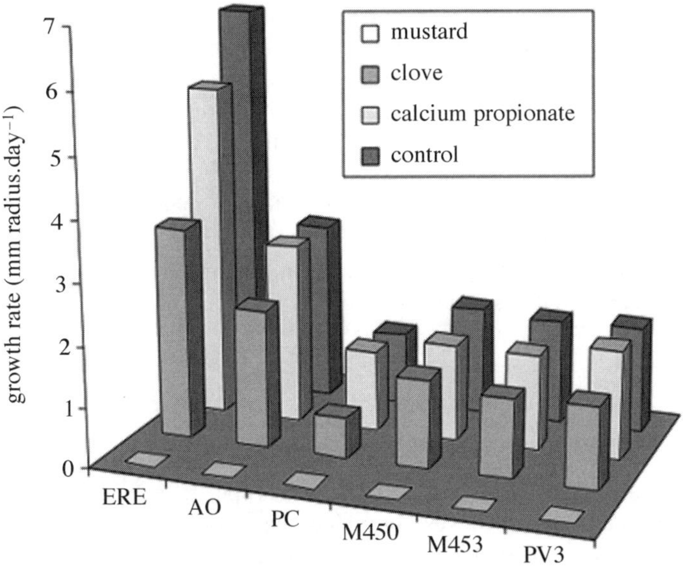

Fig. 14.1 Radial growth rate (mm/day) of *Eurotium repens* (ERE), *Aspergillus ochraceus* (AO), *Penicillium corylophilum* (PC) and strains M540, M453 and PV3 of *Penicillium verrucosum* on bread analogues at a_w 0.97, pH6 and in the presence of 1500 ppm of calcium propionate and essential oils of clove and mustard.

noticeable. In some products, this aroma/flavour may be desirable. For example, in tomato-based bakery products, basil can be used both as a flavouring and as an antifungal agent (Lachowicz *et al.*, 1998). This has particular application in novel bread products where herbs and spices are often incorporated into the bread. Eliciting in-pack volatiles may be a more effective way of using low concentrations of EOs in conjunction with modified atmosphere packaging systems. This needs to be investigated further as an alternative approach to substitute for organic acids. Few studies have examined combinations of low concentrations of existing organic acid type preservatives with EOs or combinations of EOs in bakery products. This is an area where more studies are required to examine whether these may act additively or synergistically.

The efficacy of antioxidants against spoilage moulds in some bakery products has been examined. The concentrations required for inhibition of growth in foods were generally higher than those shown to be effective *in vitro*. For example, propyl paraben had little effect on growth of four spoilage moulds on bread analogues at a_w 0.97 and pH 4.5, even at 1000 ppm (Arroyo and Magan, unpublished results). This suggests that either the antioxidants are bound by ingredients in the foodstuff and thus become less effective, or that there is less dispersion in the product providing less direct contact with the contaminant moulds.

Although ethanol is not permitted for use in bakery products in the UK, it is allowed in some other countries and could be considered as a natural antimicrobial compound. It has been demonstrated that 0.5–3.5% ethanol added after baking and cooling can significantly inhibit mould growth in bread and increase shelf-life (Vora and Sidhu, 1987; Seiler and Russell, 1991). Seiler (1984) showed that an increase of more than 50% in the mould-free shelf-life of bread was achievable by the addition of 0.5% ethanol, with 1% giving more than a 100% increase in shelf-life. Similar results have been obtained by other workers, suggesting that it could be a very effective mould inhibitor for bread and perhaps other bakery products. Ethanol has the added advantage of preventing staling in bread. This is critical in mass-produced, sliced, wrapped bread and bakery products where shelf-life is very short unless preservatives are used.

14.4 Consumer acceptance

Consumers are enthusiastic, particularly in Europe, to have bakery products containing natural preservatives, provided that mould-free shelf-life can be maintained. Some essential oils may be appropriate to use in some bakery products where they enhance product flavour and aroma. However, in some products, particularly traditional bread products, this strong odour may not be as acceptable. Another important consideration is the economics of use, which will also have a significant impact on consumer acceptance. If the costs of bakery products are significantly higher because of the use of natural preservatives, the consumer may be more reluctant to accept such products. Calcium propionate is used in a wide range of bakery products because it is effective and the costs are relatively low at the maximum permitted concentrations (approximately 2000 mg/kg). By contrast, EOs such as mustard or clove oil would cost at least twice as much if added at a level of 250 mg/kg. In cakes, potassium sorbate is effective at similar concentrations. Using cinnamon oil would be about 25% cheaper at 250 mg/kg, while bay oil at the same concentration would cost almost five times more. Antioxidants are cheaper than EOs and are similar to, if not cheaper than, organic acids. Provided they are effective in a bakery product they would be acceptable, although some have been reported to be carcinogenic to animals although not to humans. There is a direct link between consumer acceptance, economics of use and actual efficacy of preservatives, antioxidants and EOs.

14.5 Future prospects

When one looks at the legislation on food additives there are clear regulations with regard to the use of organic acids and some antioxidants in bread and bakery products (89/107/EEC; 95/2/EC; 97/77/EC; SI 1995 No. 3187; see also Chapter 15). However, at the present time, with the exception of the use of EOs as food flavourings, there is no legislation with regard to their use as preservatives. To a large extent, the use of new preservatives and preservation techniques will be driven by consumer pressure and legislation. Potential does exist for using EOs, especially where they can be combined with other abiotic factors such as packaging and modified atmospheres to improve efficacy and prolong the shelf-life of bakery products.

14.6 References

AZZOUZ, M.A. & BULLERMAN, L.B. (1982). Comparative antimycotic effects of selected herbs, spices, plant components and commercial antifungal agents. *J Food Prot* **45**, 1298–1301.

BASILICO, M.Z. & BASILICO, J.C. (1999). Inhibitory effects of some spice essential oil on *Aspergillus ochraceus* NRRL 3174 growth and ochratoxin production. *Lett Appl Microbiol* **29**, 238–241.

BILGRAMI, K.S.; SINHA, K.K. & SINHA, A.K. (1992). Inhibition of aflatoxin production & growth of *Aspergillus flavus* by eugenol & onion & garlic extracts. *Indian J Med Res* **96**, 171–175.

CAIRNS, V. & MAGAN, N. (2003). Impact of essential oils on growth and ochratoxin A production by *Penicillium verrucosum* and *Aspergillus ochraceus* on a wheat-based substrate. In *Proc. of Stored Product Protection Symposium*, E. Highley (ed.), Clarus Design, Canberra. In press.

DWIVEDI, S.K. & DUBEY, N.K. (1993). Potential use of the essential oil of *Trachyspemum ammi* against seed borne fungi of Guar. *Mycopathology,* **121**, 101–104.

EARLE, M.D. & PUTT, G.J. (1984). Microbial spoilage and use of sorbate in bakery products. *Food Technol. N.Z.* **19**, 25–36.

FRENCH, R.C. (1985). The bio-regulatory action of flavour compounds on fungal spores and other propagules. *Ann Review Phytopath* **23**, 173–199.

HOPE, R. & MAGAN, N. (2003). Multi-target environmental approach for control of growth and toxin production by *Fusarium culmorum* using essential oils and antioxidants. In *Proc. Int. Stored Prod. Protect. Symp.*, E. Highley (ed.), Clarus Design, Canberra. In press.

KUDO. L. & LEE, S.H. (1998). Potentiation of antifungal activity of sorbic acid. *J. Agric Food Chem.* **46**, 4052–4055.

LACHOWICZ, K.J., JONES, G.P., BRIGGS, D.R., BIENVENU, F.E., WAN, J., WILCOCK, A. & COVENTRY, M.J. (1998). The synergistic preservative effects of the essential oils of sweet basil (*Ocimum basilicum* L.) against acid-tolerant food microflora. *Lett Appl Microbiol* **26**, 209–214.

LEGAN, J.D. (1993). Mould spoilage of bread: the problem and some solutions. *Int Biodet Biodegr* **32**, 35–53.

MAGAN, N. (1997). Fungi in extreme environments In: *The Mycota IV, Environmental and Microbial Relationships*, Wicklow/Söderström (eds.), Springer, Berlin Heidelberg, 99–114

MAHMOUD, A.L.E. (1994). Antifungal action and antiaflatoxigenic properties of some essential oil constituents. *Lett Appl Microbiol* **19**, 111–113.

MALKKI, Y. & RAUHA, O. (1978). Mould inhibition by aerosols. *Baker's Dig* **52**, 47.

MCNEIL, V.I. & SCHMIDT, K.A. (1993). Vanillin interaction with milk protein isolates in sweetened drinks. *J Food Sci* **58**, 1142–1147.

NIELSEN, P.V. & RIOS, R. (2001). Inhibition of fungal growth on bread by volatile components from spices and herbs, and possible application in active packaging, with special emphasis on mustard essential oil. *Int J Food Microbiol* **60**, 219–229.

ÖZCAN, M. (1998). Inhibitory effects of spice extracts on the growth of *Aspergillus parasiticus* NRRL2999 strain. *Zeits Lebens-Unters Forsch* **207**, 253–255.

PASTER, N., MENASHEROV, M., RAVID, U. & JUVEN, B. (1995). Antifungal activity of oregano and thyme essential oils applied as fumigants against fungi attacking stored grain. *J Food Prot* **58**(1), 81–85.

PATKAR, K.L. USHA, C.M., SHETTY, H.S., PASTER, N. & LACEY, J. (1993). Effect of spice oil treatment of rice on moulding and mycotoxin contamination. *Lett Appl Microbiol* **17**, 49–51.

REYNOSO, M.M., TORRES, A.M., RAMIREZ, M.L., RODRIGUEZ, M.I., CHULZE, S. & MAGAN, N. (2002). Efficacy of antioxidant mixtures on growth, fumonisin production and hydrolytic enzyme production by *Fusarium verticillioides* and *F. proliferatum in vitro* on maize-based media. *Mycological Res* **106**, 1093–1099.

SAXENA, J. & MATHELA, C.S. (1996). Antifungal activity of new compounds from Nepeta leucophylla and Nepeta clarkei. *Appl Environ Microbiol* **62**(2), 702–704.

SALMERON, J., JORDANO, R. & POZO, R. (1990). Antimycotic and Antiaflatoxigenic activity of oregano and thyme. *J Food Prot* **53**(8), 697–700.

SEILER, D.A.L. (1984). Preservation of bakery products. *Institute of Food Science and Technology, Proceedings* **17**, 31–39.

SEILER, D.A.L. & RUSSELL, N.J. (1991). Ethanol as a food preservative. In: *Food Preservatives*, N.J. Russell and G.W. Gould (eds), Blackie, Glasgow, pp. 153–171.

SINHA, G.K. & GULAT, B.C. (1990). Antibacterial and antifungal study of some essential oils and some of their constituents. *Indian Perfumes* **34**, 126–129.

SMID, E.J. & GORRIS, L.G.M. (1999). Natural antimicrobials for food preservation In: *Handbook of Food Preservation*, M. Shafiur Rahman (ed.), Marcel Dekker, Inc., New York.

VORA, H.M. & SIDHU, J.S. (1987). Effect of varying concentrations of ethyl alcohol and carbon dioxide on the shelf-life of bread. *Chemie, Mikrobiologie, Technologie der Lebensmittel* **11**, 56–59.

15

Regulations: new food additives, ingredients and processes

P. Berry Ottaway, Berry Ottaway & Associates Limited, UK

15.1 Introduction

The addition of substances to help with food preservation can be traced back over millennia. Salt was valued in early civilisations, not only as a condiment but also for its ability to increase the storage life of meat and fish. Various forms of pickling go back a long way, as does the use of high sugar concentrations in preserving fruit products.

Preservatives are defined as substances that prolong the shelf-life of foods by protecting them against deterioration caused by microorganisms. This definition forms the basis of categorisation in the laws of most countries.

15.2 Natural antimicrobials: food ingredients or food additives?

The definition of a food component is, by necessity, closely linked to a definition of food. Unfortunately, there is no close agreement between the food laws of a number of the developed countries on the actual definition of food. In 1966, in the early days of the Codex Alimentarius Commission (Codex), there was an attempt to define food as:

> Food means any substance, whether processed, semi-processed or raw, which is intended for human consumption and includes drink, chewing gum, and any substance which has been used in the manufacture, preparation or treatment of food, but does not include cosmetics or tobacco or substances used only as drugs.[1]

This definition can be found in many modified forms in the legislation of different countries. Although the harmonisation of European food legislation has been in progress for over 40 years, it was only in early 2002 that the definition of a food was formally introduced into European law.[2] This definition is essentially the same as that of Codex but it also includes water and excludes live animals, residues and contaminants.

The Codex Alimentarius Commission defines a food additive as:

> Any substance not normally consumed as a food by itself and not normally used as a typical ingredient of the food, whether or not it has nutritive value, the intentional addition of which to food for a technological (including organoleptic) purpose in the manufacture, processing, preparation, treatment, packing, packaging, transport or holding of such food results, or may reasonably be expected to result (directly or indirectly) in it or its by-products becoming a component of or otherwise affecting the characteristics of such food. The term does not include contaminants or substances added to food for maintaining or improving nutritional qualities.

In the legislation of the European Union (EU), the definition of a food additive is contained in European Council Directive 89/107/EEC.[3] This definition is again very similar to the Codex definition and is given as:

> For the purpose of this directive 'Food Additive' means any substance not normally consumed as a food itself and not normally used as a characteristic ingredient of food whether or not it has nutritive value, the intentional addition of which to food for a technological purpose in the manufacture, processing, preparation, treatment, packaging, transport or storage of such food results, or may be reasonably expected to result, in its by-products becoming directly or indirectly a component of such foods.
>
> The Directive specifically excludes:

- processing aids;
- substances used in the protection of plants and plant products in conformity with Community rules relating to plant health;
- flavourings for use in foodstuffs, falling within the scope of Council Directive 88/388/EEC;
- substances added to foodstuffs as nutrients (e.g. minerals, trace minerals or vitamins).

Extraction solvents are also not considered to be food additives in the EU and are subject to specific legislation both on their use and residual levels.

A processing aid is further defined in European law as:

> any substance not consumed as a food ingredient by itself, intentionally used in the processing of raw materials, foods or their ingredients, to

> fulfil a certain technological purpose during treatment or processing and which may result in the unintentional but technically unavoidable presence of residues of the substance or its derivatives in the final product, provided that these residues do not present any health risk and do not have any technological effects on the finished product.

In European law one of the key differences between a food additive and a processing aid is that the latter must not have any technological effect in the finished product. For example, residues of a release agent used for a moulded confectionery product, such as a fancy chocolate, are unlikely to have a technological effect on the final product, whereas a preservative or antioxidant added to protect ingredients during processing could still exert a technological function if carried over into the finished product. The former would be considered a processing aid while the latter could be considered a technological additive. This is an important distinction when considering the status of substances that have antibacterial activity.

In the USA, the Food Additives Amendment 1958 to the Federal Food, Drug and Cosmetics Act[4] contains a more complex definition:

> The term food additive means any substance the intended use of which results or may reasonably be expected to result, directly or indirectly, in its becoming a component or otherwise affecting the characteristics of any food (including any substance intended for use in producing, manufacturing, packing, processing, preparing, treating, packaging, transporting, or holding food, and including any source of radiation intended for any such use), if such substance is not generally recognised, among experts qualified by scientific training and experience to evaluate its safety, as having been adequately shown through scientific procedures (or, in the case of a substance used in food prior to January 1, 1958, through either scientific procedures or experience based on common use in food) to be safe under the condition of intended use.

Pesticides on raw agricultural products and food colour additives are excluded from this legal definition as they are covered by other legislation.

This amendment categorised food chemicals as:

- those Generally Recognised As Safe (GRAS);
- those with prior sanction; and
- food additives.

In the legislation of most countries, chemical substances added as antimicrobials are regarded as food additives if the primary purpose of the substance is the preservation of the food to prolong shelf-life. While the details may differ, almost all the food additive legislation worldwide places strict controls on the preservatives that can be used in foods and drinks.

In attempting to define the legal status of an antimicrobial substance, the nature and function of the substance must be taken into consideration. Both the

Codex and European Union definition of a food additive refer to substances not normally consumed as a food by itself and not normally used as characteristic ingredients of a food and for which the intentional addition to the food is to serve a technological purpose. Thus, if a component is added to a food and the primary reason for the addition is neither for its nutritive value nor for its unique organoleptic properties but for a technological reason such as an antimicrobial property, it is most likely to be considered to be a food additive.

Although this interpretation covers most substances added entirely for their antimicrobial properties, there are anomalies. Vinegar can be used in foods both as a preservative and also as a condiment to impart acidity and its characteristic taste to foods such as sauces, mayonnaise and pickles. In the USA vinegar and acetic acid are generally recognised as safe for use in foods and are classified as GRAS under the Federal Regulations. In the EU acetic acid is considered a food additive that can be used in a relatively wide range of foods following the *quantum satis* principle when used as a preservative. Some countries outside the EU and USA specifically exclude vinegar from the categories of food additives and preservatives and it is considered to be a food ingredient. For more information on the antimicrobial properties of organic acids including acetic acid, see Chapter 6.

In the EU, the primary function of vinegar in the food would define its status. If it is used only to impart desired organoleptic properties, or as a condiment added to food before consumption, it can be considered to be a food ingredient. However, if the primary purpose is to enhance the shelf-life of the food (pickling) then it has to be classified as a food additive.

While under most circumstances it is usually clear when a substance is used in a technological capacity, and therefore as an additive, there have been cases where an ingredient, or combination of ingredients, added for nutritive or culinary purposes have exhibited antimicrobial activity in a food. An unusual example of an ingredient exhibiting such properties was found during shelf-life studies on a product containing a garlic macerate. The oil released during the processing of the garlic very significantly reduced the microbial content of the product during storage and prevented further growth. Garlic oil has been known for at least one and a half centuries to have an antimicrobial activity on a broad spectrum of bacteria and fungi, and was used to impregnate field dressings in the First World War. In this particular case the garlic macerate was used for culinary purposes and the effect of the oil on the microbiological stability of the product was unexpected. Under such circumstances the garlic could only be considered to be a food ingredient. For more information on the antimicrobial properties of garlic and other plant extracts, see Chapter 9.

Other food ingredients, such as cranberry juice and some fatty acids, also have antimicrobial properties.[5] It has also been demonstrated that the caffeine present in coffee and cocoa can inhibit the growth of a range of mycotoxigenic organisms *in vitro* and that theophylline in tea and theobromine in cocoa exhibit antimicrobial effects. The presence of these ingredients in foods or beverages is primarily for nutritional or organoleptic purposes and thus such substances would be considered to be ingredients and not food additives. In other cases, ingredients have been added in specific quantities or combinations to reduce the water activity and hence growth of microorganisms in food (e.g. sugar in jams and preserves).

15.3 The legislation on food preservatives

The general principles for the use of food additives, which are laid down in the Codex General Standard for Food Additives[6] are:

- only those food additives should be used which so far as can be judged on the evidence presently available, present no risk to the health of the consumer at the levels of use proposed, and
- the evaluation of a food additive should have taken into account any Acceptable Daily Intake, or equivalent assessment, established for the additive and its probable daily intake from all sources, and
- where the food additive is to be used in foods eaten by special groups of consumers, account should be taken of the probable daily intake by consumers in those groups.

The Codex standard also states that the use of food additives is only justified when such a use has an advantage, does not present a hazard to health, does not mislead the consumer and serves one or more of the technological functions set out by Codex. The relevant functions given in the Standard are:

- to enhance the keeping quality or stability of a food or to improve its organoleptic properties, provided that this does not change the nature, substance or quality of the food so as to deceive the consumer; and
- to preserve the nutritional quality of the food.

Food additives should be used only where the above objectives cannot be achieved by other means which are economically and technologically practicable.

Worldwide, food preservatives are one of the most regulated categories of food additives, and probably the most complex are the laws currently in force in the European Union. In EU food law, the control of preservatives is included in the directive on food additives other than colours and sweeteners (Directive 95/2/EC). This directive harmonised the legislation on food additives across the EU and came fully into effect in March 1997.[7] Within this legislation, food preservatives are controlled by three means:

1 by the permitted substance and its chemical forms;
2 by the categories of foods in which the specified substances are permitted;
3 by maximum levels in the product as consumed, calculated as g or mg/kg or g or mg/l as appropriate.

Food additives permitted for use as preservatives in the EU are very restricted and are classified into three categories:

1 sorbates, benzoates and *p*-hydroxybenzoates;
2 sulphur dioxide and sulphites;
3 other preservatives.

The category of 'other preservatives' contains 13 substances or groups of closely related substances, many of which are permitted for very specific and limited applications. For example, boric acid and sodium tetraborate can only be used with sturgeon eggs (caviar) to a maximum of 4 g/kg as boric acid. Hexamethylene tetramine is only allowed in Provolone cheese and lysozyme is only permitted for use in ripened cheese (see Chapter 7 for more information on lysozyme).

In all three categories, the permitted uses are detailed in the legislation and a permitted preservative cannot be used unless the food in which it is intended to be used is on the list for that preservative. Thus, in category 1, there are foods that can only use sorbates; a few that can only use benzoates; some that can use combinations of sorbates and benzoates; two groups of foods that can use combinations of sorbates and *p*-hydroxybenzoates, and three applications where a mixture of sorbates, benzoates and *p*-hydroxybenzoates are permitted. In each case, a maximum level on input for each of the preservatives is specified.

Although not as complex as the EU legislation, the laws on preservatives in the USA restrict the usage of antimicrobial food additives to prescribed categories of foods. In some cases a maximum level has been laid down for the preservative in specified foods and drinks, such as a maximum of 0.1% for sodium benzoate, *p*-hydroxybenzoates or combinations of both.

Before using a preservative in the USA, it is important to check whether the product is subject to a Federal Standard of Identity, a number of which have been assigned to specific foods. Where a Standard of Identity exists, the use of a preservative may either not be allowed or be specified with a maximum level of input. In the case of sorbic acid and potassium sorbate, the substances can be used with no upper levels in foods that do not have Federal Standards Identity, but for those that do, strict controls are applied.

Within the US Federal Regulations, there are a number of very specific conditions. For example, up to 0.2% of sodium benzoate may be used in orange juice destined for further manufacturing, while orange juice not intended for manufacturing and other pure fruit juices having Federal Standards of Identity may not contain a preservative. Similar situations occur in other product categories. Sodium benzoate is permitted for use in salted margarine, while potassium sorbate must be used in unsalted margarine.

In the EU and many other countries, flavourings for food use are treated separately from technological additives. However, a number of flavouring substances and plants used for flavouring purposes also possess antimicrobial activity. In such cases, the primary function in the food (i.e. as a flavour) would supersede the function as an antimicrobial agent[8] (see Chapters 9, 10, 11 and 14).

Table 15.1 Food preservatives permitted in the legislation of one or more countries worldwide

1.	Benzoic acid and sodium, potassium and calcium benzoates
2.	Parahydroxybenzoates (PHB)
	Sodium ethyl PHB
	Sodium propyl PHB
	Sodium methyl PHB
	Ethyl PHB
	Propyl PHB
	Methyl PHB
3.	Sorbic acid and potassium and calcium sorbates
4.	Sulfur dioxide
	Sodium and potassium metabisulfites
	Sodium and calcium sulfites
	Sodium, calcium and potassium hydrogen sulfites
5.	Potassium and sodium nitrites
	Potassium and sodium nitrates
6.	Propionic acid
	Sodium, calcium and potassium propionates
7.	Biphenyl and diphenyl
8.	Orthophenyl phenol
	Sodium orthophenyl phenol
9.	Thiabendazole
10.	Nisin
11.	Natamycin
12.	Hexamethylene tetramine
13.	Dimethyl dicarbonate
14.	Boric acid and sodium tetraborate
15.	Lysozyme
16.	Acetic acid
	Sodium and calcium acetates
	Sodium diocetate
17.	Diethyl pyrocarbonate

A list of food preservatives permitted in the legislation of one or more countries is shown in Table 15.1.

15.4 Authorisation of new preservatives

Over the past few years the safety evaluation of new additives has taken on a global perspective. With the advent of the World Trade Organization (WTO) in 1995,[9,10] Codex Alimentarius has assumed a greater international significance as the WTO recognises Codex as the preferred international organisation for the arbitration and settlement of disputes related to international food trade. As a consequence of the WTO and also the involvement of Codex in setting international standards for foods and commodities, the Codex Alimentarius Commission now has a major influence on food additive legislation worldwide.

Impartial advice on the safety of food additives is provided to Codex by the Joint FAO/WHO Expert Committee on Food Additives (JECFA). Since it was established in 1955, JECFA has played an international role in providing a review and opinion on food additives, particularly on safety in use. JECFA was originally set up to consider the chemical, toxicological and other aspects of contaminants and residues of veterinary drugs found in foods intended for human consumption. The committee is composed of an international group of independent scientists appointed in their own right for their expertise in food additives, veterinary drug residues and contaminants. The members are not appointed as government representatives. The committee, which is administered jointly by the Food and Agriculture Organization (FAO) and the World Health Organization (WHO), provides scientific advice directly to member countries of FAO and the WHO and also to two committees of the Codex Alimentarius Commission (Codex); the Codex Committee on Food Additives and Contaminants (CCFAC) and the Codex Committee on Residues of Veterinary Drugs in Foods (CCVDF).

The Codex committees identify the additives, contaminants and residues that should receive priority evaluation and refer them to JECFA for assessment before considering them for incorporation into Codex Standards. The JECFA reports on their toxicological evaluations are published so that the information becomes widely available. In its interaction with the Codex Committees, JECFA is independent of the Codex Alimentarius Commission and is therefore able to provide impartial and independent advice without political influence. The JECFA would be the main international body involved in the safety assessment of a new preservative.

As part of the assessment of an additive, the JECFA examines the available toxicological data and chemical specifications of the additive and establishes an Acceptable Daily Intake (ADI). The ADI is the estimate of the amount that can be taken daily in a human diet over a lifetime without appreciable health risk (based on the 'standard man' of 60 kg). Unless stated to the contrary, ADIs are expressed as mg/kg body weight.

The ADI is normally derived from the 'no observed adverse effect' level (NOAEL) determined from long-term animal (*in vivo*) studies. The ADI is calculated by applying a safety or uncertainty factor, which is commonly 100, to the NOAEL obtained from the most sensitive test species. The 100-fold safety factor is based on the need to take into account both the differences in species and differences in toxicokinetics and toxicodynamics. Although most commonly used, the factor of 100 may be varied depending on the nature of the toxic effect and the availability of relevant toxicity data.

Approval for the use of an additive is given on the understanding that the additive will be kept under surveillance and re-evaluated as necessary in the light of changing conditions of use or new scientific evidence. The ADI value may be qualified in five different ways:

1 'Not limited', which means that there is no explicit indication of an upper limit and is assigned to substances of very low toxicity (this category has now been superseded by 'not specified').
2 'Not specified', which means that on the basis of available data the total daily intake required to achieve the desired additive effect does not represent a hazard to health. For this reason the establishment of an ADI is not deemed necessary. However, the usage of an additive in this category must conform to good manufacturing practice (GMP).
3 'Unconditional' status is allocated to those substances for which the biological data includes favourable results from appropriate long- and short-term toxicological studies and/or biochemical and metabolic studies.
4 'Conditional' is allocated when the committee has considered that the quality of the available data is inadequate for an unconditional ADI and further work is required, or if there are other reasons such as those arising from dietary requirements. The reasons and the restrictions imposed on the use of the substance are stated in the evaluation.
5 'Temporary' acceptance means that there are insufficient data to establish whether or not the substance is toxic and further evidence must be submitted within a stated period of time. Details of the requirements are included in the evaluation.

After the JECFA have completed their assessment of an additive and issued their opinion, the Codex Alimentarius Commission is responsible for the formal approval or disapproval. When the Commission gives approval for the inclusion of the substance in the list of additives, in many cases the approval is limited, as far as possible, to specific foods for specific purposes at the lowest effective level of use. Any judgments are determined on the basis of the ADI of the additive and its probable intake from all dietary sources.[11]

In the EU, evaluations and approvals of additives have been carried out by the Scientific Committee on Food of the European Commission (SCF). From 2002, there has been a transfer of responsibilities from the SCF to the European Food Safety Authority. In July 2001 the SCF published guidance notes on the preparation of submissions for food additive evaluations. This is a detailed document that lays down the requirements for the scientific data, including the toxicological studies and other studies considered necessary (Table 15.2).[12]

An important requirement under European Law is that the applicant must provide a persuasive 'case of need' for the new additive. This must include reasons why the functions cannot be met by additives already approved. Acceptable reasons would be a better safety profile or reduced usage levels giving the same effect as existing additives. Purely commercial needs (e.g. commercial competition with other sources) are not considered sufficient justification for approval.

Those substances that do not have GRAS status in the USA are subject to a similar evaluation and approval system by the Food and Drug Administration (FDA). If the JECFA have already given an opinion and established an ADI for

Table 15.2 Requirements for safety evaluation of a food additive in the European Union[12]

Core toxicological studies*
(a) Metabolism and toxicokinetics
(b) Subchronic toxicity
(c) Genetic toxicity
(d) Chronic toxicity and carcinogenicity
(e) Reproduction and developmental toxicity
Other studies*
(a) Immunotoxicity
(b) Allergenicity
(c) Food intolerance
(d) Neurotoxicity
(e) Acute toxicity
(f) Skin and eye irritation sensitisation

* Core studies are those normally required to support an application. Other studies may be necessary to establish safety of certain substances.

the substance this would now be taken into consideration by other reviewing bodies, such as the SCF and FDA.

15.5 Genetic modification

If a food additive or food ingredient is derived from a process or source in which genetic modification of an organism has taken place, the legal status of the substance will change significantly in many countries, including those in the EU. A large number of additives and ingredients are now being produced by fermentation processes in which the fermentative organism (bacterium, yeast or mould) has been genetically modified. Additives and ingredients known to be produced in this way include gums, vitamins, colours, flavours and food acids.

A number of organic acids used in food preservation, such as citric, fumaric, malic, acetic and lactic acids (see also Chapter 6), are being, or have the potential to be, produced from genetically modified organisms. The preservative nisin is a bacteriocin obtained from *Lactococcus* (formally *Streptococcus*) *lactis*. Nisin is used to delay spoilage in a number of dairy products such as cheese and clotted cream (Chapters 2 and 3). This, along with similar antimicrobial substances, is produced by a biotechnological process that could probably benefit from the use of genetically modified microorganisms.

Legislation to control the use of genetic modification in the production of food additives and food ingredients is rapidly increasing and the potential controls and authorisation procedures will have to be taken into consideration during either the development of new food antimicrobials or when changing the method of production of existing ones.

15.6 Processes and packaging

15.6.1 Irradiation

The use of irradiation for the preservation of food goes back some decades. Developments have been hampered by adverse public opinion on its use and, in many countries, by stringent legislation. Commercial food irradiation is normally carried out using gamma rays emitted from the element Cobalt-60 or electron beams generated by machines (ionising radiation). In general, foods are treated by the following sources of ionising radiation:

- gamma rays from radionuclides ^{60}Co or ^{137}Cs;
- x-rays generated from machine sources operated at or below a nominal energy (maximum quantum energy) level of 5 MeV;
- electrons generated from machine sources operated at or below a nominal energy (maximum quantum energy) level of 10 MeV.

When food is exposed to ionising radiation it absorbs energy and the amount of energy absorbed is called the 'absorbed dose', which is measured in units called Gray (Gy) units. The energy absorbed by the food causes the formation of short-lived free radicals that kill many microorganisms and also interact with the food. It is the action of the free radicals, rather than the irradiation, that is responsible for the preservative effects (for further details see Chapter 13).

Irradiation is subject to detailed legislation in many countries, including the member states of the EU. Most legislation specifies the categories of foods that are permitted to be irradiated (e.g. herbs, spices, fruit, vegetables and poultry) and also the maximum dosage in kGy.

When a food or ingredient has been irradiated, it is a legal requirement in many countries to have a label statement to this effect. The Codex Alimentarius Standard on Food Labelling requires that when an irradiated product is used as an ingredient in another food, the irradiation should be declared in the list of ingredients. Where a single ingredient product is prepared from a raw material that has been irradiated, the label of the product should contain a statement indicating that the treatment has taken place.[13] The label statements most commonly required by legislation are 'irradiated' or 'treated with ionising radiation'. The Codex Standard also gives the international food irradiation symbol, which is optional, but when used should be in close proximity to the name of the food.

15.6.2 Impregnated packaging

It is technically feasible to impregnate packaging in contact with food with antimicrobial substances. However, the legal situation surrounding such a packaging material can be complex, with the primary concern being the potential migration of a substance from the packaging into the food.

Within European food law there is specific legislation on materials and articles which come into contact with food.[14] The main requirement of this legislation is that, under foreseeable conditions of use, packaging material in contact with food must not transfer any of their constituents to foods in quantities that could:

- endanger human health;
- bring about an unacceptable change in the composition of the food or a deterioration in the organoleptic characteristics of the food.

This means that any antimicrobial used in packaging would most likely have to be in compliance with the preservative legislation for the countries in which the packaging is intended to be used. In addition, good data on potential migration from the packaging to the food would be necessary to demonstrate that permitted levels of the substance were not exceeded in the food. As the migration into solid or semi-solid food is often only a small distance in from the point of contact, interpretation of the data in terms of legal compliance can be complex.

15.7 New or novel ingredients and processes

15.7.1 Novel ingredients

As discussed earlier, there may be some cases where an antimicrobial substance could be classified as a food ingredient rather than as a food additive. There are a rapidly decreasing number of countries where new or novel ingredients can be introduced into the food chain without some form of prior notification or approval. From May 1997 the EU has had legislation requiring the official approval of all novel foods and ingredients.[15] A novel food or ingredient is defined in the European Regulation No. 258/97 as a food or ingredient:

> which has not hitherto been used for human consumption to a significant degree within the (European) Community and which falls into one of the following categories.

The first two of the six categories listed cover foods and ingredients produced from, or are by-products of, genetically modified organisms. The third includes all foods and ingredients with a new or intentionally modified primary molecular structure, while the fourth includes all those consisting of or isolated from microorganisms, fungi or algae. The fifth category covers all foods and ingredients consisting of or isolated from plants and food ingredients isolated from animals except for those obtained by traditional propagating or breeding practices and having a safe history of food use. The last category covers food and ingredients from a new or novel production process where the process gives rise to significant changes in the composition or structure of the foods or ingredients which affect their nutritional value, metabolism or level of undesirable substances.

The categories, when considered carefully, are such that almost all novel foods and ingredients are included within the scope of the regulation. The regulation does not apply to food additives, food flavourings or extraction solvents as there is European legislation in force covering the introduction of new substances into these three categories.

The regulation requires that a full dossier on the substance or process must be reviewed by a competent authority in the EU country of intended first sale. Requirements for the scientific data (mainly on safety) and the format for the dossier are given in detail in European Commission Recommendation 97/618/EC of 29 July 1997.[16] The initial assessment should be completed in 90 days. Summaries of the application have to be sent to the European Commission who is required to notify the other 14 member states. The other member states have a 60 day period in which to object to the assessment, make comments or request clarification from the applicant. In theory, the total time should be no longer than 150 days. In practice, however, applications are, in effect, being reviewed in detail by the competent authorities in all or most of the member states, resulting in up to 15 reviews of the same data. In addition, many of the applications have been referred by the European Commission to other EU committees, such as the SCF. The SCF is not bound by law to a time limit in which it has to give its opinion and delays by the SCF have added considerably to the length of time taken to process applications; in many cases, over two years.

The costs of obtaining the data to support a novel food application, most of which relates to safety studies, are substantial and are considered to be prohibitive by many small to medium-sized innovative companies. In the first five years after the regulation came into effect, there were only 37 applications, of which 11 were for genetically modified plants such as maize and rape (canola oil) intended to be used as sources of food ingredients. Of the total applications, there were only six approvals, of which one was for fruit preparations produced using high-pressure processing. Another approval was for the very specific use of phytosterol esters in fat spreads. Two applications were refused and three withdrawn by their applicants.

As a result of this legislation, there have been only five new ingredients officially entering the European market in over five years. These are:

- yellow fat spreads with added phytosterol esters;
- phospholipids derived from egg yolk;
- dextran produced from a bacterium;
- trehalose (sweetener);
- coagulated potato protein.

Approval for the phytosterols in fat spreads is very specific and does not give approval for other combinations of phytosterol esters or other uses.

In June 2002, the European Commission published a discussion paper, which considered some of the major issues that had emerged during the first five years of the operation of the regulation. A number of options to amend the regulation were given in the paper and comments elicited.[17]

A similar control on novel ingredients was introduced in Canada and amended in October 1999 to introduce categories almost identical to those in the

EU regulation. For the first few years of its existence, the Canadian law has primarily been used for the authorisation of foods and ingredients derived from genetic modification.

The situation in the USA with regard to new ingredients is more complex than in the EU and Canada. This has been further complicated since the introduction of the Dietary Supplement and Health Education Act (DSHEA) in 1995.[18] Under DSHEA, the safety of ingredients destined for use in dietary supplements is the responsibility of the manufacturer/supplier and not the FDA. This is the antithesis of the requirements for the evaluation of ingredients for other food categories that is carried out by the FDA. As the definition of a dietary supplement is being very loosely interpreted, a number of new ingredients have entered the American market under this category, including a number of herbs and herbal extracts considered to be medicinal in other countries.

15.7.2 Novel processes

The EU Regulation 258/97 includes within its scope processes that are not currently used and that give rise to significant changes in the composition or structure of the foods or ingredients that affect their nutritional value, metabolism or level of undesirable substances. The interpretation of what falls into this definition is very unclear and has been the subject of debate. In the first five years after the regulation came into force, the only process put forward for evaluation was for the use of a high-pressure treatment process for the pasteurisation of fruit preparations. Although the application was made, and approved, it was questionable whether the application was required. The process method (high pressure) does not introduce such significant changes to the food, in this case fruit preparations, that the nutritional value, metabolism, or level of undesirable substances are affected.

The key aspect of the requirement is the effect of the process on the food or ingredient. Thus, if it can be demonstrated that a new process does not have a measurable effect on the nutritional quality of the food, nor introduces contaminants or by-products of degradation, the process may not fall into the scope of the regulation.

15.8 Borderline between food and medicine

A complex area that appears in all national legislative systems, but with varying interpretations, is the borderline between food and medicine. This has to be taken into consideration in assessing the status of a new substance, and it is particularly relevant in respect to herbs and other substances of botanical origin.

Within the EU, a substance can be considered medicinal if it is capable of restoring, correcting or modifying physiological functions in human beings or animals. This means that if a substance, or combination of substances, can exert a physiological function on the body that is not normally associated with nutrient function, the substance or combination may fall into the classification of a medicine. With the increasing knowledge of the physiological effects that can be brought about by common components of the diet, such as phytoestrogens in soya products, there has been a proposal that the EU definition of a medicine be amended to 'by modifying physiological functions and by exerting a pharmacological action'.

Even with the proposed changes to the definition there will be a large number of herbs and their extracts that are considered medicinal because of their function.

15.9 Future prospects for natural antimicrobials

As can be seen from this review, the rapidly increasing legislative environment worldwide is making it increasingly more difficult to introduce new substances into the food chain. New entries will have to be supported by a significant amount of safety data irrespective of whether they are intended as food additives or ingredients. The costs of supporting applications for approval are becoming prohibitive for the smaller innovative companies and the commercial future of the substance (i.e. the pay-back) is the prime consideration of even the larger companies. To be even considered for commercial use, a new antimicrobial substance or process must show very considerable advantages over those already in existence.

15.10 References

1. Codex Alimentarius Commission. http://www.fao.org/es*/esn/codex
2. European Parliament and Council Regulation (EC) No. 178/2002. O.J. of the EC L31/1 of 1 February 2002.
3. European Council Directive 89/107/EEC. O.J. of the EC L40/27 of 11 February 1989.
4. Furia TE (Ed.) (1972). *The Handbook of Food Additives*, CRC Press, Boca Raton, USA.
5. Beuchat LR and Golden DA (1989). 'Antimicrobials occurring naturally in foods', Food Tech **43** (1), 134–143.
6. Codex Alimentarius Commission (1999). 'General Standard For Food Additives Codex Stan 192-1995 (Rev 2-1999). Codex Alimentarius Commission, Rome.
7. European Parliament and Council Directive 95/2/EC. O.J. of the EC L61/1 of 18 March 1995.
8. Jay, JM and Rivers GM (1984). 'Antimicrobial activity of some food flavoring compounds', *J. Fd Safety* **6**, 129–139.
9. World Trade Organisation (1994a). *Agreement Establishing the World Trade Organisation*. World Trade Organisation, Geneva
10. World Trade Organisation (1994b). *Understanding on Rules and Procedures*

Governing the Settlement of Disputes. World Trade Organization, Geneva.
11. Joint FAO/WHO Expert Committee on Food Additives (JECFA) (2001). *Compendium of Food Additive Specifications*. http://www.codexalimentarius.net/CONTACT/Htframe.htm
12. Scientific Committee on Food of the European Commission (2001). 'Guidance on Submissions for Food Additive Evaluations'. SCF/CS/ADD/Gen/26 Final of 12 July 2001, Brussels.
13. Codex Alimentarius Commission (2000). '*Food Labelling – Complete Texts,* Revised 1999'. Codex Alimentarius Commission, Rome.
14. European Council Directive 89/109/EEC. O.J. of EC L40/38 of 11 February 1989.
15. European Parliament and Council Regulation (EC) No. 258/97. O.J. of EC L43/1 of 14 February 1997.
16. European Commission Recommendation 97/618/EC. O.J. of EC L253/1 of 16 September 1997.
17. European Commission (2002). 'Discussion Paper: Implementation of Regulation (EC) No. 258/97 Concerning Novel Foods and Novel Food Ingredients'. SANCO D4 of the European Commission, July 2002, Brussels.
18. Congress of the United States of America (1994). 'The Dietary Supplement and Health Education Act of 1994'. Public Law 103-417, 103d Congress, Washington DC.

Appendix

Useful web sites

Topic or organisation	Web site address
Government sites (UK):	
Food Standards Agency	www.food.gov.uk
Public Health Laboratory Service	www.phls.co.uk
Government sites (USA):	
Centers for Disease Control	www.cdc.gov
Department of Agriculture	www.usda.gov
Food and Drug Administration	www.fda.gov
FDA Center for Food Safety and Applied Nutrition	www.cfsan.fda.gov
FDA Antimicrobial Additives Guidelines	http://vm.cfsan.fda.gov/~dms/opa-antg.html
Food Safety and Inspection Service	www.fsis.usda.gov
National Toxicology Program	http://ntp-server.niehs.nih.gov
International organizations	
Food and Agriculture Organization	www.fao.org
Codex Alimentarius Commission	www.codexalimentarius.net
European Commission Food Safety	www.europa.eu.int/comm/food/
European Food Safety Authority	www.efsa.eu.int
Joint FAO/WHO Expert Committee on Food Additives	www.fao.org/es/esn/jecfa/database/cover.htm
World Health Organization	www.who.int
Manufacturers of food additives:	
Nisin producers	www.dsm.com/dfs/
	www.danisco.com
	www.aplin-barrett.co.uk
Chitosan producers	www.randburg.com/is/genis.html
	www.vanson.com/plainSite/index.html

Miscellaneous:	
Activated lactoferrin	http://activatedlactoferrin.com
Aromatic and medicinal plants	www.hort.purdue.edu/newcrop/med-aro/toc.html
Edible films	www.ediblefilms.org
European Antimicrobial Resistance Surveillance System	www.earss.rivm.nl
Regulations:	
Code of Federal Regulations (USA)	www.access.gpo.gov/nara/cfr/cfr-table-search.html
Official Journal of the European Communities (where European Directives are published)	http://europa.eu.int/eur-lex/en/oj/index.html

Index